Integrated Nanophotonic Resonators

Integrated Nanophotonic Resonators

Fundamentals, Devices, and Applications

edited by

Ya Sha Yi

PAN STANFORD PUBLISHING

Published by

Pan Stanford Publishing Pte. Ltd.
Penthouse Level, Suntec Tower 3
8 Temasek Boulevard
Singapore 038988

Email: editorial@panstanford.com
Web: www.panstanford.com

British Library Cataloguing-in-Publication Data
A catalogue record for this book is available from the British Library.

Integrated Nanophotonic Resonators: Fundamentals, Devices, and Applications

ISBN 978-981-4613-78-1 (Hardcover)
ISBN 978-981-4613-79-8 (eBook)

Printed in the USA

Contents

Preface

The emerging development of integrated nanoelectronics and nanophotonics has been driven considerably by miniaturization. Following the path taken in integrated semiconductor electronics of reducing devices to their ultimately fundamental scale, for instance, few-nanometer transistor, now integrated nanophotonic devices have also been scaled down, creating the increasingly active research fields of integrated and coupled photonic systems. Recently, the integrated resonator (micro/nanophotonic/electronic/mechanical resonator) has seen widespread applications in biomedicine, telecommunications, sensing/detection, security, solid-state lighting, and the renewable energy field. The interactions between coupled integrated micro- and nanostructures can provide us the fundamental understanding and engineering of the complex systems for a variety of applications.

This book aims to bring to the readers the latest development in this rapidly emerging field. It compiles state-of-the-art research by the interdisciplinary team from around the world, who are leading experts in this rapidly advancing field. In this book, we introduce the readers to the fundamental knowledge of coupled integrated photonic/electronic/mechanical micro- and nanoresonators and their interactions, as well as cutting-edge development along this exciting direction.

Key features of this book include

- wide coverage on coupled integrated photonic systems, including fundamental principles, design, simulation, fabrication and device characterization, passive and active system, and widespread applications in electronics/photonics/mechanics;
- contributions by internationally well-known experts who have been working actively in this emerging field for the past decade;
- excellent figures and references to accompany each chapter.

In preparation of this book, I would like to thank the editors at Pan Stanford Publishing for their hard work and tireless efforts, Stanford Chong for initiating the idea of editing this book, and Arvind Kanswal, who helped me turn the manuscript into a production-ready book.

Last but not the least, I am indebted to my family for their endless love and support, my wife, Lilly Tan, my daughter, Grace Yi, and my son, Alexander Yi.

I hope you will enjoy reading the book!

Ya Sha Yi
July 27, 2015

Chapter 1

Hybrid and Coupled Photonic System between Nanoparticle and Integrated Microresonator

Ya Sha Yi[a,b]

[a]*University of Michigan, MI, USA*
[b]*Massachusetts Institute of Technology, MA, USA*

yashayi@umich.edu, yys@alum.mit.edu

The rapid advancement of integrated optoelectronics has been driven considerably by miniaturization. Following the path taken in electronics of reducing devices to their ultimately fundamental forms, for instance, single-electron transistors, now optical devices have also been scaled down, creating the increasingly active research fields of integrated nanophotonics.[1–5] A fundamental nanophotonic device, involving single elementary excitations such as photons and spins, offers various advantages over a macroscopic object: its state can be controlled, initialized and read out with a precision at the fundamental quantum mechanical limit; its operational fidelity based on the interconversion of single excitations can approach 100%; and its functionality may be based on coherent rather than incoherent dynamics, allowing for devices that maintain quantum coherence, which is a crucial

Integrated Nanophotonic Resonators: Fundamentals, Devices, and Applications
Edited by Ya Sha Yi

ISBN 978-981-4613-78-1 (Hardcover), 978-981-4613-79-8 (eBook)
www.panstanford.com

requirement for applications in quantum information processing. Thus, a fundamental device could outperform traditional devices in terms of operation speed, integrability, and energy efficiency. On the fundamental level, a device consists of various coupled and/or integrated entities, such as single emitters, optical microresonators, optical waveguides, and photon-to-electron or photon-to-spin interfaces (Fig. 1.1). If different-material systems are involved (for example a combination of organic and inorganic units), then the resulting device might be appropriately described as "hybrid." However, this term can also be used in a broader sense when new functionality is gained from a combination of different physical effects. Examples are the fusion of bioassays and micro- or nanoelectromechanical systems and that of microfluidics and optics. A merging of quantum optics and nanomechanics, photonics and atomic physics, or quantum optics and plasmonics is likely to provide new fundamental insight as well as novel applications such as quantum coherent devices or quantum-limited sensors.[6–13] For example, a quantum computer, by contrast with its classical counterpart, needs to maintain coherent superpositions of discrete quantum states, acting as so-called quantum bits. A similar challenge occurs in quantum sensors, where exploitation of entangled quantum states allows for improved precision over classical equivalents. Whereas long storage times of quantum states may be possible, for example in atomic systems, fast read-out and information transfer using integrated optics, that is, an atomic–photonic hybrid system, would be desirable. The quantum mechanical decoherence and energy transfer processes in these hybrid systems are, however, not fully understood. This broader class of hybrid system is the motivation of this chapter, many of above applications will benefit from the understanding of the fundamental coupled integrated photonic system. This chapter emphasizes model systems combining metallic and dielectric nanophotonic constituents by scanning probe manipulation.

A typical example of a hybrid nanophotonic system is photonic molecule. With the rapid progress of nanotechnology, many nanoscale photonic devices as small as 30 nm have been realized, which are very promising to achieve manipulation of photons at chip scale and having broad applications in renewable energy (photovoltaic cells, solid state lighting), telecommunications and bio medical field.[14,15] Recently, it was found that the electromagnetic modes of certain photonic devices are very similar to the electronic wave function of molecules[16] (Figs. 1.2a–f). One of the most interesting examples is microresonators, especially when we arrange two or more microresonators together within optical coupling regime, the electromagnetic modes of the whole structure are very similar to the bonding (symmetric) or antibonding (antisymmetric) electronic wave function modes formed in molecules.[17–20] It is interesting to study the photonic molecule of various structures using optical techniques and it may further improve our understanding of more complex photonic structures.

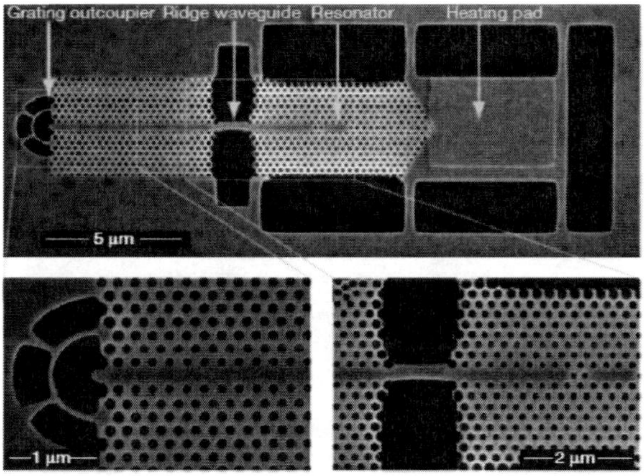

Figure 1.1 Nanophotonic functional elements. An integrated photonic structure. The structure includes a grating coupler (left; also shown magnified) and a quantum dot inside a photonic crystal resonator coupled to a photonic crystal waveguide[13] (also shown magnified).

Scanning probe microscopy (SPM) tip (Fig. 1.2g) has played a key role in nanophotonics and has found many applications in medicine, drug delivery, solar cells, and sensors. It is also an important tool for the study of many nanoscale structures and is used to interact with nanoscale devices, as critical information could be obtained to understand their characteristics. Recently, SPM tips have been utilized to simulate various nanoparticles, as the tip size is typically in the order of 10 to 100 nm size, which are perfect match to the nanoparticles. Nanoparticles have been heavily used in photonics, such as optical detection and sensing area, as fast, non-invasive, and potentially label-free techniques are becoming more important for biosensing, gas sensing, and chemical sensing. Single nanoparticle detection is one of the ultimate goals for detection and sensing device and represents sensing at the extreme. In recent years, many novel methods have been utilized to realize nanoparticle detection. For example, metal nanoparticles are used as contrast agents in biomolecule sensing, semiconductor nanoparticles are used as single-photon emitters in quantum information processing and as fluorescent markers for biological processes, nanoshells with special engineering methods are used for cancer therapies and photothermal tumor ablation, and polymer nanoparticles are employed as calibration standards and probes in biological imaging in functionalized form.[21–24]

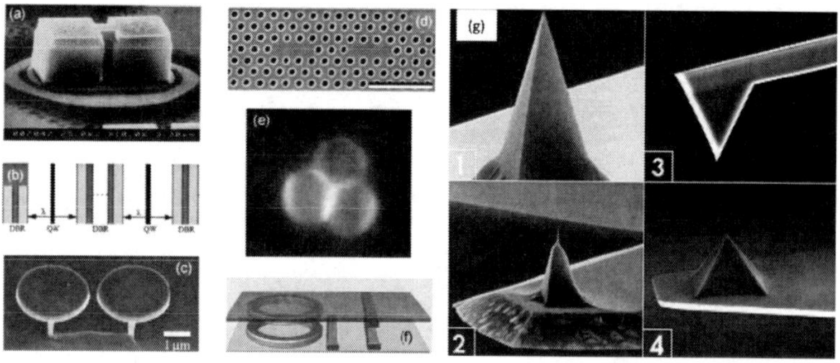

Figure 1.2 Typical configurations of photonic molecules: (a) square-shape photonic dots coupled via a semiconductor bridge, (b) planar Fabry–Pérot cavities coupled through a partially-transparent Bragg mirror, (c) whispering-gallery mode microdisks side-coupled via an airgap, (d) closely located defects in a photonic crystal membrane, (e) triangular photonic molecule composed of touching microspheres, (f) vertically coupled microrings. (g) various types of SPM tips.

The synergy between the hybrid photonic systems and SPM tips provides us with a unique opportunity, as we can utilize the special hybrid photonic modes to interact with the SPM tip to achieve more understanding of external perturbation, inversely, we can also utilize SPM tip to study hybrid photonic systems and their properties. To further explore the hybrid and integrated photonic systems, this work will employ SPM tips with integrated active optical components. For example, the SPM tip can be coated with various dielectric, metallic and magnetic materials, including inorganic, organic and biological molecule (DNA, RNA, protein, etc.) attached materials. Multiple layer materials can also be coated on the tip to form an effective core-shell or multi-shell nanoparticle structures. Active optoelectronic materials (quantum dots, etc.) can be attached to the SPM tips. These active probes will be controllably scanned in proximity to metal or semiconductor nanoscale structures, essentially creating a highly tunable and versatile hybrid system. I will begin by highlighting some key proposed works in this chapter utilizing these hybrid nanophotonic functionalities and then briefly introduce our several preliminary and ongoing works for these hybrid nanophotonic systems using active SPM tips. As scanning probe techniques at present offer the greatest versatility, the successful application of such techniques for the assembly of hybrid nanophotonic architectures will be critical for our next generation complex integrated nanophotonic devices and systems.

The chapter describes three representative directions:

(a) *Integrated nanophotonics for multiple metallic nanoparticle sensing and detection.* We have utilized the unique optical properties of the integrated microresonator mode to enhance the light signal from potentially biomolecule tagged metallic nanoparticles. Extremely low concentration biomolecule sensing and detection can be achieved.

(b) *Photonic resonator with nanoscale notch for high-sensitivity label-free biomolecule quantification, single-bacteria detection, and single-virus sensing.* The basic working mechanism of photonic resonator and the optical properties of whispering-gallery-mode will be explained; the future lab-on-chip single biomolecule sensing using integrated photonic resonator will also be described.

(c) *Nanoscale electromagnetic field and mode profile mapping and measurement.* We have demonstrated higher-order optical resonance node detection by using an integrated disk microresonator from the Through port of the coupling bus waveguide. In addition to the fundamental mode, the disk resonator has higher-order whispering gallery modes. The excited second-order higher-order mode has a node at the position where the electromagnetic energy of the fundamental mode is close to a maximum. This high-resolution measurement of optical resonance mode profile has a variety of applications for optical sensing and detection. The self-referencing characteristics of the two optical resonance modes have potential to achieve optical detection independent of external perturbation, such as temperature change.

(d) *Back reflection caused by optically interacting nanoparticle.* We have investigated theoretically and experimentally a new approach using optical ring and disk micro-resonators as nanoparticle detection. This approach is based on the observation that when a nanoparticle having a large index contrast from the medium surrounding the microresonator is brought within the evanescent field outside the resonator, some of the light circulating in the resonator is reflected and circulates in the opposite direction. When the resonator is tuned to a resonant frequency for the forward propagating light, the backscattered light is also in resonance with the cavity. Therefore particle-induced coupling between the forward and backward propagating resonant modes results in a buildup of significant intensity in the backward propagating modes. This principle applies to all the forward-propagating resonant modes of the resonator, thus with broadband excitation of the forward modes, many backward propagating modes can be excited.

1.1 Metallic Nanoparticle on Microring Resonator for Bio Optical Detection and Sensing

1.1.1 Introduction

Optical resonator has generated wide interests in the detection and sensing field. For the relatively high-Q microresonators, a small change on the refractive index can be detected from the shift of resonance wavelength. Recently, the splitting of the resonance modes has been observed, which is caused as a consequence of clockwise and anticlockwise propagating mode coupling. This phenomenon has been proposed for various applications, such as photonic molecules.[15,18–20,25–27] Nanoparticles have been heavily used in the optical detection and sensor area, as fast, non-invasive, and potentially label-free techniques are becoming more important for biosensing, gas sensing, chemical sensing, and nanomedicine. For example, metal nanoparticles are used as contrast agents in biomolecule sensing. Semiconductor nanoparticles are used as single-photon emitters in quantum information processing, and as fluorescent markers for biological processes. Nanoshells with special engineering methods are used for cancer therapies and photothermal tumor ablation. Polymer nanoparticles are employed as calibration standards and probes in biological imaging in functionalized form.[21–24] The synergy between microresonator and nanoparticle is becoming more important with the rapid progress of nanophotonics field.

The influence on microresonators by *dielectric* nanoparticles has been intensively studied recently; fiber tip is used to study the resonance mode profile, especially from the splitting of resonance mode. The mean resonance mode wavelength shift, splitting bandwidth, and their dependence on dielectric nanoparticle size and position have also been studied by many groups.[28,29]

Optical sensors (including biosensor, chemical sensor, and gas sensor) based on the whispering gallery mode (WGM) micro-resonator have generated worldwide interest in this emerging filed involving dielectric nanoparticles or dielectric bio layer.[30–39] However, the microsphere is made manually, which makes the large-scale manufacture of optical sensor devices very challenging. The integrated microring resonator does not have this limit as we can use the semiconductor microelectronics process and millions of devices can be fabricated and integrated on a single chip with nanoscale precision.

Metallic nanoparticles have also been used intensively on detection and sensing, especially in the form of surface plasmon polariton (SPP), as the surface mode generated at the interface between the dielectric surface and metallic surface is strongly confined at the interface, which can be utilized for many potential applications.[40] Recently, in the conventional four-port micro-resonator configuration, it was found that metallic nanoparticles can be used as a strong scatters on the microring resonators, which induces large reflection signals at an output port, which is normally dark.[41,42] The interaction between *metallic* nanoparticles and microring resonators has revealed many interesting phenomena and has not been emphasized much in previous studies as *dielectric* nanoparticles.

1.1.2 On-Chip Microring Resonator Device Structure and Simulation Method

Here we have numerically demonstrated a unique result by Au nanoparticles, when it is adsorbed at the edge of microring resonator. Compared to the resonance position without any Au nanoparticle, it was found that there is a *blue* shift for the resonance peak, which is opposite to the resonance wavelength shift direction when the dielectric nanoparticles are adsorbed onto the microring resonator. Owing to the unique refractive index properties of Au, the number and position effects are also appealing and investigated in detail in this chapter.

We used the conventional four-port microring resonator configuration on a chip, as illustrated in Fig. 1.3. The microring resonator is 4 μm in diameter, and the ring waveguide width is 200 nm as a single mode waveguide, the thickness of the ring resonator and bus waveguide is 250 nm. The two bus waveguides are evanescently coupled to the microring resonator, with the coupling gap 100 nm. In this work, we simulated the Si ring resonator and coupled waveguide system, with SiO_2 as the bottom cladding and air as the top cladding. The Au nanoparticle was placed at the outside edge of the microring resonator. The refractive indexes of Si and SiO_2 are is 3.48 and 1.45, respectively, and the dispersion relation of Au around wavelength 1.55 μm is used.[43]

We used finite difference time domain (FDTD) method in three dimensions to simulate the four-port microring resonator with/without Au nanoparticles.[44] Due to the small size of nanoparticles, fine grid size as small as 2 nm and sufficient long evolution time steps are used to check the reliability of the simulation until the optimized grid size and time steps were found to reduce the required memory and simulation time. Perfect matched layer (PML) absorbing boundary condition is used for the entire simulation window (10 μm × 10 μm). The bus waveguide was excited with a Gaussian pulse, which covers the wavelength window around 1.55 μm, the detected signal at Drop port was Fourier transformed to obtain the Drop port vs. wavelength information.

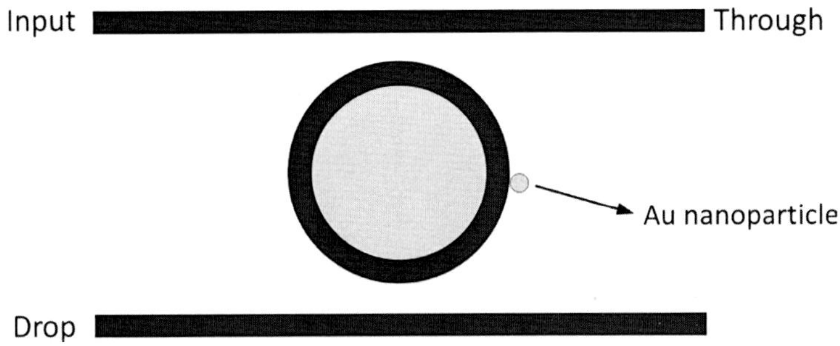

Figure 1.3 The on-chip four-port microring resonator configuration. The core of the microring resonator and the two bus waveguides is Si (n = 3.48), with bottom cladding SiO_2 (n = 1.45) and air cladding on top. The waveguide width is 200 nm and satisfies the single mode condition. The Au nanoparticle is adsorbed on the microring resonator.

1.1.3 Multiple Au Nanoparticle Effects on Microring Resonator and Simulation Results

Similar to the dielectric nanoparticles, the metallic nanoparticles will make the resonance wavelength shift and broadening of the splitting bandwidth. As illustrated in Fig. 1.4, the main difference of the Au nanoparticle is its smaller real refractive index than air and very large imaginary part (absorption part), which leads to the *blue* shift of the resonance mode wavelength position. As the nanoparticle numbers reach a certain number, the splitting of the resonance began to appear, within our simulation resolution. For the 80 nm Au nanoparticle on the microring resonator with the nanoparticle number increasing from one to two, it is observed that both splitting modes are *blue* shifted. This phenomenon is unique as it provides us a very convenient approach to distinguish the dielectric nanoparticles and the Au nanoparticles, both are used extensively for sensing and nanomedicine field. Furthermore, the intensity of Drop port is reduced rapidly with the increasing number of Au nanoparticles (30 in this case), which represents its large absorption characteristics at this wavelength. For sensing applications using Au nanoparticles or other metallic nanoparticles, it is inferred from this work that there is a limit for the number of metallic nanoparticles adsorbed on the ring resonator, as there is normally large absorption for metallic nanoparticles. When the number reaches a certain point—critical number (30 Au nanoparticles in this case), the interaction between metallic nanoparticles and microring resonator becomes so strong that they completely degrade the resonance—the Q is strongly degraded and the intensity at Drop port approaches zero.

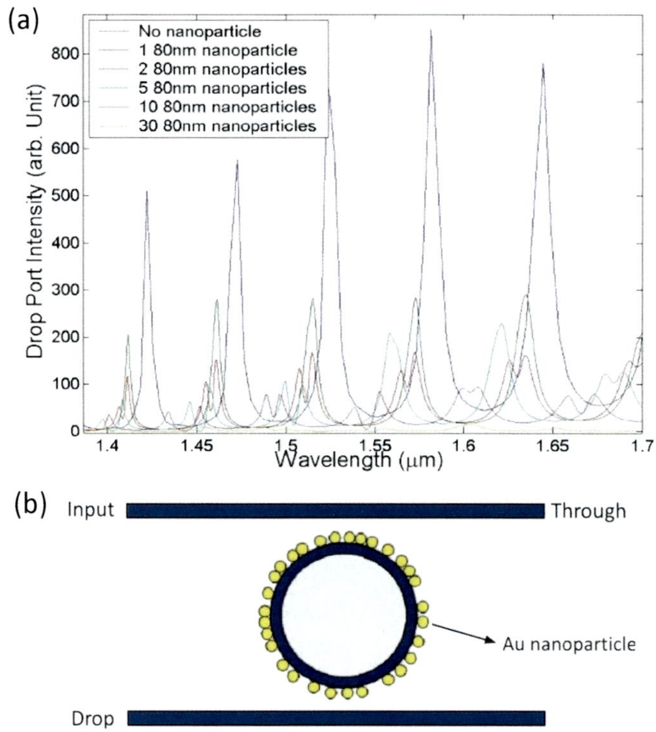

Figure 1.4 (a) The Drop port intensity with different number of 80 nm size Au nanoparticles adsorbed on the microring resonator. (b) Example of multiple Au nanoparticles randomly distributed on the microring resonator.

For using Au nanoparticle and Drop port as a detection mechanism, the dependence on position is also important as the Au nanoparticles are possible to be adsorbed randomly on the microring resonator, the relatively position independence is necessary. To this purpose, for 30 Au nanoparticles 80 nm in size, we have randomly distributed the 30 Au nanoparticles on the microring resonator and compared the Drop intensity in Fig. 1.5. It is shown that the intensity at Drop port for three random positions is at the same order and this result demonstrates the relatively independence of the Au nanoparticle position on the microring resonator.

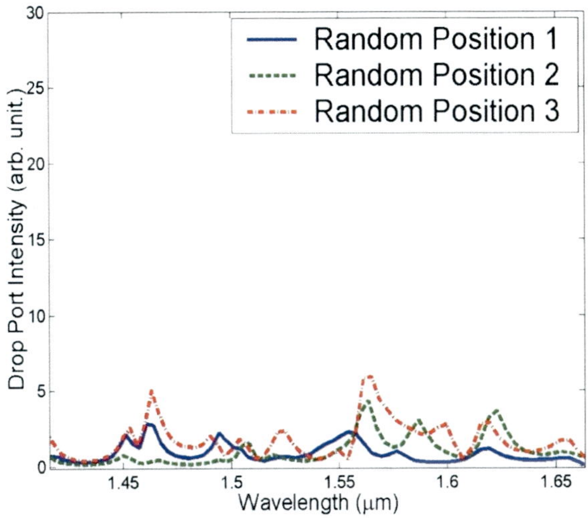

Figure 1.5 The Drop port intensity with three random positions of 30 Au nanoparticles with 80 nm in size.

Metallic nanoparticle size uniformity is very important for practical sensing and detection, as the nanoparticle size normally has a distribution around the target nanoparticle size, which we would like to use. To study the effect of uniformity of the size of the Au nanoparticles on the performance of the integrated microring resonator, for 5 Au nanoparticle case with 80 nm size, we have randomly chosen the nanoparticle size that has certain distribution around 80 nm. Figure 1.6 is the comparison between the uniform size nanoparticles and nanoparticles with certain distribution, the overall signal from drop port is almost the same although there is some small difference. The result demonstrates the robustness of our sensing mechanism using Au nanoparticles, which can tolerate certain non-uniformity of Au nanoparticles.

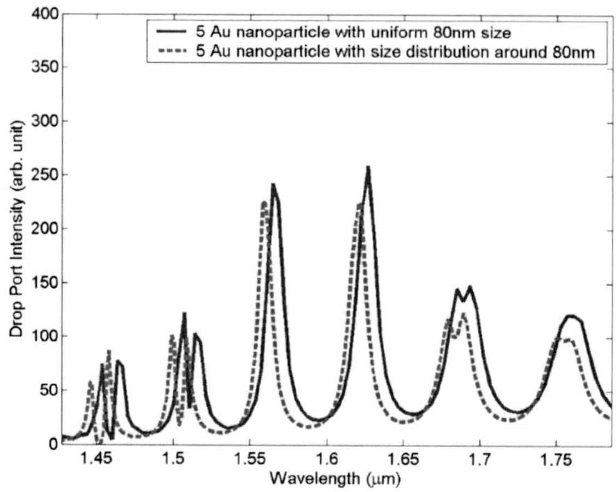

Figure 1.6 The Drop port intensity with Au nanoparticle size distribution, 5 Au nanoparticles with uniform 80 nm in diameter (solid line) and 5 Au nanoparticles with random size distribution around 80 nm in size.

We also studied the dependence on Au nanoparticle size, which is illustrated in Fig. 1.7; 10 nm, 40 nm, and 80 nm Au nanoparticle are compared at Drop port with the different number of Au nanoparticles. The intensity at Drop port for 10 nm size Au nanoparticle is reduced in a much slower pace than that for the 40 nm and 80 nm Au nanoparticle. The reduction of the intensity is mainly caused by the strong absorption of Au nanoparticles, around 1.55 μm; the penetration depth of Au is about 45 nm. The dependence on nanoparticle size for the Drop port reveals the correlation between the penetration depths, nanoparticle size and resonance mode evanescent tail length, which might be utilized to measure the Au nanoparticle size. For optical sensing and detection purpose, the optimized Au nanoparticle size should meet two requirements: One is relatively large shift when the nanoparticle is adsorbed to the ring resonator; the other is the slow Q degradation ratio when more Au nanoparticles are adsorbed on the microring resonator. Based on the results in Figs. 1.4 and 1.7, the optimized size of Au nanoparticle is estimated as 40 nm, which is around the penetration depth.

In summary, we have numerically demonstrated the unique optical response behavior for Au nanoparticle in the four-port microring resonator configuration, which can be utilized for single nanoparticle detection and related applications. The *blue* shift of the resonance mode position at Drop port due to the Au nanoparticle is drastically different from that of dielectric nanoparticles, which are widely used for various applications in detection, sensing and biomedical field. For sensing, the unwanted nanoparticle adsorbed on the microring resonator is mostly dielectric nanoparticles, which may be mixed with true signal, the unique *blue* shift by Au nanoparticle could be utilized to differentiate from the dielectric nanoparticles, therefore, it can be utilized to increase the signal-to-noise ratio. Due to large absorption of Au, the number of nanoparticles will reach a critical number before the microring resonator can still maintain an effective optical resonator device. The results on position and size dependence suggest the robustness to use Au nanoparticle for future applications in the fields of detection, sensing, and nanomedicine.

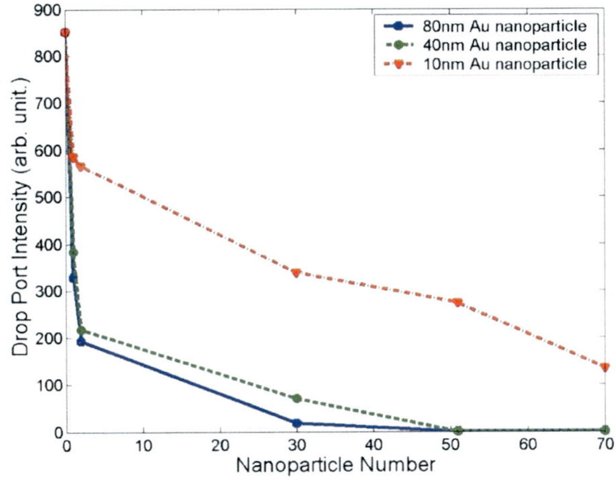

Figure 1.7 The Drop intensity vs. nanoparticle number for different size of Au nanoparticles.

1.2 Strong Coupling between On-Chip Notched Ring Resonator and Nanoparticle

With the rapid progress of nanotechnology, many nanoscale photonic devices as small as 30 nm have been realized, which are very promising to achieve manipulation of photons at chip scale and having broad applications in renewable energy (photovoltaic cells, solid state lighting), telecommunications and bio medical field.[14] Recently, it was found that the electromagnetic modes of certain photonic devices are very much similar to the electronic wave function of molecules.[15,16] One of the most interesting examples is microring resonators, especially when we arrange two or more microresonators together within optical coupling regime, the electromagnetic modes of the whole structure are very similar to the bonding (symmetric) or antibonding (antisymmetric) electronic wave function modes formed in molecules.[17–20] It is interesting to study the photonic molecule of various structures using optical techniques and it may further improve our understanding of the real molecular structures. The nanoparticle has played a key role in nanophotonics and has found many applications in medicine, drug delivery, solar cells, and sensors. It is also an important tool for the study of many nanoscale structures and is used to interact with nanoscale devices, as critical information could be obtained to understand their characteristics. Recently, nanoparticles have been heavily used in the optical sensor area, as fast, non-invasive, and potentially label-free techniques are becoming more important for biosensing, gas sensing, and chemical sensing. Single nanoparticle detection is one of the ultimate goals for a sensing device and represents sensing at the extreme. In recent years, many novel methods have been utilized to realize nanoparticle detection. For example, metal nanoparticles are used as contrast agents in biomolecule sensing, semiconductor nanoparticles are used as single-photon emitters in quantum information processing and as fluorescent markers for biological processes, nanoshells with special engineering methods are used for cancer therapies and photothermal tumor ablation, and polymer nanoparticles are employed as calibration standards and probes in biological imaging in functionalized form.[21–24] The synergy between the photonic molecule and nanoparticle provides us with a unique opportunity, as we can utilize the special photonic molecular modes to interact with the nanoparticle to achieve single nanoparticle detection, inversely, we can also utilize nanoparticles to study photonic molecules and their properties.

In this work, we used on-chip notched microresonator as a photonic molecular example, as a variety of types of optical micro-resonators have been investigated and it is a natural photonic molecule to use (Fig. 1.8a); for the nanoparticle, we used the atomic force microscope (AFM) tip to simulate a single nanoparticle, where the small tip can be either dielectric materials (Si, GaAs, Si_3N_4, etc.) or metallic materials (Au, Ag, Al, etc.). We have achieved, for the first time, strong coupling between an on-chip notched microring resonator and a single nanoparticle. Specifically, we have used the nanoscale notch (~100 nm) in the microring resonator with diameter around 4 μm to strongly interact with the AFM tip. The intentionally created notch in the ring resonator cause the splitting of the original ring resonance mode and formation of bonding (symmetric) photonic states and antibonding (antisymmetric) states. The AFM tip can be positioned inside the notch. The strong coupling between core electromagnetic modes in the notch and the tip cause the bonding photonic modes to shift at the nanometer scale, while there is almost no shift for the antibonding photonic modes. It confirms the photonic molecular modes characteristics generated by the notched microresonator. The result suggests the potential to deeply study the photonic molecular mode characteristics. Furthermore, we have found the unique and very different shift behavior of the splitting modes from the dielectric Si tip and the metallic Au tip, which can be a critical detection and sorting mechanism for the different type of nanoparticle systems.

In the conventional approach to study the splitting modes using microspheres, because the interaction between the fiber tip (or nanoparticle) and the evanescent mode of the microsphere is very small, it leads to a very small change in the effective refractive index of the sphere, which shifts the wavelength position of the peaks and causes small splitting in the resonator transmission (or reflection) spectrum; these changes are typically on the order of picometer in size.[35] In order to detect such small shifts, one must normally use an expensive tunable narrow-linewidth laser source to scan the relevant spectral region of the resonator output spectrum. Furthermore, the resonator itself must be designed to yield a very narrow linewidth, so that the small peak shifts and splitting can be detected. This requires a high finesse (free spectral range divided by linewidth), or equivalently high quality factor (operating wavelength divided by linewidth $\sim 10^8$), which translates to low-loss waveguides in the resonator and weak coupling between the resonator and fiber tip. Here, we demonstrate strong coupling between on-chip notched ring resonator and nanoparticle, where the nanoparticle can be placed inside the notch. Compared to the ring resonator without notch, as illustrated in Fig. 1.8b, where the interaction strength between the nanoparticle and the resonator mode field is relatively weak and only a small portion of the field (evanescent tail) is interacting with the nanoparticle, the notch provides access to the peak of the electromagnetic field localized in the core, so that when a nanoparticle is placed there, the strong core field, rather than the weak evanescent cladding field, overlaps the nanoparticle and thereby produces an enhanced response, as shown in Fig. 1.8c.

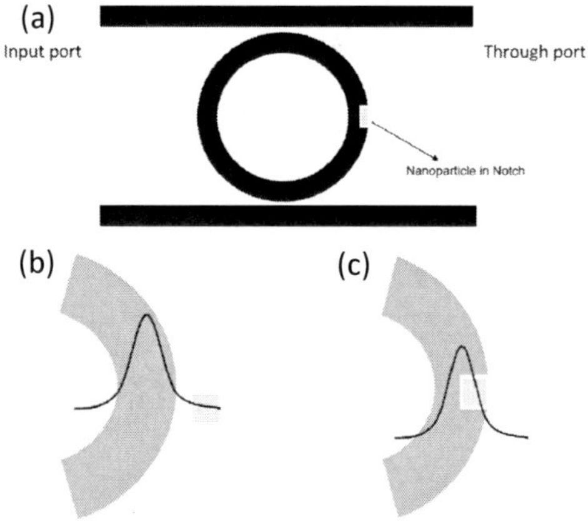

Figure 1.8 (a) The notched microring resonator and the coupling to the bus waveguide with input and output ports. The bus and ring waveguide width is 200 nm. The notch size is around 100 nm × 100 nm. (b) The evanescent coupling between the resonance mode and the nanoparticle. (c) The notch in the microring resonator. The nanoparticle is in the notch and coupling to the core of the resonance mode, the black curve represents the single mode behavior for the ring waveguide. The ring waveguide is a sing mode waveguide at 1.53 μm.

The on-chip photonic device configuration of a microresonator with a notch, nanoparticle and integrated with two bus waveguide has been fabricated.[34,41,45–47] The notch of 100 nm size in the ring was fabricated by e-beam lithography. We have analyzed the case of a 100 nm long notch with a 20 nm diameter dielectric nanoparticle Si tip and 20 nm diameter metallic Au particle inside the notch. The thickness of Si waveguide is

220 nm. Figure 1.9a is the image of the 100 nm sized notch in the microresonator using portable SEM (Hitachi TM-1000). The ring resonator is 4.0 μm in diameter with the waveguide width of 200 nm. The core material of the ring is Si with refractive index 3.48 at around 1.53 μm wavelength, and the bottom cladding material is SiO_2 with refractive index 1.46. We used SOI wafer for the small ring resonator fabrication, and an e-beam is utilized to fabricate the 100 nm × 100 nm notch size at the edge of the Si microring resonator. The refractive index of the Au nanoparticle tip is 0.54 + 9.58*i* at 1.53 μm. Figure 1.9b is the SEM image within the coupling gap area, which shows the clear 100 nm gap between the bus waveguide and the notched ring resonator. A tunable laser from 1480 nm to 1580 nm is used to couple the light from tapered optical fiber to the Si waveguide, a Ge detector is put at other end to collect the Through port signal. Figure 1.10 is the measurement result for the "Through" port of bus waveguide for the ring resonator with notch at the edge. We can clearly see the splitting of the original resonance modes at around 1.53 μm, which represents the bonding photonic mode at shorter wavelength (1.534 μm) and the antibonding photonic mode at longer wavelength (1.537 μm). The splitting width is almost 3 nm; this nanometer-scale splitting bandwidth is much larger than the picometer splitting bandwidth we normally see for the fiber tip close to the microsphere resonator.

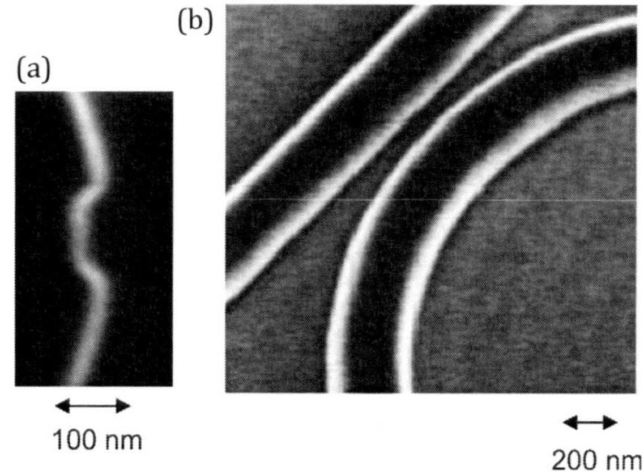

Figure 1.9 (a) The image of the 100 nm size notch by portable SEM. (b) The SEM image with clear 100 nm coupling gap between the notched ring resonator and the single mode bus waveguide.

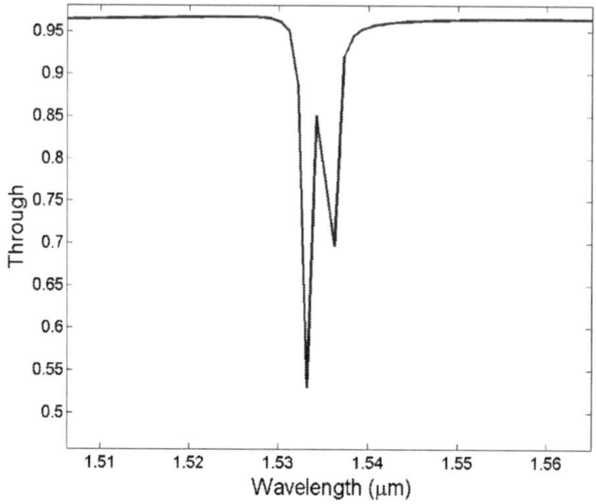

Figure 1.10 The transmission at Through port around 1.53 μm, the splitting is around 3 nm, the bonding photonic mode is at shorter wavelength 1.534 μm and antibonding photonic mode is at longer wavelength 1.537 μm.

We used a Si nanoparticle tip with 20 nm in diameter and put it inside the center of 100 nm size notch of the ring resonator using portable AFM, transmission at the "Through" port of the coupled waveguide is shown in Fig. 1.11a. It is observed that the bonding mode is ***red*** shifted by 1.1 nm, while the position of the antibonding mode is almost unchanged, leading to a smaller splitting bandwidth due to the red shift of the bonding mode. Next, the Si nanoparticle tip is replaced with a 20 nm Au nanoparticle tip and also put inside the notch center of the microring resonator. We see the dramatically different shift with the Au nanoparticle for the bonding state, a 1 nm ***blue*** shift is observed instead and the antibonding state remains unchanged, causing larger splitting bandwidth than the original splitting due to the notch with just air, as shown in Fig. 1.11b. The results reveal the drastically different photonic mode properties between the symmetric and antisymmetric states and suggests that we can utilize the nanoparticle to study the photonic molecule characteristics. We can understand the red shift induced by the Si nanoparticle by considering an effective index increase in the notch due to the substitution of air with high index Si; the same reason for Au nanoparticle, as the real part is less than that of air. We can also understand the fact that nanoparticles have little effect on the anti-bonding mode by considering that they are placed at the center of the notch and close to the zero field node of the anti-bonding mode. It can be clearly seen that the strong coupling effect of a nanoparticle placed in a notch is fundamentally different from the evanescent coupling effect of a nanoparticle placed in the evanescent field of the resonator, which high-Q microspheres are often used.

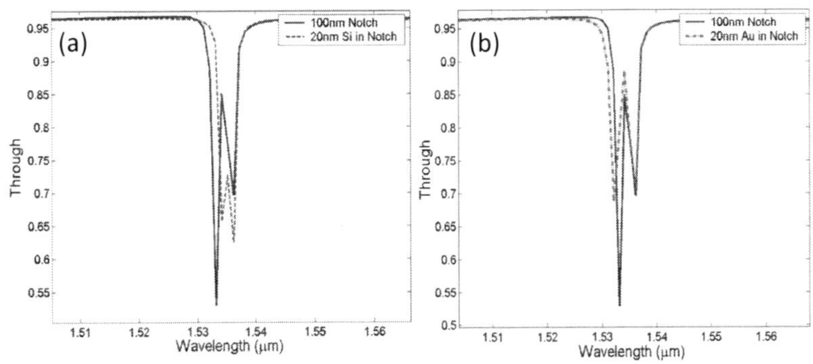

Figure 1.11 (a) Si nanoparticle tip with 20 nm in diameter is put inside the 100 nm notch size of the ring resonator, the Through port from the coupled waveguide is shown, it is observed that the bonding mode wavelength (resonance at shorter wavelength) is ***red*** shifted with 1.1 nm, while the wavelength of the antibonding mode (resonance at longer wavelength) is almost unchanged, the splitting bandwidth is smaller due to the red shift of the bonding mode. (b) 20 nm Au nanoparticle tip is put inside the notch size of the microring resonator, we see the dramatic different shift from the Au nanoparticle for the bonding state (resonance at shorter wavelength), the 1.0 nm ***blue*** shift is observed instead and the antibonding state (resonance at longer wavelength) is still unchanged, the splitting bandwidth is larger than the original splitting due to the notch with air.

The different wavelength shift between a dielectric Si nanoparticle and a metallic Au nanoparticle, as well as the splitting bandwidth narrowing with the dielectric Si compared to the widening with the metallic Au nanoparticle, provides us with a unique, *self-referencing* mechanism to distinguish these different types of nanoparticles. This is very important for the biosensor area utilizing nanoparticles, as Au nanoparticles are often used for tagging, but other dielectric nanoparticles other than the analyte may be present and cause misleading sensing signals; the Au nanoparticle's unique effects on the shift direction and the splitting bandwidth widening will enable us to distinguish the real signal from the background dielectric nanoparticle noise and greatly enhance the signal-to-noise ratio. The intentionally fabricated notch in the microring resonator provides us with a localized position to trap the nanoparticle. Based on the strong splitting and wavelength shift when the nanoparticle is localized in the notch, we can identify whether the nanoparticle stays within the fabricated notch and differentiate the nanoparticle position. The future study on the electromagnetic force applied upon the nanoparticle close to the notch will be interesting to help us to understand the mechanic response from the notch and how to deliver the nanoparticle to the notch.

In summary, we have experimentally demonstrated the strong coupling between on-chip nanoscale notched ring resonators and nanoparticle, in which a notch is introduced in the resonator to provide access to the core field, which is drastically different from previous studies using evanescent coupling to microsphere. Placing a nanoparticle in the notch produces a much stronger response than simply placing the nanoparticle in contact with the exterior of the core. In the exemplary case of a dielectric silicon and metallic gold nanoparticle placed in a notch, we have demonstrated that nanoparticle induces a large wavelength splitting (~nm) and very different shift in the resonant modes of the resonator. This is a significant improvement over the smaller wavelength shifts and splitting (~pm) observed in earlier experiments where the particle was placed outside the core of a conventional microsphere resonator, and lowers the requirement for very high-Q resonator devices. Note that the utility of this approach is not limited to ring resonators used as examples here but can be extended to other types of resonator geometries, such as racetracks, and polygons. This work provides us a unique way to achieve single nanoparticle detection and sorting with thousands of times signal enhancement. The nature of the on-chip microresonators will also make large-scale integration on a sing chip possible.

1.3 Nanoscale Electromagnetic Field and Mode Profile Mapping and Measurement

We have demonstrated higher-order optical resonance node detection by using an integrated disk microresonator from the Through port of the coupling bus waveguide. In addition to the fundamental mode, the disk resonator has higher-order whispering gallery modes. The excited second-order higher-order mode has a node at the position where the electromagnetic energy of the fundamental mode is close to a maximum. This high-resolution measurement of optical resonance mode profile has a variety of applications for optical sensing and detection. The self-referencing characteristics of the two optical resonance modes have potential to achieve optical detection independent of external perturbation, such as temperature change.

The integrated disk microresonator device is composed of a disk resonator and one or two coupled bus waveguides for light injection and signal extraction. A wavelength tunable laser was placed at the input port of the first bus waveguide to launch the waveguide mode; the photo detector was placed at the Through port of the bus waveguide to receive the output signal after the coupling of the bus waveguide and the disk microresonator (Fig. 1.12a). A detection of the excited WGMs event occurs when an AFM tip couples with the mode-field on the disk, causing changes in the frequencies of the disk resonant modes and coupling between the disk modes. As illustrated in Fig. 1.12b, the solid curve is the Through port signal without an AFM tip close to the disk microresonator, while the dashed curve is the Through port signal when a gold AFM tip is staged close to the surface of the disk resonator, where the tip position is at a point that is close to the maximum electromagnetic field intensity of the fundamental mode, while almost at the nodal point of the second-order mode. We observed about 0.4 nm resonance wavelength shift of the Through port signal for the fundamental WGM mode around 1.55 μm (solid curve in Fig. 1.12b). For the higher-order WGM mode, the change induced by the attached gold AFM tip is almost zero for the Through port, which can be clearly seen in Fig. 1.12b (compared to the shift between the solid curve and the dashed curve). The differential effect for multiple WGMs indicates the advantages of the disk resonator—we can make it a self-referencing detection and sensing device, which can reference out frequency shifts induced by effects other than the nanoparticle interaction, e.g. changes in the refractive index of the medium surrounding the resonator, such as the fluctuation of temperature. Although we only use a gold AFM tip as an example, other nanoparticles with high refractive index and large absorption, such as Si, Al, and Ag nanoparticles, should possesses similar characteristics. The mode profiles were simulated using the finite element method. Figure 1.12c illustrates the mode profile of the fundamental mode and the second-order mode and it clearly shows the drastic

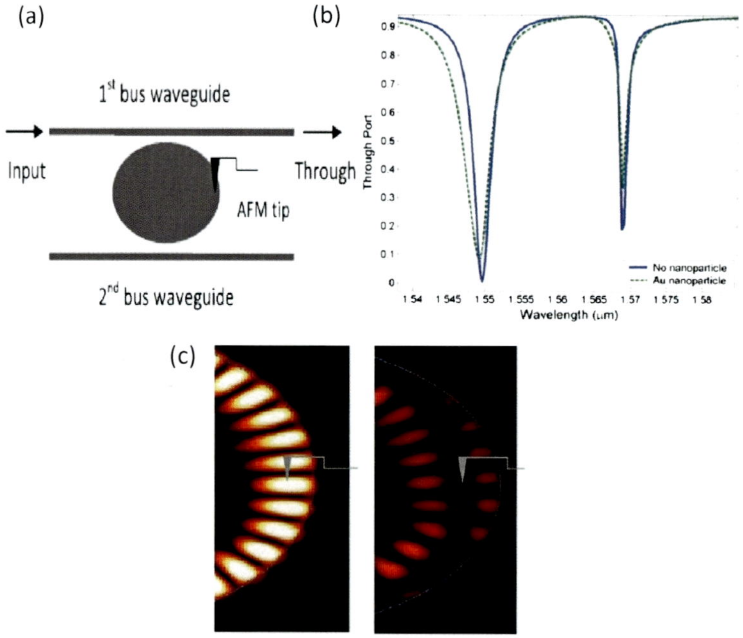

Figure 1.12 (a) Integrated disk microresonator coupled with two bus waveguides. (b) Higher-order optical resonance node detection. Fundamental mode at 1549.6 nm, second-order mode at 1569 nm, solid curve—without a gold AFM tip, dashed curve—when the tip is close to the nodal position of the second-order mode. (c) Mode profile (electromagnetic field intensity) of the fundamental optical resonance mode and second-order optical resonance mode, excited by the coupling bus waveguide.

different mode characteristics of these two excited modes. The different mode profile also explains our observations why the resonance wavelength shifts of the two modes are so different when the AFM tip is placed at the nodal point of the second-order resonance mode. At the position where the fundamental mode has close to maximum electromagnetic energy, because there is almost no electromagnetic energy (node) at the position for the higher-order mode, placing an AFM tip close to the point causes the relevant large resonance wavelength shift of the fundamental mode and almost no shift for the second-order mode. The higher-order mode nodal detection is potential for making ultra sensitive integrated micro- and nanoscale devices with many applications in biomedicine, food, and security areas. As nanotechnology has enabled us to make various nanoscale objects with the dimension from a few nanometers to several hundred nanometers, how to make extremely sensitive and small size detection devices is critical for us to have a better understanding and further control the nanoscale objects.

1.4 Back Reflection Caused by Optically Interacting Nanoparticle

We have investigated theoretically and experimentally a new approach using optical ring and disk micro-resonators as nanoparticle detection. This approach is based on the observation that when a nanoparticle having a large index contrast from the medium surrounding the microresonator is brought within the evanescent field outside the resonator, some of the light circulating in the resonator is reflected and circulates in the opposite direction. When the resonator is tuned to a resonant frequency for the forward propagating light, the backscattered light is also in resonance with the cavity. Therefore, particle-induced coupling between the forward and backward propagating resonant modes results in a buildup of significant intensity in the backward propagating modes. This principle applies to all the forward-propagating resonant modes of the resonator, thus with broadband excitation of the forward modes, many backward propagating modes can be excited. As will be shown below, these backward propagating modes can be detected using a broadband detector.

This provides an intensity-based signature of the presence of the nanoparticle. The basic principle of reflection mode detection is illustrated in Fig. 1.13. In this case, the principle is demonstrated using a conventional integrated optical Add/Drop filter configuration, in which a micro-ring or micro-disk resonator is evanescently coupled to two bus waveguides, yielding a four-port device.

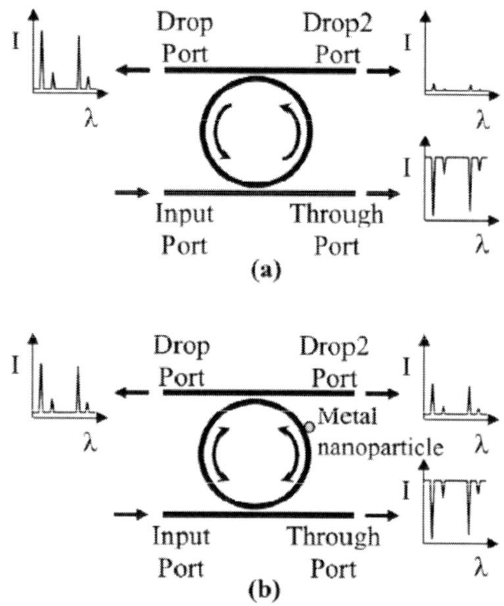

Figure 1.13 (a) Diagram showing port outputs for an Add/Drop device without a scattering nanoparticle present. (b) Diagram showing port outputs for an Add/Drop device with a scattering nanoparticle optically coupled to the resonator.

Figure 1.13a is a schematic diagram of the conventional operation of the device as an Add/Drop filter. In this case, light entering the Input port couples into the resonator and then exits one or both of the Through or Drop ports, depending on the wavelength of the light and the strength of coupling between the resonator and bus waveguides.[9] With appropriate design, at resonant wavelengths of the cavity, the input light is almost entirely routed to the Drop port. Note that the port in the upper right of the diagram, the port that would be the "Add" port in Add/

Drop filter operation, receives essentially none of the light from the input waveguide. This is because the coupling of the input bus to the cavity excites only traveling wave resonances, which propagate in the counter-clockwise direction. Any light exiting the Add port in this configuration is a result of backscattering from fabrication defects and is expected to be very weak and constant over time.

Figure 1.13b shows the proposed operation of the same device as reflection mode detection. With the nanoparticle inducing backscattering, light builds up in the clockwise propagating traveling wave modes. This clockwise propagating light is strongly coupled to the upper right port (and also back into the input port) and has essentially the same frequency spectrum as the light exiting the Drop port. Because of the similarity between the characteristics of the upper right port to the Drop port, we have renamed it the "Drop2" port and call this approach to sensing "Drop2 detection."

Quantitative validation of the feasibility of Drop2 detection was demonstrated via finite different time domain (FDTD) simulations using both Fortran code and commercial software (Photon Design). For convenience, simulations were performed as two-dimensional calculations, using effective indexes appropriate to devices having silicon nitride waveguides with silica cladding. Figure 1.14a shows the result of such a simulation for the case of a 30 micron diameter ring resonator fabricated with waveguides 2.5 micron wide and 400 nm thick, with air upper cladding. The dotted trace shows the Drop2 port spectral intensity (for unit input) in the absence of nanoparticle coupling; the weak peak structure is an artifact related to "roughness" corresponding to the grid size used in the numerical calculation. The solid trace is the Drop2 spectrum when a 60 nm diameter platinum nanoparticle (n = 5.17, k = 7.0 at 155 nm) is placed in contact with the resonator. Note the dramatic increase in Drop2 intensity at the wide range of resonant frequencies. Note also that the solid curve (with a 60 nm platinum nanoparticle) has been vertically shifted by 0.003 to make both curves more clearly visible.

Drop2 detection has been experimentally demonstrated using a variety of resonator designs and nanoparticle types. In one experiment, an Add/Drop type resonator as shown schematically in Fig. 1.14b was employed. The device was realized using silicon nitride waveguides with silica lower cladding, fabricated on a silicon wafer using plasma-enhanced chemical vapor deposition, contact lithography and reactive ion etching. Typical devices consisted of a 30 micron diameter ring, based on a silicon nitride channel waveguide 400 nm thick and either 1, 1.6, or 2.5 micron wide. Rings were vertically coupled to two buried silicon nitride ridge waveguides, 250 nm thick and 1, 1.2, 1.4, or 1.6 micron wide, with a ridge height of 150 nm. The coupling gap between the top of the bus waveguide and the bottom of the ring waveguide was 100 nm. Low insertion loss in the waveguides was achieved by thermal reflow of the patterned photo-resist before etching. An optical micrograph of a typical device is shown in Fig. 1.14b.

In order to facilitate controlled study of the interaction of the resonator with a nanoparticle scattering center, the nanoparticle was simulated by the probe tip of an AFM. This allowed the "particle" to be repeatedly placed in contact with the resonator and removed, as well as to be scanned across different regions of the resonator (Fig. 1.14c). By using tips made of different materials, or coated with different materials, nanoparticles of different composition can be simulated. Because the spatial extent of the evanescent field outside the resonator is roughly 100 nm, it is expected that for tips with radii near 100 nm, the response of the resonator to the AFM tip will be a reasonable approximation to the response of the device to a particle of similar radius. Figure 1.14d shows a trace of the power recorded from the Drop2 port versus time, while the AFM tip slowly scanned a small region (approximately 1 micron scan length) on the periphery of the ring. We observed increases in Drop2 power as high as 3.5 dB. Note that the fiber-to-fiber received power of a typical straight waveguide in this configuration was approximately −20 dBm, so that the total additional optical loss seen at the Drop2 port is approximately 40 dB. Note also that this device had a relatively-low quality factor Q of approximately 300. Our initial results seem to suggest that better Drop2 sensitivity (larger change in output power due to a nanoparticle) results from devices with lower Q, possibly because with broadband excitation this results in a higher amount of total optical power circulating within the device.

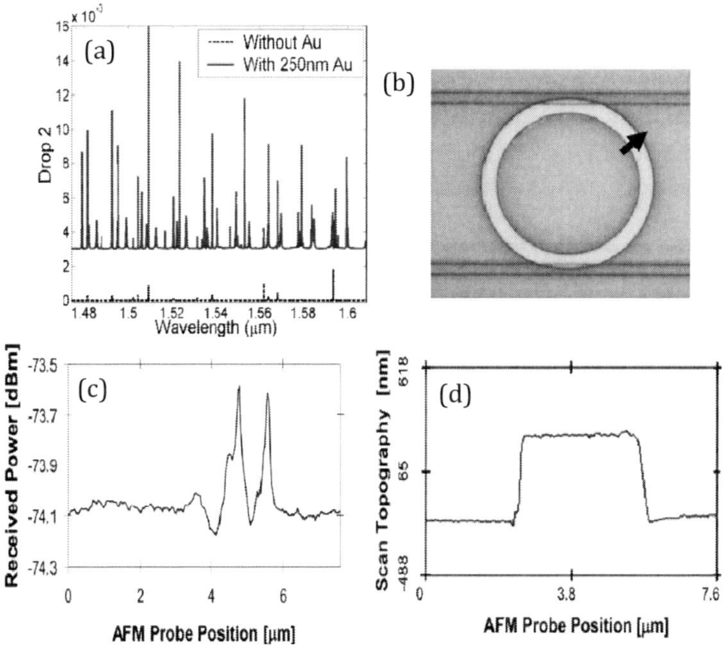

Figure 1.14 (a) Simulated Drop2 intensity-relative to input excitation-with (solid curve) and without (dotted curve) a 250 nm gold nanoparticle present. Note that the solid curve (with a gold nanoparticle) has been vertically shifted by 0.003 for clarity. (b) Optical micrograph of a typical Add/Drop device. (c) Measured Drop2 power while scanning a 250 nm diameter gold probe tip over the ring waveguide. (*Note*: Black arrow in (a) shows position and trajectory of scan for (b); total scan time was 120 sec.) (d) Topography recorded by the AFM during the scan. (*Note*: The *x* axes for (b) and (c) are identical for location of power within the waveguide.)

1.5 Summary

The works on hybrid and coupled photonics with experimental demonstration at the nanometer scale represent the future scientific trend in these forefront areas. Utilizing the current nanophotonics technology (design, simulation, fabrication and measurement techniques), it is possible to study these novel coupled photonic nanomaterials and devices in ways that were unimaginable a decade or two ago.

Acknowledgements

We thank the support from the University of Michigan, Microsystems Technology Laboratory, and Center for Materials Science and Engineering at the MIT for the access to the facility and measurement equipment, and 3M central research lab.

References

1. H. Pothier, et al., *Physica B*, **169**, 573 (1991).
2. K. K. Likharev, *Proc. IEEE*, **87**, 606 (1999).
3. D. L. Klein, R. Roth, A. K. L. Lim, A. P. Alivisatos, and P. L. McEuen, *Nature*, **389**, 699 (1997).
4. T. Dietl, D. D. Awschalom, M. Kaminska, and H. Ohno (eds.) *Spintronics, Semiconductors and Semimetals*, **82**, Elsevier (2008).
5. P. N. Prasad, *Nanophotonics*, Wiley (2004).
6. M. A. Nielsen, and I. L. Chuang, *Quantum Computation and Quantum Information*, Cambridge University Press (2000).
7. T. J. Huang, *MRS Bull.*, **33**, 226 (2008).
8. D. Psaltis, S. R. Quake, and C. Yang, *Nature*, **442**, 381 (2006).
9. T. J. Kippenberg, and K. J. Vahala, *Opt. Express*, **15**, 17172 (2007).
10. R. Folman, P. Krueger, J. Schmiedmayer, J. Denschlag, and C. Henkel, C., *Adv. At. Mol. Opt. Phys.*, **48**, 263 (2002).
11. A. V. Akimov, et al., *Nature*, **450**, 402 (2007).
12. J. I. Cirac and P. A. Zoller, *Nature*, **404**, 579 (2000).
13. A. Faraon, et al., *Opt. Express*, **16**, 12154 (2008).
14. W. D. Li, S. Y. Chou, *Opt. Express*, **18**, 931 (2010).
15. B. E. Little, S. T. Chu, and H. A. Haus, *Opt. Lett.*, **23**, 1570 (1998).
16. S. V. Boriskina, *J. Opt. Soc. Am. B*, **23**, 1565 (2006).
17. A. Yariv, Y. Xu, R. K. Lee, and A. Scherer, *Opt. Lett.*, **24**, 711–713 (1999).
18. Q. Song, H. Cao, S. T. Ho, and G. S. Solomon, *Appl. Phys. Lett.*, **94**, 061109 (2009).
19. A. Francois and M. Himmelhausa, *Appl. Phys. Lett.*, **92**, 141107 (2008).
20. M. L. Gorodetsky, A. D. Pryamikov, and V. S. Ilchenko, *J. Opt. Soc. Am. B*, **17**, 1051 (2000).

21. Z. Yuan, B. E. Kardynal, R. M. Stevenson, A. J. Shields, C. J. Lobo, K. Cooper, N. S. Beattie, D. A. Ritchie, and M. Pepper, *Science*, **295**, 102 (2002).

22. M. Bruchez, M. Moronne, P. Gin, S. Weiss, and A. Paul Alivisatos, *Science*, **281**, 2013 (1998).

23. C. Loo, A. Lin, L. Hirsch, M. Lee, J. Barton, N. Halas, J. West, and R. Drezek, *Technol. Cancer Res. Treat.*, **3**, 33 (2004).

24. R. Wiese, *Luminescence*, **18**, 25 (2003).

25. M. Bayer, T. Gutbrod, J. P. Reithmaier, A. Forchel, T. L. Reinecke, P. A. Knipp, A. A. Dremin, and V. D. Kulakovskii, *Phys. Rev. Lett.*, **81**, 2582 (1998).

26. A. Forchel, M. Bayer, J. P. Reithmaier, T.L. Reinecke, and V.D. Kulakovskii, *Physica E.*, **7**, 616 (2000).

27. K. J. Vahala, Optical microcavities. *Nature*, **424**, 839–846 (2003).

28. A. Mazzei, S. Gotzinger, L. de S. Menezes, G. Zumofen, O. Benson, and V. Sandoghdar, *Phys. Rev. Lett.*, **99**, 173603 (2007).

29. M. Borselli, T. J. Johnson, and O. Painter, *Opt. Express*, **13**, 1516 (2005).

30. F. Vollmer and S. Arnold, *Nat. Methods*, **5**(7), 591–596 (2008).

31. X. Fan, I. M. White, S. I. Shopova, H. Zhu, J. D. Suter, and Y. Sun, *Anal. Chim. Acta*, **620**(1–2), 8–26 (2008).

32. S. Arnold, M. Khoshsima, I. Teraoka, S. Holler, and F. Vollmer, *Opt. Lett.*, **28**(4), 272–274 (2003).

33. A. M. Armani, R. P. Kulkarni, S. E. Fraser, R. C. Flagan, and K. J. Vahala, *Science*, **317**(5839), 783–787 (2007).

34. M. Loncar, *Nat. Photonics*, **1**(10), 565–567 (2007).

35. D. Evanko, *Nat. Methods*, **4**(9), 683 (2007).

36. F. Vollmer, S. Arnold, and D. Keng, *Proc. Natl. Acad. Sci. U.S.A.*, **105**(52), 20701–20704 (2008).

37. S. A. Wise, and R. A. Watters, Bovine serum albumin (7% Solution) (SRM 927d), NIST Gaithersburg, MD (2006).

38. W. E. Moerner, and D. P. Fromm, *Rev. Sci. Instrum.*, **74**(8), 3597–3619 (2003).

39. J. B. Jensen, L. H. Pedersen, P. E. Hoiby, L. B. Nielsen, T. P. Hansen, J. R. Folkenberg, J. Riishede, D. Noordegraaf, K. Nielsen, A. Carlsen, and A. Bjarklev, *Opt. Lett.*, **29**(17), 1974–1976 (2004).

40. A. Polman and H. A. Atwater, *Mater. Today*, January, **56** (2005).

41. B. Koch, Y. Yi, J. Zhang, S. Znameroski, and T. Smith, *Appl. Phys. Lett.*, **95**, 201111 (2009).
42. B. Koch, L. Carson, C. Guo, C. Lee, Y. Yi, J. Zhang, M. Zin, S. Znameroski, and T. Smith, *Sens. Actuators B*, **147**, 573–580 (2010).
43. E. D. Palik, in *Handbook of Optical Constants of Solids* (ed. E. D. Palik), Academic, Orlando, FL (1985).
44. A. Taflove and S. C. Hagness, *Computational Electrodynamics: The Finite Difference Time Domain Method*, Artech House, Inc. (2005).
45. A. Gondarenko, J. S. Levy, and M. Lipson, *Opt. Express* **17**, 11366 (2009).
46. E. S. Hosseini, S. Yegnanarayanan, A. H. Atabaki, M. Soltani, and A. Adibi, *Opt. Express*, **17**, 14543 (2009).
47. T. Barwicz, M. A. Popovic, P. T. Rakich, M. R. Watts, H. A. Haus, E. P. Ippen, and H. I. Smith, *Opt. Express*, **12**, 1437 (2004).

Chapter 2

Coupled-Mode Theory and Its Applications on Computational Nanophotonics

Ya Sha Yi[a,b] **and Jianwei Mu**[b]

[a]*University of Michigan, MI, USA*
[b]*Massachusetts Institute of Technology, MA, USA*

yashayi@umich.edu, yys@alum.mit.edu

Coupled-mode theory (CMT) has been widely utilized in the analysis of electromagnetic wave coupling and conversions due to its mathematical simplicity and physical intuitiveness. The early version of CMT was proposed by Pierce and Miller in 1950s to study the microwaves, and was mathematically formulated by Schelkunoff using mode expansion, and by Haus with variational principles.[1–5] The CMT was later introduced to the investigation of optical waveguides by Marcuse,[6] Snyder,[7] and Kogelnik[8] in the early 1970s. Since then, a series of formulations and applications in optical waveguide has been proposed and studied. The physical model of CMT is expanding the total field inside and optical waveguide in terms of the field of a reference waveguide structure, by applying the orthogonality condition, a set of ordinary differential equations are obtained. Generally, the CMT focused on

Integrated Nanophotonic Resonators: Fundamentals, Devices, and Applications
Edited by Ya Sha Yi

ISBN 978-981-4613-78-1 (Hardcover), 978-981-4613-79-8 (eBook)
www.panstanford.com

the guided modes with the assumption that only a limited number of guided modes (usually one or two) close to phase matching play significant roles in the interaction of the modal fields. In situations of applications involving the radiation mode coupling, the application of CMT becomes cumbersome due to the continuous spectrum of radiation modes. One possible solution to circumvent the problem of radiation modes is to introduce leaky modes to approximate the radiation modes.[9–20] The leaky modes are, however, not orthogonal and normalizable in real domain. For this reason, it is difficult to deal with leaky mode formulations analytically and even more so numerically for practical applications.

Recently, a new computation model was introduced to the mode-matching method in which the waveguide structure is enclosed by a perfectly matched layer (PML) terminated by a perfectly reflecting boundary conditions (PRB).[21–23] This seemingly paradoxical combination of PML and PRB leads to a somewhat unexpected yet remarkable result: It creates an open and reflectionless environment in a close and finite computation domain. A set of complex modes can be derived from this waveguide model that are well behaved in terms of orthogonality and normalization and can be readily solved by standard analytical and numerical techniques. By utilizing the complex modes as an orthogonal base functions to represent the radiation fields, the CMT can be applied as if all the modes are discrete and guided. The complex CMT was subsequently applied to simulation and analysis of slab, circular, and channel optical waveguide structures and shown to be highly accurate and versatile.

2.1 Modal Analysis with Perfectly Matching Layers

We make the following assumptions: The medium in the waveguide structure is lossless, linear, and isotropic. The permittivity and permeability of vacuum are denoted as ε_0 and μ_0, respectively. The permeability μ in the medium is equal to the free space value μ_0 throughout this chapter. The time dependency is expressed as $\exp(j\omega t)$. The wave is propagating along z, and the z dependency is expressed as $\exp(-j\gamma z)$, which refers to the propagation in the positive z direction, or $\exp(-j\gamma z)$ in the negative z direction.

ω and γ are the angular frequency and the propagation constant, respectively.

Considering a waveguide structure where the transverse index profile $n(x, y)$ is arbitrary and defined in the Cartesian coordinate system, if we enclose the optical waveguide by an anisotropic perfectly matching layers (PML) terminated by a perfectly reflection boundary, the Maxwell's equations can be written as

$$\Delta \times E = -j\omega\mu_0[\Lambda]H \tag{2.1}$$

$$\Delta \times H = -j\omega\varepsilon_0 n^2[\Lambda]E \tag{2.2}$$

The tensor $[\Lambda]$ accounts for the PML and is given by

$$[\Lambda] = \begin{bmatrix} S_y/S_x & 0 & 0 \\ 0 & S_x/S_y & 0 \\ 0 & 0 & S_x S_y \end{bmatrix} \tag{2.3}$$

where S_x and S_y are called the coordinate stretching factors and given by

$$S_x = \kappa_x - j\frac{\sigma_x}{\omega\varepsilon} \tag{2.4}$$

$$S_y = \kappa_y - j\frac{\sigma_y}{\omega\varepsilon} \tag{2.5}$$

and $\kappa_x(\kappa_y)$ and $\sigma_x(\sigma_y)$ are the parameters to control the phase shift and absorption of the travelling waves in the PML along $x(y)$. For non-PML media, we have $\kappa_x = \kappa_y = 1$ and $\sigma_x = \sigma_y = 0$.

The vector wave equation for the electric field is derived from Eqs. (2.1) and (2.2)

$$\nabla \times ([\Lambda]^{-1}\nabla \times E) = n^2k^2[\nabla]E, \tag{2.6}$$

where $\kappa = \omega\sqrt{\ \varepsilon_0\mu_0}$ is the wave number in free space. Further, we have the transverse wave equation as

$$\nabla_t \times \left(\frac{1}{s_x sLy}\nabla_t \times E_t\right) + \hat{z} \times ([\Lambda]_t^{-1}\nabla_t \times (-j\gamma E_z\hat{z})) - \gamma^2\hat{z} \times ([\Lambda]_t^{-1}\hat{z} \times E_t) = n^2k^2[\Lambda]_t E_t, \tag{2.7}$$

where

$$[\Lambda]_t = \begin{pmatrix} s_y/s_x & 0 \\ 0 & s_x/s_y \end{pmatrix}. \tag{2.8}$$

Using $\nabla\cdot(\varepsilon_0 n^2[\Lambda]E) = 0$, we have

$$E_z = (\nabla_t\cdot(\varepsilon_0 n^2[\Lambda]_t E_t))/\, j\gamma n^2 s_x s_y. \tag{2.9}$$

One can derive the vectorial wave equation for the transverse electric field,

$$\nabla_t\times\left(\frac{1}{s_x s_y}\nabla_t\times E_t\right) - \hat{z}\times\left\{[\Lambda]_t^{-1}\nabla_t\times\left(\frac{\nabla_t\cdot(n^2[\Lambda]_t E_t)}{n^2 s_x s_y}\hat{z}\right)\right\} - \gamma^2\hat{z}\times\{[\Lambda]_t^{-1}\hat{z}\times E_t\} = n^2k^2[\Lambda]_t E_t, \tag{2.10}$$

Equation (2.10) can be written in matrix form

$$\begin{bmatrix} P_{xx} & P_{xy} \\ P_{yx} & P_{yy} \end{bmatrix}\begin{bmatrix} E_x \\ E_y \end{bmatrix} = \gamma^2\begin{bmatrix} E_x \\ E_y \end{bmatrix} \tag{2.11}$$

where the differential operators are defined as

$$P_{xx}E_x = \frac{\partial}{\partial x}\left[\frac{1}{n^2}\frac{1}{s_x}\frac{\partial}{\partial x}\left(n^2\frac{1}{s_x}E_x\right)\right] + \frac{s_x}{s_y}\frac{\partial}{\partial y}\left[\frac{1}{s_y}\frac{\partial}{\partial y}\left(\frac{1}{s_x}E_x\right)\right] + n^2k^2E_x \tag{2.12}$$

$$P_{xy}E_y = \frac{s_y}{s_x}\frac{\partial}{\partial x}\left[\frac{1}{n^2}\frac{1}{s_y}\frac{\partial}{\partial y}\left(n^2\frac{1}{s_y}E_y\right)\right] - \frac{\partial}{\partial y}\left[\frac{1}{s_x}\frac{\partial}{\partial x}\left(\frac{1}{s_y}E_y\right)\right] \tag{2.13}$$

$$P_{yy}E_y = \frac{\partial}{\partial y}\left[\frac{1}{n^2}\frac{1}{s_y}\frac{\partial}{\partial y}\left(n^2\frac{1}{s_y}E_y\right)\right] + \frac{s_y}{s_x}\frac{\partial}{\partial x}\left[\frac{1}{s_x}\frac{\partial}{\partial x}\left(\frac{1}{s_y}E_y\right)\right] + n^2k^2E_y \tag{2.14}$$

$$P_{yx}E_x = \frac{s_x}{s_y}\frac{\partial}{\partial y}\left[\frac{1}{n^2}\frac{1}{s_x}\frac{\partial}{\partial x}\left(n^2\frac{1}{s_x}E_x\right)\right] - \frac{\partial}{\partial x}\left[\frac{1}{s_y}\frac{\partial}{\partial y}\left(\frac{1}{s_x}E_x\right)\right] \tag{2.15}$$

Equation (2.11) is a full vectorial equation. All the vectorial properties of the electromagnetic field are included. $P_{xx} \neq P_{yy}$ causes the polarization dependence, whereas $P_{xx} \neq 0$ and $P_{yy} \neq 0$ induces the polarization coupling between the two components E_x and E_y. If the coupling between the two polarization is weak and negligible, by neglecting the cross terms P_{xy} and P_{yx}, the full vectorial (2.11) reduces to two decoupled equations:

$$P_{xx}E_x = \gamma^2 E_x \tag{2.16}$$

$$P_{yy}E_y = \gamma^2 E_y \tag{2.17}$$

There is only one transverse field component in Eqs. (2.16) and (2.17), and the eigenmodes associated with those equations are commonly named as quasi-TE modes and quasi-TM modes. The equations are referred to as semivectorial vector equations. If the structures are weakly-guiding, even the polarization dependence may be neglected. Equation (2.11) reduces to

$$P_{xx} = P_{yy} = \frac{\partial^2}{\partial x^2} + \frac{\partial^2}{\partial y^2} + n^2k^2. \tag{2.18}$$

Since the polarization difference is negligible, we call this scalar approximation. Equations (2.16) and (2.17) are then replaced with a single equation:

$$P\Psi = \gamma^2\Psi \tag{2.19}$$

If we have the transverse electric field components obtained from Eq. (2.11), we can find the transverse magnetic transverse field components by

$$H_x = -Y_0 N_{\text{eff}} \frac{s_x}{s_y} E_y + \frac{Y_0}{N_{\text{eff}} k^2} \frac{s_x}{s_y} \left\{ \frac{\partial}{\partial y} \left[\frac{1}{n^2} \frac{1}{s_y} \frac{\partial}{\partial y} \left(n^2 \frac{1}{s_y} E_y \right) \right] + \frac{\partial}{\partial y} \left[\frac{1}{n^2} \frac{1}{s_x} \frac{\partial}{\partial x} \left(n^2 \frac{1}{s_x} E_x \right) \right] \right\} \tag{2.20}$$

$$H_y = +Y_0 N_{\text{eff}} \frac{s_y}{s_x} E_x - \frac{Y_0}{N_{\text{eff}} k^2} \frac{s_y}{s_x} \left\{ \frac{\partial}{\partial x} \left[\frac{1}{n^2} \frac{1}{s_x} \frac{\partial}{\partial x} \left(n^2 \frac{1}{s_x} E_x \right) \right] + \frac{\partial}{\partial x} \left[\frac{1}{n^2} \frac{1}{s_y} \frac{\partial}{\partial y} \left(n^2 \frac{1}{s_y} E_y \right) \right] \right\}, \tag{2.21}$$

where $Y_0 = \sqrt{\varepsilon_0/\mu_0}$ is the admittance of light in vacuum, and $N_{\text{eff}} = \gamma/\kappa_0$ is the effective modal index.

For waveguides with the weakly polarization coupling

$$\frac{\partial}{\partial y}\left[\frac{1}{n^2}\frac{1}{s_x}\frac{\partial}{\partial x}\left(n^2\frac{1}{s_x}E_x\right)\right]\approx\frac{\partial}{\partial x}\left[\frac{1}{n^2}\frac{1}{s_y}\frac{\partial}{\partial y}\left(n^2\frac{1}{s_y}E_y\right)\right]\approx 0,$$

the semivector solutions are given by

$$H_x = -Y_0 N_{\text{eff}}\frac{s_x}{s_y}E_y + \frac{Y_0}{N_{\text{eff}}k^2}\frac{s_x}{s_y}\frac{\partial}{\partial y}\left[\frac{1}{n^2}\frac{1}{s_y}\frac{\partial}{\partial y}\left(n^2\frac{1}{s_y}E_y\right)\right] \tag{2.22}$$

$$H_y = +Y_0 N_{\text{eff}}\frac{s_y}{s_x}E_x - \frac{Y_0}{N_{\text{eff}}k^2}\frac{s_y}{s_x}\frac{\partial}{\partial x}\left[\frac{1}{n^2}\frac{1}{s_x}\frac{\partial}{\partial x}\left(n^2\frac{1}{s_x}E_x\right)\right] \tag{2.23}$$

For very weakly guided waveguide structures with $\frac{\partial n^2}{\partial x}\approx\frac{\partial n^2}{\partial y}\approx 0$, we have the scalar approximation:

$$\tilde{H} = \pm N_{\text{eff}}Y_0\tilde{E}, \tag{2.24}$$

where $\tilde{H}$ stands for (H_x, H_y), and $\tilde{E}$ stands for $\left(\frac{s_x}{s_y}E_y, \frac{s_y}{s_x}E_x\right)$.

For general Media, the time average power for each guided mode is real and finite and can be normalized as

$$\frac{1}{4}\iint_{\text{Entire region}}(E_{tn}\times H_{tn}^{*} + E_{tn}^{*}\times H_{tn})\cdot\hat{z}dA = 1, \tag{2.25}$$

where t denotes transverse components, and n denotes the mode number. The expression reduces to

$$\frac{1}{2}\iint_{\text{Entire region}}(E_{tn}\times H_{tn})\cdot\hat{z}dA = 1 \tag{2.26}$$

for lossless media. However, it is noted for general media

$$\frac{1}{2}\iint_{\text{Entire region}}(E_{tn}\times H_{tn})\cdot\hat{z}dA = N_n, \tag{2.27}$$

where N_n may not be equal to unity and may even be complex!

Let m and n denote the modal indices of two distinct guided modes with corresponding propagation constants represented by β_m and β_n, the orthogonal relation is given by

$$\iint_{\text{Entire region}} (E_{tn} \times H_{tm}) \cdot \hat{z} da = 0. \tag{2.28}$$

This orthogonality relation is valid even for active or lossy media. In particular, if the media are passive and lossless, the orthogonality relation maybe rewritten as

$$\iint_{\text{Entire region}} (E_{tm}^{*} \times H_{tn}) \cdot \hat{z} da = 0. \tag{2.29}$$

In practice, we normally set the phase-shift parameter κ_x and κ_y to unity since they are significant only for dealing with evanescent fields. On the other hand, the profiles of the absorption coefficients σ_x and σ_y are critical for effectively reducing the reflections from the perfectly reflecting boundary. A commonly used expression for the absorption profile is

$$\sigma = \sigma_{\max} \left(\frac{\rho}{T_{\text{PML}}} \right)^{m}, \tag{2.30}$$

where T_{PML} is the thickness of the PML layer and ρ is the distance measured from the starting position of the PML. A good measure for the effectiveness of the PML is the reflection coefficient defined by[61]

$$R_{\text{PML}} = \exp\left\{ -\frac{2\sigma_{\max}}{n\sqrt{\varepsilon_0/\mu_0}} \int_0^{T_{\text{PML}}} \left(\frac{\rho}{T_{\text{PML}}} \right)^{m} d\rho \right\}. \tag{2.31}$$

In terms of the PML reflection coefficient, the PML coefficient can be conveniently expressed as

$$S = \kappa - j\frac{\lambda}{4\pi n T_{\text{PML}}} \left[(m+1) \ln\left(\frac{1}{R_{\text{PML}}} \right) \right] \left(\frac{\rho}{T_{\text{PML}}} \right)^{m}. \tag{2.32}$$

It was shown that when $m = 2$, the PML seems to be the most effective.

2.2 Complex Coupled-Mode Theory Based on Normal Modes

2.2.1 Derivation of Complex Coupled-Mode Equations Based on Normal Modes

To derive the coupled-mode equations based on the complex modes described in Section 2.1, we assume that the permittivity function distribution along the waveguide with perturbations can be expressed as

$$\tilde{\varepsilon}(x,y,z) = \varepsilon(x,y) + \Delta\varepsilon(x,y,z), \tag{2.33}$$

where the index perturbation $\Delta\varepsilon$ is defined as the difference between the index profiles of the practical waveguides under investigation and the reference waveguides for which the complex modes are known. Maxwell's equations for the perturbed waveguides are

$$\nabla \times \mathbf{E}(x,y,z) = -j\omega\mu_0[\Lambda]\mathbf{H}(x,y,z). \tag{2.34}$$

$$\nabla \times \mathbf{H}(x,y,z) = +j\omega\tilde{\varepsilon}(x,y,z)[\Lambda]\mathbf{E}(x,y,z). \tag{2.35}$$

Suppose that the difference between the perturbed and the reference waveguides is sufficiently small so that we can expand the unknown transverse electromagnetic fields of the perturbed waveguides in terms of the transverse modal fields of the reference waveguides, i.e.,

$$\mathbf{E}_t(x,y,z) = \sum_{n=1}[a_n(z) + b_n(z)]\mathbf{e}_{tn}(x,y) \tag{2.36}$$

$$\mathbf{H}_t(x,y,z) = \sum_{n=1}[a_n(z) - b_n(z)]\mathbf{h}_{tn}(x,y) \tag{2.37}$$

The functions $a_n(z)$ and $b_n(z)$ are the mode amplitudes for the forward and backward propagating waves, respectively. The longitudinal fields can be expressed in terms of transverse components as

$$\mathbf{E}_z(x,y,z) = \sum_{n=1}[a_n(z) - b_n(z)]\frac{\varepsilon}{\tilde{\varepsilon}}\mathbf{e}_{2n}(x,y) \tag{2.38}$$

$$\mathbf{H}_z(x,y,z) = \sum_{n=1}[a_n(z)+b_n(z)]\mathbf{h}_{2n}(x,y) \tag{2.39}$$

The next step is to derive the coupled-mode equations governing the mode amplitudes. To do so, we simply substitute Eqs. (2.36)–(2.39) into (2.34) and (2.35). After mathematical manipulations, we derive the following coupled equations:

$$N_m \frac{da_m}{dz} + j\gamma_m a_m = -j\sum_{n=1}\kappa_{mn}a_n - j\sum_{n=1}\chi_{mn}b_n \tag{2.40}$$

$$N_m \frac{db_m}{dz} - j\gamma_m b_m = +j\sum_{n=1}\kappa_{mn}b_n + j\sum_{n=1}\chi_{mn}a_n \tag{2.41}$$

in which the coupling coefficients are given by

$$\kappa_{mn} = \frac{\omega\varepsilon_0}{4}\int_A(\tilde{n}^2 - n^2)\left(\mathbf{e}_{tn}\cdot\mathbf{e}_{tm} - \frac{n^2}{\tilde{n}^2}e_{2n}\cdot e_{2m}\right)da \tag{2.42}$$

$$\chi_{mn} = \frac{\omega\varepsilon_o}{4}\int_A(\tilde{n}^2 - n^2)\left(\mathbf{e}_{tn}\cdot\mathbf{e}_{tm} + \frac{n^2}{\tilde{n}^2}e_{2n}\cdot e_{2m}\right)da, \tag{2.43}$$

where the refractive indices are used to replace the permittivity functions, i.e., $\varepsilon = n^2\varepsilon_0$ and $\tilde{\varepsilon} = \tilde{n}^2\varepsilon_0$. The coefficient N_m is defined as

$$N_m = \frac{1}{2}\int\int(\mathbf{e}_{tm}\times\mathbf{h}_{tm})\cdot\tilde{\mathbf{z}}da. \tag{2.44}$$

Under the normalization condition, $N_1 = 1$ for the real core guided modes and $N_m \approx 1$ for the complex cladding leaky modes with relatively small mode losses. In general, however, N_m may not be equal to unity and can be complex.

The coupled-mode equations (2.40) and (2.41) and the expressions for the coupling coefficients (2.42) and (2.43) are formally identical to those derived for guided modes in waveguides made of reciprocal media in presence of loss and/or gain and hence can be solved by the same analytical and numerical techniques as previously done in literature. In the classical CMT, to deal with radiation fields based on the integral of the continuous radiation modes is possible but extremely cumbersome. On the other hand, however, both guided and radiation fields are considered in the new complex CMT in a unified fashion. Also, it is noted that the coupling coefficients for the co- and contra-coupling modes are symmetrical in the sense that

$$\kappa_{mn} = \kappa_{nm} \tag{2.45}$$

$$\chi_{mn} = \chi_{nm} \tag{2.46}$$

Further, we may rewrite the mode amplitudes by separating the slowly varying envelopes with the fast oscillating carriers so as to

$$a_n = A_n \exp(-j\gamma_n z) \tag{2.47}$$

$$b_n = B_n \exp(+j\gamma_n z) \tag{2.48}$$

On substitution of Eq. (2.19) by Eq. (2.23), we derive

$$N_m \frac{dA_m}{dz} = -j\sum_{n=1} \kappa_{mn} A_n \exp[-j(\gamma_n - \gamma_m)z] - j\sum_{n=1} \chi_{mn} B_n \exp[+j(\gamma_n + \gamma_m)z] \tag{2.49}$$

$$N_m \frac{dB_m}{dz} = +j\sum_{n=1} \kappa_{mn} B_n \exp[+j(\gamma_n - \gamma_m)z] + j\sum_{n=1} \chi_{mn} A_n \exp[-j(\gamma_n + \gamma_m)z] \tag{2.50}$$

For grating structures whose index perturbations are periodic along the waveguide axis, the coupling coefficients Eqs. (2.42) and (2.43) are also periodic functions with the same period. We may expand the coupling coefficients in terms of Fourier series as

$$\kappa_{mn} = \sum_{l=-\infty}^{+\infty} D_{mn}^{(l)} \exp\left(jl\frac{2\pi}{\Lambda}z\right) \tag{2.51}$$

$$\chi_{mn} = \sum_{l=-\infty}^{+\infty} C_{mn}^{(l)} \exp\left(jl\frac{2\pi}{\Lambda}z\right) \tag{2.52}$$

Substituting Eqs. (2.51) and (2.52) into Eqs. (2.49) and (2.50), we derive

$$\begin{aligned} N_m \frac{dA_m}{dz} = &-j\sum_{n=1} A_n \sum_{l=-\infty}^{+\infty} D_{mn}^{(l)} \exp\left[-j\left(\gamma_n - \gamma_m - l\frac{2\pi}{\Lambda}\right)z\right] \\ &-j\sum_{n=1} B_n \sum_{l=-\infty}^{+\infty} C_{mn}^{(l)} \exp\left[+j\left(\gamma_n + \gamma_m + l\frac{2\pi}{\Lambda}\right)z\right] \end{aligned} \tag{2.53}$$

$$N_m \frac{dB_m}{dz} = +j \sum_{n=1} B_n \sum_{l=-\infty}^{+\infty} D_{mn}^{(l)} \exp\left[+j\left(\gamma_n - \gamma_m + l\frac{2\pi}{\Lambda}\right)z\right]$$
$$+j \sum_{n=1} A_n \sum_{l=-\infty}^{+\infty} C_{mn}^{(l)} \exp\left[-j\left(\gamma_n + \gamma_m - l\frac{2\pi}{\Lambda}\right)z\right]. \tag{2.54}$$

It can be shown that the phase factors $\gamma_n - \gamma_m \pm l2\pi/\Lambda$ and $\gamma_n + \gamma_m \pm l2\pi/\Lambda$ in the exponential terms in Eqs. (2.53) and (2.54) are most critical in determining the strength of interactions between the different modes over distance. Only when these factors are close to zero there will be appreciable power exchange between the coupled modes, a condition referred to as the phase-matching conditions. In fact, the role of grating is to facilitate the phase matching between two propagation modes with different propagation constants by providing a grating space harmonic component related to the grating period and profile. Note that the phase matching conditions for the co- and contra-propagating modes are quite distinct and may not be readily realized by the same grating. In practice, we normally design the grating to assist coupling for either contra-propagating (e.g., Bragg gratings) or co-propagating (e.g., long-period gratings) waves only. Under this assumption, the coupled-mode equations are further reduced to two groups as described below.

For the contra-propagation waves, the coupled-mode equations reduce to

$$N_m \frac{dA_m}{dz} = -j \sum_{n=1} B_n \sum_{l=-\infty}^{+\infty} C_{mn}^{(l)} \exp\left[+j\left(\gamma_n + \gamma_m + l\frac{2\pi}{\Lambda}\right)z\right] \tag{2.55}$$

$$N_m \frac{dB_m}{dz} = +j \sum_{n=1} A_n \sum_{l=-\infty}^{+\infty} C_{mn}^{(l)} \exp\left[-j\left(\gamma_n + \gamma_m - l\frac{2\pi}{\Lambda}\right)z\right] \tag{2.56}$$

and for the co-directional propagation waves, we have

$$N_m \frac{dA_m}{dz} = -j \sum_{n=1} A_n \sum_{l=-\infty}^{+\infty} D_{mn}^{(l)} \exp\left[-j\left(\gamma_n - \gamma_m - l\frac{2\pi}{\Lambda}\right)z\right] \tag{2.57}$$

$$N_m \frac{dB_m}{dz} = +j \sum_{n=1} B_n \sum_{l=-\infty}^{+\infty} D_{mn}^{(l)} \exp\left[+j\left(\gamma_n - \gamma_m + l\frac{2\pi}{\Lambda}\right)z\right] \tag{2.58}$$

Further, for a given grating, the phase matching condition is a function of wavelength through wavelength dependence of the mode propagation constants, i.e., $\gamma_n(\lambda)$ and $\gamma_m(\lambda)$. It also depends on the mode index (m,n = 1,2,...) and the order of space harmonics in the Fourier expansion. Any combination of these parameters that leads to a phase matching condition will likely yield a distinct resonant signature in the mode coupling as illustrated later in the transmission and reflection spectra in this chapter. By identifying these phase matching conditions in the coupled-mode equations, we can greatly simplify the solutions and also gain great insight into the salient features underlying the interaction of the modes in presence of the gratings.

First of all, we may consider only the largest Fourier expansion coefficient, i.e., $l = \pm 1$ or the first-order grating effect and ignore all these other high-order space harmonics. Investigation of the higher-order gratings can be performed in the similar fashion, but will not be pursued further in this work. The coupled-mode Eqs. (2.55)–(2.58) are subsequently decoupled into two separate sets such that

$$N_m \frac{dA_m}{dz} = -j \sum_{n=1} B_n C_{mn}^{(-1)} \exp\left[+j\left(\gamma_n + \gamma_m - \frac{2\pi}{\Lambda} \right) z \right] \tag{2.59}$$

$$N_m \frac{dB_m}{dz} = +j \sum_{n=1} A_n C_{mn}^{(+1)} \exp\left[-j\left(\gamma_n + \gamma_m - \frac{2\pi}{\Lambda} \right) z \right] \tag{2.60}$$

for the contra-directional propagation waves and

$$N_m \frac{dA_m}{dz} = -j \sum_{n=1} A_n D_{mn}^{(-1)} \exp\left[-j\left(\gamma_n - \gamma_m + \frac{2\pi}{\Lambda} \right) z \right] \tag{2.61}$$

$$N_m \frac{dB_m}{dz} = +j \sum_{n=1} B_n D_{mn}^{(+1)} \exp\left[+j\left(\gamma_n - \gamma_m + \frac{2\pi}{\Lambda} \right) z \right] \tag{2.62}$$

for the co-directional propagation waves.

In many practical situations, the phase matching conditions can only be realized at a distinct wavelength (i.e., λ_{mn}) for a given pair of modes (i.e., m and n). For the sake of simplicity, we suppose that the m-th mode is the forward propagating fundamental guided mode with largest real propagation constant, i.e., $m = 1$ and $\gamma_m = \beta_1$ (real). In the proximity of the phase-matching wavelength $\lambda \cong \lambda_{1n}$, we may consider only the two modes that are close to the phase match (whenever it is possible!) so that Eqs. (2.59)–(2.62) are simplified to

$$N_1 \frac{dA_1}{dz} = -jC_{1n}^{(-1)} B_n \exp\left[+j\left(\gamma_n + \beta_1 - \frac{2\pi}{\Lambda} \right) z \right] \tag{2.63}$$

$$N_n \frac{dB_n}{dz} = +jC_{n1}^{(+1)} A_1 \exp\left[-j\left(\gamma_n + \beta_1 - \frac{2\pi}{\Lambda} \right) z \right] \tag{2.64}$$

for the contra-directional propagation modes, and

$$N_1 \frac{dA_1}{dz} = -jD_{1n}^{(-1)} A_n \exp\left[-j\left(\gamma_n - \beta_1 + \frac{2\pi}{\Lambda} \right) z \right] \tag{2.65}$$

$$N_n \frac{dA_n}{dz} = -jD_{n1}^{(+1)} A_1 \exp\left[+j\left(\gamma_n - \beta_1 + \frac{2\pi}{\Lambda} \right) z \right] \tag{2.66}$$

for the co-directional propagating modes. Note that $\gamma_n = \beta_n - j\alpha_n$ and we define the phase detuning factors such that

$$\Delta\beta_n = \frac{1}{2}\left(\gamma_1 + \gamma_n - \frac{2\pi}{\Lambda} \right) \tag{2.67}$$

for the contra-directional propagation modes and

$$\Delta\beta_n = \frac{1}{2}\left(\gamma_1 - \gamma_n - \frac{2\pi}{\Lambda} \right) \tag{2.68}$$

for the co-directional propagation modes. The phase matching conditions happened as

$$\Re(\Delta\beta_n) = 0 \tag{2.69}$$

so that the grating period for the phase matching conditions for the contra- and co-directional modes are

$$\Lambda = \frac{2\pi}{\Re(\gamma_1 + \gamma_n)} \tag{2.70}$$

and

$$\Lambda = \frac{2\pi}{\Re(\gamma_1 - \gamma_n)} \tag{2.71}$$

We may recast Eqs. (2.65) and (2.66) into more revealing forms as follows:

$$N_1 \frac{dA_1}{dz} = -jC_{1n}^{(-1)} B_n \exp[+j(2\Delta\beta_n)z] \tag{2.72}$$

$$N_n \frac{dB_n}{dz} = +jC_{n1}^{(+1)} A_1 \exp[-j(2\Delta\beta_n)z] \tag{2.73}$$

for the contra-directional modes and

$$N_1 \frac{dA_1}{dz} = -jD_{1n}^{(-1)} A_n \exp[-j(2\Delta\beta_n)z] \tag{2.74}$$

$$N_n \frac{dA_n}{dz} = -jD_{n1}^{(+1)} A_1 \exp[+j(2\Delta\beta_n)z] \tag{2.75}$$

for the co-directional modes.

2.2.2 Solutions of Complex Coupled-Mode Equations

Computationally, we may solve the full coupled-mode Eqs. (2.59)–(2.62) (referred to as the full CMT) or the reduced Eqs. (2.72)–(2.75) (referred to as the reduced CMT). The former can be readily carried out by a standard numerical algorithm such as the Runge–Kutta method, whereas the latter can be solved to yield simple analytical formulas.

Suppose that all the power is initially launched in the forward-propagating fundamental mode, i.e., $a_1(0) = 1$. For the contra-directional modes, we assume that no power is associated with the backward-propagating modes at the other side of the grating, i.e., $b_n(L) = 0$. The analytical solutions are

$$a_1(z) = a_1(0) \frac{\Delta\beta_n \sinh S(z-L) - jS\cosh S(z-L)}{-\Delta\beta_n \sinh SL - jS\cosh SL} e^{-j(\beta_1 - \Delta\beta_n)z} \tag{2.76}$$

$$b_n(z) = -a_1(0)\frac{C_{n1}^{(+1)}}{N_n}\frac{\sinh S(z-L)}{\Delta\beta_n \sinh SL + jS\cosh SL}e^{j(\gamma_n-\Delta\beta_n)z}, \tag{2.77}$$

where

$$S = \sqrt{\kappa_n^2 - (\Delta\beta_n)^2} \tag{2.78}$$

$$\kappa_n = \sqrt{\frac{C_{1n}^{(-1)}C_{n1}^{(+1)}}{N_1 N_n}}. \tag{2.79}$$

For the co-directional modes, we assume that no power is associated with the secondary mode at the starting point of the grating, i.e., $a_n(0) = 0 (n \neq 1)$. The solutions are

$$a_1(z) = a_1(0)\frac{j\Delta\beta\sin(Q_n z) + Q_n\cos(Q_n z)}{Q_n}e^{-j(\beta_1-\Delta\beta_n)z} \tag{2.80}$$

$$a_n(z) = -je^{j\Delta\beta_n z}\frac{D_{n1}^{(+1)}}{N_n}\frac{\sin(Q_n z)}{Q_n}a_1(0)e^{-j(\gamma_1+\Delta\beta_n)z}, \tag{2.81}$$

where

$$Q_n = \sqrt{\chi_n^2 + (\Delta\beta_n)^2} \tag{2.82}$$

$$\chi_n = \sqrt{\frac{D_{n1}^{(+1)}D_{1n}^{(-1)}}{N_n N_1}}. \tag{2.83}$$

Note that the analytical solutions of the reduced coupled-mode equations are formally identical to those derived previously for real modes, except that the effective coupling coefficients κ_n (or γ_n) and the equivalent phase detuning factor $\Delta\beta_n$ may become complex.

Once we obtain the mode amplitudes, we will be able to calculate the guided powers carried by each of the modes. In general, the power flow along the waveguide is given as

$$P(z) = \frac{1}{2}\Re\int_A [\mathbf{E}_t(x,y,z)\times\mathbf{H}_t^*(x,y,z)]\cdot\hat{z}da \tag{2.84}$$

By substitution of the field expansions in terms of the complex modes into Eq. (2.84), we obtain

$$P(z)=\sum_{m=1}\sum_{n=1}M_{mn}[a_m a_n^* - b_m b_n^*]-\sum_{m=1}\sum_{n=1}N_{mn}[a_m b_n^* - b_m a_n^*], \tag{2.85}$$

where the first summation is for the power associated with the co-directional propagating modes in the forward and backward directions, whereas the second summation is related to the power of the contra-directional propagating modes. The cross-power coefficients are defined as

$$M_{mn}=\frac{1}{4}\iint_A(\mathbf{e}_{tm}\times\mathbf{h}_{tn}^*+\mathbf{e}_{tn}^*\times\mathbf{h}_{tm})\cdot\hat{z}da \tag{2.86}$$

$$N_{mn}=\frac{1}{4}\iint_A(\mathbf{e}_{tm}\times\mathbf{h}_{tn}^*-\mathbf{e}_{tn}^*\times\mathbf{h}_{tm})\cdot\hat{z}da. \tag{2.87}$$

Normally, the cross-power associated with the contra-directional modes is negligible in comparison with that with the co-directional modes and hence the second summation in Eq. (2.77) may be ignored so that

$$P(z)\cong\sum_{m=1}\sum_{n=1}M_{mn}[a_m a_n^* - b_m b_n^*]. \tag{2.88}$$

For the complex modes that are significant in the mode coupling process, the transverse fields are almost real and hence these modes are almost power orthogonal so that we may further approximate Eq. (2.88) as

$$P(z)\cong\sum_{n=1}[|a_n|^2-|b_n|^2] \tag{2.89}$$

Assume that the total power is launched into the forward propagating fundamental mode at the input. For the contra-directional mode coupling in Bragg grating, the total power at the input of the grating is therefore

$$P(0)=1-R(\lambda), \tag{2.90}$$

in which the total reflected power is expressed as

$$R(\lambda) \cong \sum_{m=1}\sum_{n=1} M_{mn} b_m(0) b_n^*(0) \tag{2.91}$$

and the mode indices m and n cannot be equal to unity simultaneously. If we further neglect the cross-power associated with the co-directional propagating modes, i.e., $M_{mn} \cong 0$ and $M_{mn} \cong 1$, we have

$$R(\lambda) \cong \sum_{n=1} | b_m(0) |^2. \tag{2.92}$$

Similarly, if assume that at the output of the grating, no backward propagating modes exist. The total power at the output of the grating is

$$P(L) = |a_1(L)|^2 = T(\lambda). \tag{2.93}$$

The power conservation of the waveguide gratings calls for $P(0) = P(L)$, i.e.,

$$T(\lambda) = 1 - R(\lambda). \tag{2.94}$$

2.2.3 Complex Coupled-Mode Equations Based on Local Modes

The coupled-mode equations based on local modes are as follows:

$$\frac{d}{dz} a_m = -j\beta_m a_m - \sum \kappa_{mn} a_n - \sum \chi_{mn} b_n \tag{2.95}$$

$$\frac{d}{dz} b_m = j\beta_m a_m - \sum \kappa_{mn} b_n - \sum \chi_{mn} a_n, \tag{2.96}$$

where

$$\kappa_{mn} = \frac{1}{4N_m} \iint \left(\frac{\partial e_{tn}}{\partial z} \times h_{tm} + e_{tm} \times \frac{\partial h_{tn}}{\partial z} \right) \cdot \hat{z} da \tag{2.97}$$

$$\chi_{mn} = \frac{1}{4N_m} \iint \left(\frac{\partial e_{tn}}{\partial z} \times h_{tm} + e_{tm} \times \frac{\partial h_{tn}}{\partial z} \right) \cdot \hat{z} da \tag{2.98}$$

The coupling coefficients can be further simplified as the approaches used by Marcuse[33]:

$$\kappa_{mn} = \frac{1}{4N_m}\frac{\omega\varepsilon_0}{\beta_n - \beta_m}\iint\left(\frac{\partial n^2}{\partial z}e_{tm}\cdot e_{tn}\right)\cdot\hat{z}da \tag{2.99}$$

$$\chi_{mn} = \frac{1}{4N_m}\frac{\omega\varepsilon_0}{\beta_n + \beta_m}\iint\left(\frac{\partial n^2}{\partial z}e_{tm}\cdot e_{tn}\right)\cdot\hat{z}da. \tag{2.100}$$

2.2.4 Applications of Complex Coupled-Mode Equations in Gratings

To illustrate the salient features of the complex modes theory for grating structures, we use the waveguide structure and grating structure shown in Fig. 2.1.

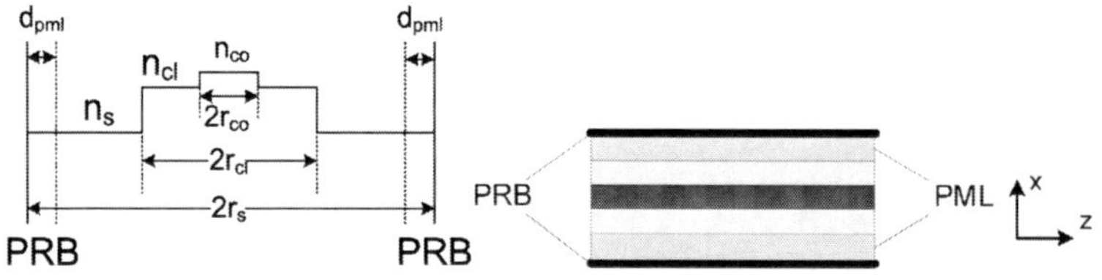

Figure 2.1 Slab waveguide and waveguide grating structure.

The step-index slab waveguide with refractive indices of the core, inner cladding, and outer cladding are denoted by n_{co}, n_{cl}, and n_s, respectively. The waveguide parameters are chosen such that the refractive indices of the core and the inner cladding are $n_{co} = 1.458$ and $n_{cl} = 1.450$, respectively. The half widths of the core and the inner cladding layers are $r_{co} = 2.5$ μm and $r_{cl} = 12.5$ μm.

2.2.4.1 Applications of complex coupled-mode equations in Bragg reflectors

For Bragg reflectors, we assume that $\Delta n_{grating} = 9 \times 10^{-4}$ and $L = 800$ μm and consider the following cases in our simulation.

2.2.4.1.1 Bragg gratings with lower index outer cladding ($n_s < n_{cl}$)

By considering the phase matching condition, we readily identify three distinct wavelengths corresponding to the Bragg conditions between the forward propagating fundamental mode and the first three backward propagating modes, respectively. Note that under this situation all modes are guided with real propagation constants and coupling coefficients so that the conventional CMT does apply. We subsequently calculated the coupling coefficients for these three pairs of mode coupling and show them in Fig. 2.2a. By applying the analytical solutions of the reduced CMT around each of the phase matching wavelengths, we obtained the transmission spectra as indicated by the solid, dotted and dash lines in Fig. 2.2a. It is observed that each of the phase-matching conditions produces a distinct dip in the transmission spectrum with magnitude proportional to the strength of the coupling coefficient $|\kappa_n|$. The entire transmission spectra may be obtained by first calculating

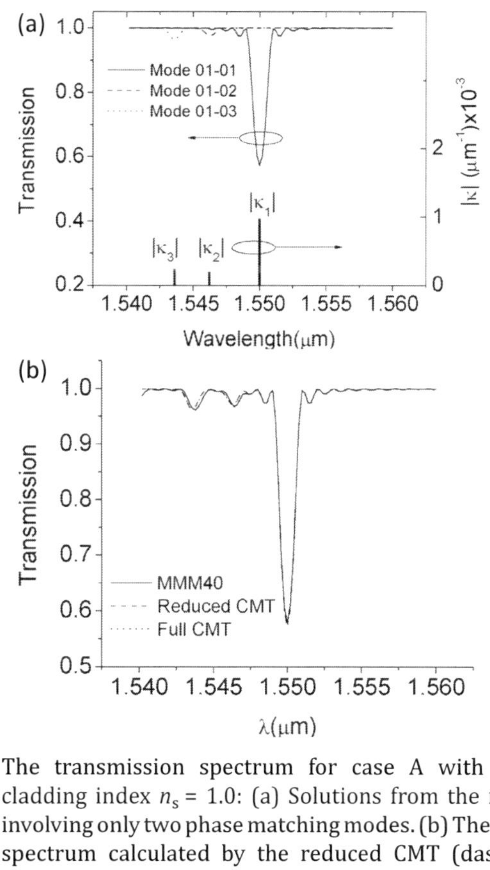

Figure 2.2 The transmission spectrum for case A with lower outer cladding index n_s = 1.0: (a) Solutions from the reduced CMT involving only two phase matching modes. (b) The transmission spectrum calculated by the reduced CMT (dash lines), the full CMT (dotted lines), and the rigorous MMM (solid lines).

the reflection spectra using the reduced CMT near each of the phase matching points. The results are shown in Fig. 2.2b as dash lines. Also shown in the same figure are the results obtained by solving the full coupled-mode equations numerically with total of three modes and also by applying the rigorous mode-matching method (MMM) with total of 40 modes. The results of the reduced CMT, the full CMT and the rigorous MMM are all in good agreement, indicating that the reduced CMT is sufficient.

2.2.4.1.2 Bragg gratings with equal index outer cladding ($n_s = n_{cl}$)

For the infinite cladding gratings ($n_s = n_{cl}$), the real, guided cladding modes evolve to complex quasi-leaky modes as well as PML modes. The latter plays negligible roles in the interactions with the grating-assisted couplings between the forward propagating guided mode and the backward propagating complex modes due to their huge mode losses and small mode overlaps with the guided modes in the core. For the quasi-leaky modes, however, the spectral spacing between them is too small and hence the phase-matching wavelengths are hardly distinguishable relative to the spectral width of each transmission dip as illustrated in Fig. 2.3a. Further, the coupling strengths $|\kappa_n|$ at these phase matching modes are similar.

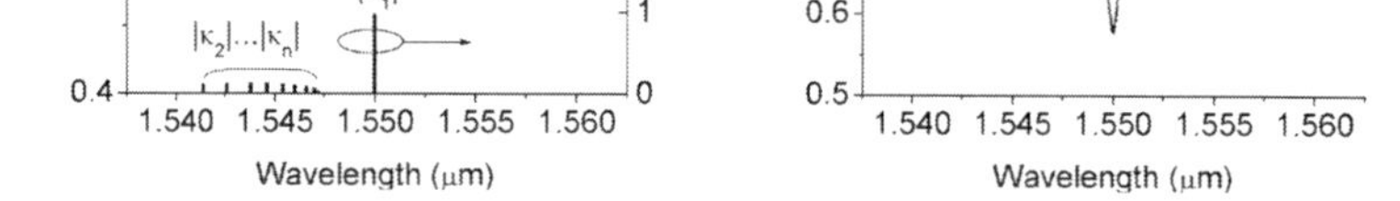

Figure 2.3 The transmission spectrum for case B with equal outer cladding index n_s = 1.450. (a) Phase matching wavelengths, corresponding coupling strengths, and the transmission spectrum predicted by the full CMT involving from 2 up to 11 modes. (b) The transmission spectrum calculated by the full CMT (dotted lines) and the rigorous MMM (solid lines).

The results by solving the full CMT with consideration for coupling from 1 to 10 backward propagating modes are illustrated in Fig. 2.3a. A flat overall drop in the transmission spectrum is predicted by considering all the relevant modes. The accuracy of the complex CMT involving 10 modes is verified by comparison with the results from the rigorous MMM with total of 60 modes as evident in Fig. 2.3b. Shown in the same figure with dotted lines are results obtained using the reduced CMT considering only the phase-matched modes near the phase matching wavelengths.

It is indeed surprising to see that the simple solutions of the reduced CMT yield remarkably accurate results even under the situation in which the phase matching wavelengths are very close to each other. In comparison with the conventional CMT, which has to resort to either continuous radiation modes or large number of box modes or tricky leaky modes, the complex CMT is much more straightforward in dealing with strong radiation fields in this case.

2.2.4.1.3 *Bragg gratings with higher index outer cladding ($n_s > n_{cl}$)*

If the refractive index of the outer cladding is higher than that of the cladding, the waveguide structure becomes leaky in the sense that no guided modes exist for mode index lower than the cladding index. We identify a total of seven (7) phase-matching modes and calculate the transmission spectrum by solving the full CMT considering coupling from 1 to 7 modes, respectively. Figure 2.4b shows the comparison between the full CMT (total of 7 modes), the reduced CMT (two modes near each phase-matching wavelengths) and the rigorous MMM (total of 60 modes), which are in excellent agreement with each other.

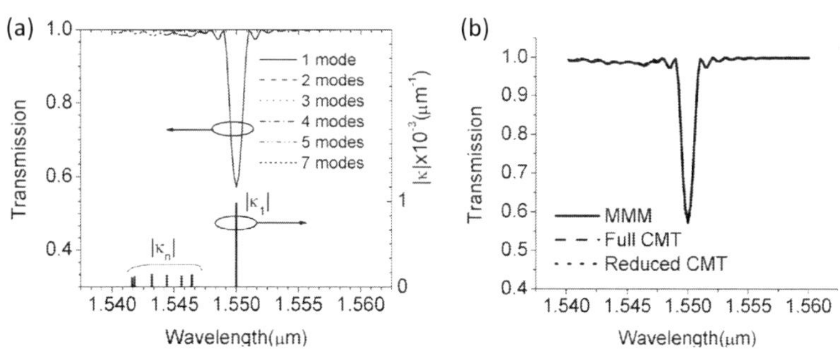

Figure 2.4 Transmission spectra with index of the outer cladding n_s slightly larger than the index of the inner cladding n_{cl} (n_s = 1.455).

2.2.4.2 Applications of complex coupled-mode equations in transmission gratings

We consider a long period grating with the following structure parameters: period Λ = 300 μm, length L = 30 mm, the grating is written in the core with the "dc" index change $\Delta n = 2 \times 10^{-4}$; core radius r_{co} = 2.5 μm, cladding radius r_{cl} = 62.5 μm; the refractive index of the core n_{co} = 1.458 and the refractive index of the inner cladding n_{cl} = 1.45. Long period gratings with three different surrounding materials (n_s) will be analyzed as follows:

2.2.4.2.1 Long period gratings with lower index outer cladding ($n_s < n_{cl}$)

In this case ($n_s < n_{cl}$), the waveguide structure supports a number of guided cladding modes. As a result, the key parameters in the coupled-mode equations, i.e., the coupling coefficients, the normalizing factors, and the detuning factors, are all real. The complex CMT therefore is the same as the conventional CMT. Through identifying the phase matching conditions, the transmission spectra can be easily obtained. We plot the transmission spectra for n_s = 1.0 and n_s = 1.44 in Fig. 2.5. A blue shift of the transmission is observed as the increase of the refractive index of the surrounding material.

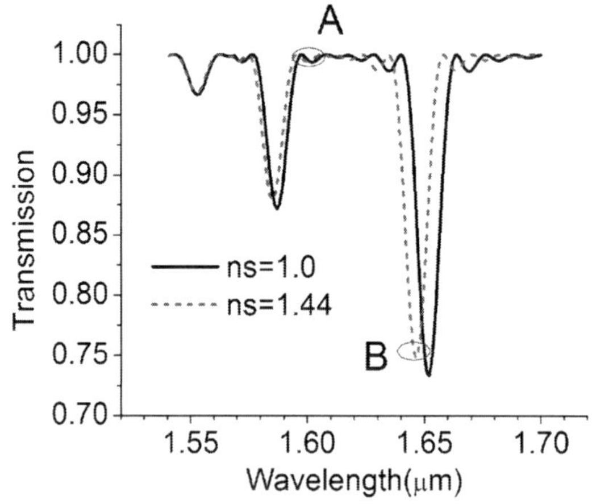

Figure 2.5 Transmission spectra of LPG with low-refractive-index outer cladding.

The power conservation is studied by assessing the power carried by the guided core mode and the cladding mode (which is the most closely phase-matched) when n_s = 1.44. The LPG is investigated at two wavelengths, i.e., 1600 nm (the out of phase point) and 1645 nm (the in phase point) as indicated in Fig. 2.5. The simulation results in Fig. 2.6 show that power exchange between the coupled modes becomes significant only at the phase matched point. Also, the power conservation along the propagation direction is observed.

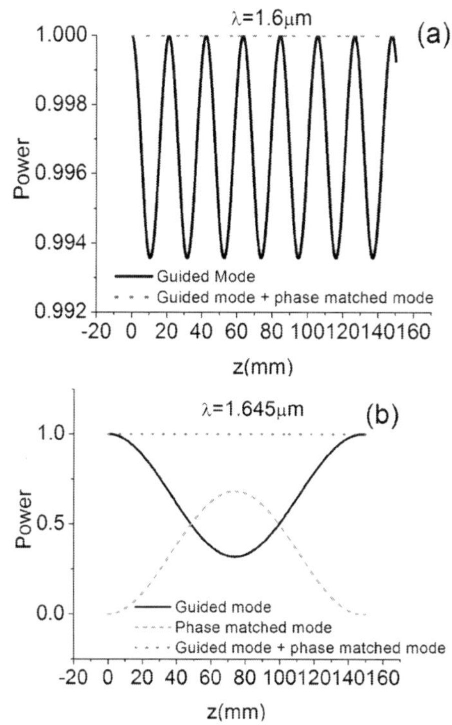

Figure 2.6 Power variation along propagation direction for LPG with lower refractive index of the outer cladding layer (n_s = 1.44): (a) Out of phase point; (b) in-phase point.

2.2.4.2.2 Long period gratings with higher refractive index outer cladding ($n_s > n_{cl}$)

If the surrounding material has a higher refractive index than that of the inner cladding, the cladding modes become complex cladding modes in which the field mainly is confined in the inner cladding while oscillated in the surrounded materials. To disclose the impact of variations of the surrounding materials, we studied the LPG with two different refractive indices of the outer cladding medium. The detuning factors as functions of wavelengths in Fig. 2.7 show that each dip in the transmission spectra corresponds to a specific phase matching point. For this reason, the reduced coupled-mode equation is valid and can be used to predict the performance of the LPG. The transmission spectra shown in Fig. 2.7b indicate that the coupling strength is enhanced with the increase of the refractive index for the outer cladding.

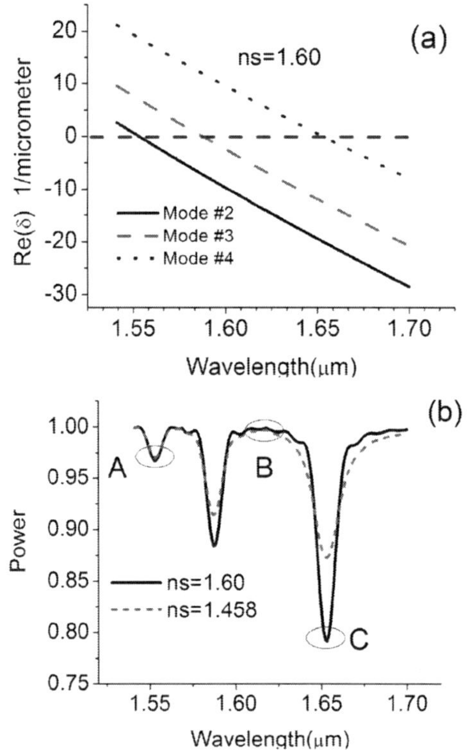

Figure 2.7 Characteristics of LPG with high-refractive-index surrounding media: (a) Detuning factor as a function of wavelengths; (b) transmission spectra.

To further illustrate the salient features of the power exchange and attenuation in the presence of radiation, we examine power carried by the guided and the complex mode along the waveguide at two distinct wavelengths corresponding to the points A and C in Fig. 2.7b. The results are shown in Figs. 2.8a,b, respectively. In both cases, the guided mode is coupled with the phase-matched complex mode with significant power exchange as indicated by the oscillation of the power along the waveguide. On the other hand, the coupling strength at the point A is much stronger than that at the point C, while the leakage loss for the corresponding complex mode is much less (The effective indices of the complex modes at the points A and C are $1.45 - j10^{-7}$ and $1.45 - j10^{-8}$, respectively. The refractive index of the core mode is 1.451). Therefore, the power damping at the point A is smaller than that at the point C. On the other side, the total power carried by the guided mode and the complex mode decays monotonically due to the radiation loss.

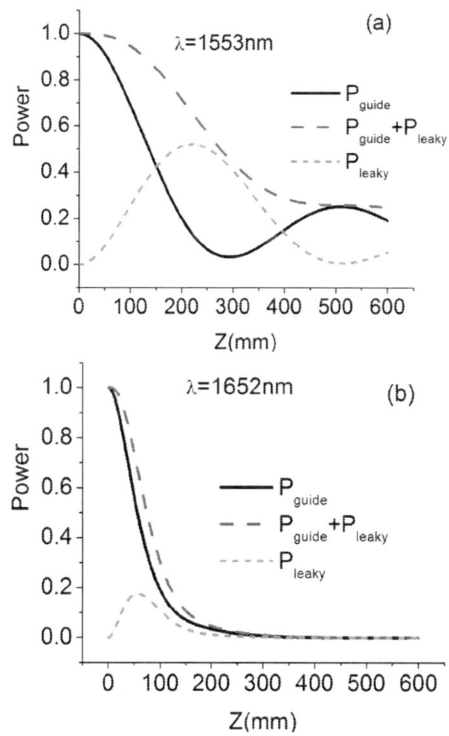

Figure 2.8 Power evolution of guided mode and complex mode as functions of propagation distance in LPG with n_s = 1.60 for phase matched wavelengths: (a) Wavelength equals to 1553 nm; (b) wavelength equals to 1652 nm.

2.2.4.2.3 Long period gratings with infinite claddings ($n_s = n_{cl}$)

Now we consider the long period grating with the refractive index of the surrounding material equals to that of the cladding, e.g., LPG with infinite cladding. This seemingly simple structure has the most challenging complexity comparing to other cases in regard of computation effort for two reasons: On one hand, the cladding modes are all continuous radiation modes, implying that a set of modes will satisfy the phase matching condition; on the other hand, special attention has to be paid on choosing the PML parameters to obtain the accurate modal indices. Large computation window is preferred to reduce the mode spacing to yield a smooth transmission spectrum. In this study, the computation window is set to r_s = 800 μm; the PML reflection is set to 1e-50 with 5.5 μm thickness on both sides. The coupling length defined by $2\pi/(\beta_{co} - \beta_v)$ is plotted in Fig. 2.9. It is shown that the many radiation modes are simultaneously close to the phase matching condition.

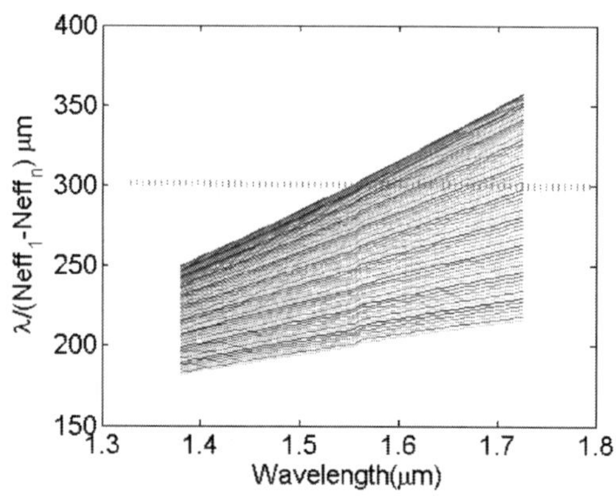

Figure 2.9 Coupling length for LPG with infinite cladding.

Strictly speaking, the reduced coupled-mode equations, which utilize two phase matched modes, are not valid, as those modes close to resonant conditions will contribute to the power exchange and cannot be neglected any more. On the other hand, by noting the fact that the interactions among the radiation modes are negligible, we may apply the approximated analytical solution as follows:

$$T = 1 - \sum_{v}^{N} (1 - |A_{co,v}|^2) \tag{2.101}$$

where $A_{co,v}$ is the amplitude of the core mode coupling to the v-th radiation mode and is obtained through Eqs. (2.75). The convergence of the core mode transmission has been investigated at wavelength equals to 1725 nm. The results in Fig. 2.10a indicate that at least 30 modes are required to obtain a satisfactory precision. The transmission spectra in Fig. 2.10b confirm that the analytical approximation agrees well with the numerical solutions from the full CMT equations.

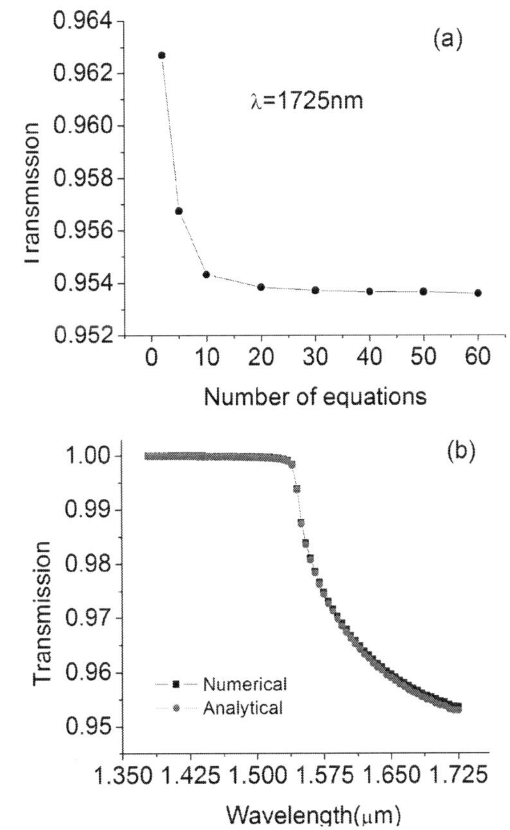

Figure 2.10 Transmission characteristics of LPG with infinite claddings. (a) Convergence of complex coupled equations; (b) transmission spectrum.

2.2.4.3 Applications of complex coupled-mode equations in waveguide taper structures

We consider a linear taper, which has been investigated in the literature.[111] The structure is represented by

$$d(z) = d_{\text{in}} + z\tan\theta \tag{2.102}$$

Here θ is the taper angle.

The structure geometry is shown in Fig. 2.11 and the physical parameters are refractive index of substrate, n_s = 1.515; refractive index of the core, n_{co} = 1.517; and the refractive index of the cladding n_{cl} = 1.0. The width of the input waveguide core d_{in} = 5 μm and the width of the output waveguide core d_{out} = 10 μm. The working wavelength is chosen to be λ = 1.32 μm. Although the ideal taper is continuous, we can use staircase approximation to evaluate the radiation loss. The number of steps (M) and grid size (Δz) along the propagation direction is determined by step height (Δd). The step height (Δd), on the other hand, is decided by the beam resolution.

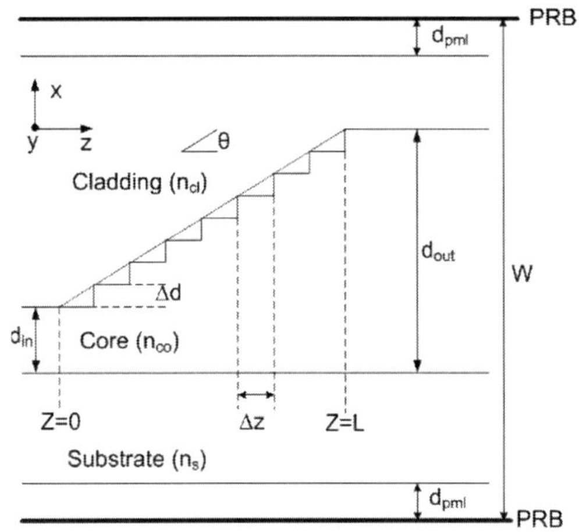

Figure 2.11 Geometry of linear taper waveguide structure.

For a linear taper shown above, the two refractive index profiles are the same everywhere except in the vicinity of the boundary between the cladding and the core. Moreover, for taper waveguide structure with small angles, the backward reflection is negligible.

Based on the analysis, we may simplify the coupling coefficients for TE modes as

$$\kappa_{mn}(z) = \frac{\omega\varepsilon_0}{2N_m} \frac{n_{\text{co}}^2 - n_{\text{el}}^2}{\beta_n - \beta_m} \tan\theta \cdot (e_{tm} \cdot e_{tn})_{x=d} \tag{2.103}$$

The computation parameters are as follows: the thickness of the substrate, d_s = 55 μm; computation window, W = 70 μm; PML thickness is 2.5 μm on both side; and the PML reflection is 1e-4. The power transmission coefficient obtained by solving the coupled equations ((2.84) and (2.85)) will be referred to as full CMT. For weakly coupled waveguide structure in which the mutual coupling among the higher order modes is negligible, we may consider the coupling between the fundamental mode and higher order modes separately, e.g., instead of using Eqs. (2.84) and (2.85), the coupled equations can be expressed as

$$\frac{d}{dz}a_1 = -j\beta_1 a_1 - \kappa_{1n} a_n \tag{2.104}$$

$$\frac{d}{dz}a_n = -j\beta_n a_n - \kappa_{1n} a_n. \tag{2.105}$$

The transmission coefficient is approximated by

$$T = 1 - \sum_n |a_n|^2, \tag{2.106}$$

The transmission obtained through (2.94) will be referred to as reduced CMT. The power transmission of the studied linear taper as a function of taper angle has been shown in Fig. 2.12. Comparing to the benchmark obtained from mode matching method, the results from full CMT overestimate the transmission, on the other hand, results of reduced CMT underestimate the power transmission. For small taper angle, full CMT and reduced CMT are good approximations. However, for large angle, full CMT is more accurate as the mutual coupling among higher modes becomes pronounced.

The convergence of the full CMT with the grid size has been investigated. The relative error of power transmission for different grid size (Δd) has been shown in Fig. 2.13. It is observed that the relative error becomes smaller with the decreasing of mesh size for different taper angles. Further, the relative error decreases with the taper angle.

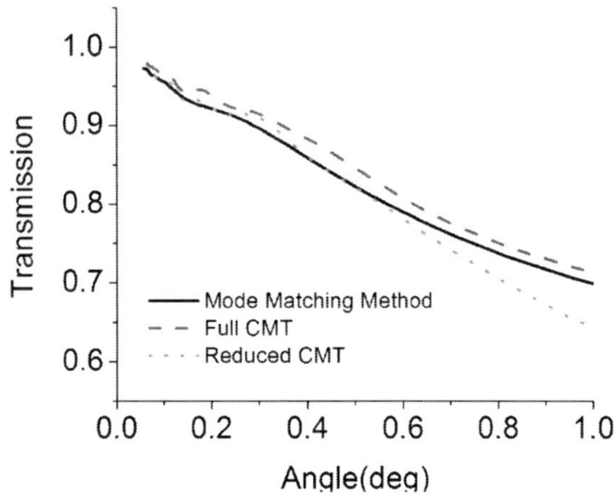

Figure 2.12 Power transmission of a linear taper.

We also studied the convergence of the full CMT with respect to the number of modes being used. It is observed from Fig. 2.14 that the relative error decreases with the number of modes. Similar to the convergence behavior of the mesh sizes, the accuracy is deteriorated with the increase of the taper angle.

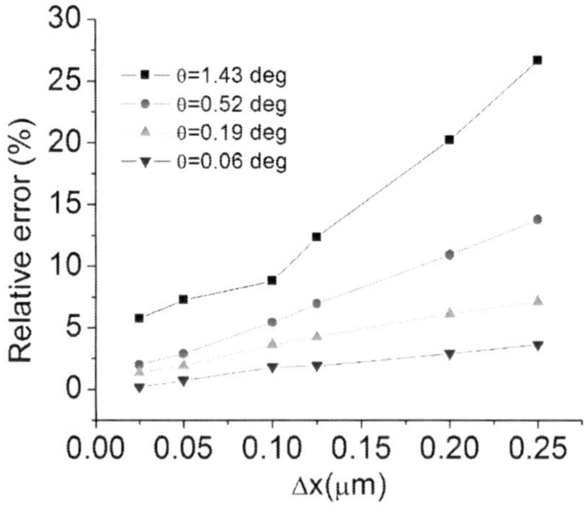

Figure 2.13 Convergence behavior of grid size (Δd).

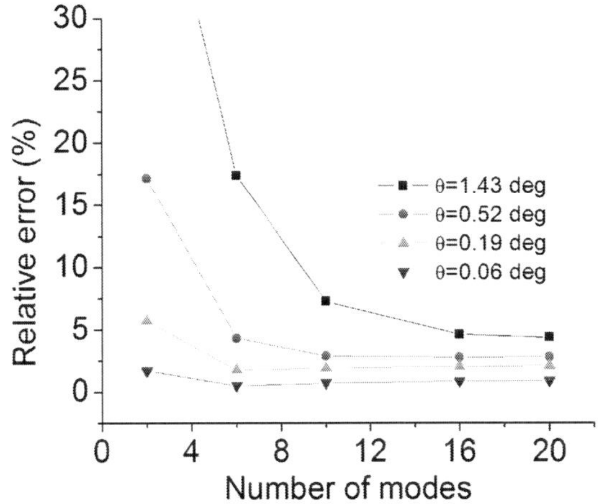

Figure 2.14 Convergence behavior of number of modes.

2.3 Summary

In this chapter, a coupled-mode theory based on complex modes has been proposed and verified through examples of short/long period gratings. Further, we formulated the complex CMT based on local mode and applied it to study the transition loss of the tapered waveguide structure. An analytical solution of radiation loss for long period grating has been derived. The effectiveness as well as the accuracy has been validated through the study of transmissive waveguide grating with different surrounding materials.

References

1. S. E. Miller, Integrated optics: An introduction, *Bell Sys. Tech. J.*, **48**(7), 2059–2069, 1969.
2. I. P. Kaminow, Optical integrated circuits: A personal perspective, *J. Lightw. Technol.*, **26**(9–12), 994–1004, 2008.
3. D. Iazikov, C. Greiner, and T. W. Mossberg, Apodizable integrated filters for coarse WDM and FTTH-type applications, *J. Lightw. Technol.*, **22**(5), 1402–1407, 2004.
4. M. Smit and Y. S. Oei, Photonic integrated circuits for advanced communication networks, *J. Optoelectron.*, **12**(1), 25–30, 1998.

5. J. J. G. M. van der Tol, Y. S. Oei, U. Khalique, R. Notzel, and M. K. Smit, InP-based photonic circuits: Comparison of monolithic integration techniques, *Prog. Quantum Electron.*, **34**(4), 135–172, 2010.

6. Y. Sakamaki, K. Hattori, Y. Nasu, T. Hashimoto, Y. Hashizume, T. Mizuno, T. Goh, and H. Takahashi, One-chip integrated polarisation-multiplexed DQPSK demodulator using silica-based planar lightwave circuit technology, *Electron. Lett.*, **46**(16), 1152–1153, 2010.

7. Y. Sakamaki, H. Yamazaki, T. Mizuno, T. Goh, Y. Nasu, T. Hashimoto, S. Kamei, K. Hattori, H. Takahashi, T. Kobayashi, and M. Ishikawa, Dual polarisation optical hybrid using silica-based planar lightwave circuit technology for digital coherent receiver, *Electron. Letters*, **46**(1), 58–59, 2010.

8. H. Fukuda, K. Yamada, T. Tsuchizawa, T. Watanabe, H. Shinojima, and S. I. Itabashi, Silicon photonic circuit with polarization diversity, *Opt. Express*, **16**(7), 4872–4880, 2008.

9. W. Bogaerts, D. Taillaert, P. Dumon, D. Van Thourhout, and R. Baets, A polarization-diversity wavelength duplexer circuit in silicon-on-insulator photonic wires, *Opt. Express*, **15**(4), 1567–1578, 2007.

10. R. Nagarajan, M. Kato, J. Pleumeekers, P. Evans, D. Lambert, A. Chen, V. Dominic, A. Mathur, P. Chavarkar, M. Missey, A. Dentai, S. Hurtt, J. Baeck, R. Muthiah, S. Murthy, R. Salvatore, S. Grubb, C. Joyner, J. Rossi, R. Schneider, M. Ziari, F. Kish, and D. Welch, Single-chip 40-channel InP transmitter photonic integrated circuit capable of aggregate data rate of 1.6 Tbit/s, *Electron. Lett.*, **42**(13), 771–773, 2006.

11. W. A. Zortman, D. C. Trotter, and M. R. Watts, Silicon photonics manufacturing, *Opt. Express*, **18**(23), 23598–23607, 2010.

12. B. Jalali, S. Yegnanarayanan, T. Yoon, T. Yoshimoto, I. Rendina, and F. Coppinger, Advances in silicon-on-insulator optoelectronics, *IEEE J. Sel. Topics Quantum Electron.*, **4**(6), 938–947, 1998.

13. A. J. Lowery, Computer-aided photonics design, *IEEE Spectr.*, **34**(4), 26–31, 1997.

14. D. Gallagher, Photonic CAD matures, in *IEEE LEOS Newsl.*, pp. 8–14, 2008.

15. R. Soref, The past, present, and future of silicon photonics, *IEEE J. Sel. Topics Quantum Electron.*, **12**(6), 1678–1687, 2006.

16. C. Xu, CAD for photonic devices and circuit, in *Asia Optical Fiber Comm & Optoelectronic Expo* Shanghai, 2006.

17. A. Taflove and S. C. Hagness, *Computational Electrodynamics: The Finite-Difference Time-Domain Method*, Boston: Artech House, 2005.

18. K. S. Yee, Numerical solution of initial boundary value problems involving Maxwell's equations in isotropic media, *IEEE Trans. Antennas Propagat.*, **16**(3), 302–307, 1966.

19. D. Marcuse, Coupled mode theory of round optical fibers, *Bell Sys. Tech. J.*, **52**(6), 817–842, 1973.

20. A. Yariv, Coupled-mode theory for guided-wave optics, *IEEE J. Quantum Electron.*, **9**(9), 919–933, 1973.

21. W. Snyder, Coupled-mode theory for optical fibers, *J. Opt. Soc. Am*, **62**(11), 1267–1277, 1972.

22. S. E. Miller, Coupled-wave theory and waveguide applications, *Bell Sys. Tech. J.*, **33**(3), 661–719, 1954.

23. M. D. Feit and J. A. Fleck, Light-Propagation in Graded-Index Optical Fibers, *Appl. Opt.*, **17**(24), 3990–3998, 1978.

24. J. Gerdes and R. Pregla, Beam-propagation algorithm based on the method of lines, *J. Opt. Soc. Am. B*, **8**(2), 389–394, 1991.

25. G. R. Hadley, Wide-angle beam propagation using Pade approximant operators, *Opt. Lett.*, **17**(20), 1426–1428, 1992.

26. H. Shigesawa and M. Tsuji, Mode propagation through a step discontinuity in dielectric planar wave-guide, *IEEE Trans. Microw. Theory Tech.*, **34**(2), 205–212, 1986.

27. T. E. Rozzi, Rigorous analysis of step discontinuity in a planar dielectric waveguide, *IEEE Trans. Microwave Theory and Techniques*, **26**(10), 738–746, 1978.

28. H. A. Jamid and M. Z. M. Khan, 3-D full-vectorial analysis of strong optical waveguide discontinuities using Pade approximants, *IEEE J. Quantum Electron.*, **43**(3–4), 343–349, 2007.

29. P. L. Ho and Y. Y. Lu, A bidirectional beam propagation method for periodic waveguides, *IEEE Photon. Technol. Lett*, **14**(3), 325–327, 2002.

30. P. L. Ho and Y. Y. Lu, A stable bidirectional propagation method based on scattering operators, *IEEE Photon Technol. Lett*, **13**(12), 1316–1318, 2001.

31. J. Ctyroky, S. Helfert, R. Pregla, P. Bienstman, R. Baets, R. De Ridder, R. Stoffer, G. Klaasse, J. Petracek, P. Lalanne, J. P. Hugonin, and R. M. De La Rue, Bragg waveguide grating as a 1D photonic band gap structure:

COST 268 modelling task, *Opt. Quantum Electron.*, **34**(5), 455–470, 2002.

32. K. Q. Le, Complex Pade approximant operators for wide-angle beam propagation, *Opt. Commun.*, **282**(7), 1252–1254, 2009.
33. D. Marcuse, *Theory of Dielectric Optical Waveguides*. New York: Academic Press, 1974.
34. H. Kogelnik, Theory of dielectric waveguides, in *Integrated Optics* (T. Tamir, Ed). New York: Springer-Verlag, 1975.
35. A. Hardy and W. Streifer, Coupled mode theory of parallel wave-guides, *J. Lightw. Technol.*, **3**(5), 1135–1146, 1985.
36. H. A. Haus, W. P. Huang, S. Kawakami, and N. A. Whitaker, Coupled-mode theory of optical wave-guides, *J. Lightw. Technol.*, **5**(1), 16–23, 1987.
37. H. A. Haus, W. P. Huang, and A. W. Snyder, Coupled-mode formulations, *Opt. Lett.*, **14**(21), 1222–1224, 1989.
38. W. P. Huang and H. A. Haus, Self-consistent vector coupled-mode theory for tapered optical wave-guides, *J. Lightw. Technol.*, **8**(6), 922–926, 1990.
39. H. A. Haus and W. P. Huang, Coupled-mode theory, *Proc. IEEE*, **79**(10), 1505–1518, 1991.
40. S. L. Chuang, A coupled mode formulation by reciprocity and a variational principle, *J. Lightw. Technol.*, **5**(1), 5–15, 1987.
41. A. W. Snyder and J. D. Love, *Optical Waveguide Theory*. Chapman and Hall: Chapman and Hall, 1983.
42. T. Erdogan, Cladding-mode resonances in short- and long-period fiber grating filters, *J. Opt. Soc. Am. A.*, **14**(8), 1760–1773, 1997.
43. T. Erdogan, Fiber grating spectra, *J. Lightw. Technol.*, 15(8), 1277–1294, 1997.
44. R. Sammut and A. W. Snyder, Leaky modes on a dielectric waveguide: Orthogonality and excitation, *Appl. Opt.*, **15**(4), 1040–1044, 1976.
45. S. L. Lee, Y. C. Chung, L. A. Coldren, and N. Dagli, On leaky mode approximations for modal expansion in multilayer open waveguides, *IEEE J. Quantum Electron.*, **31**(10), 1790–1802, 1995.
46. P. Bienstman and R. Baets, Optical modelling of photonic crystals and VCSELs using eigenmode expansion and perfectly matched layers, *Opt Quantum Electron.*, **33**(4–5), 327–341, 2001.
47. P. Bienstman, H. Derudder, R. Baets, F. Olyslager, and D. De Zutter, Analysis of cylindrical waveguide discontinuities using vectorial

eigenmodes and perfectly matched layers, *IEEE Trans. Microw. Theory Tech.*, **49**(2), 349–354, 2001.

48. H. Derudder, D. De Zutter, and F. Olyslager, Analysis of waveguide discontinuities using perfectly matched layers, *Electron. Lett.*, **34**(22), 2138–2140, 1998.

49. H. Derudder, F. Olyslager, D. De Zutter, and S. Van den Berghe, Efficient mode-matching analysis of discontinuities in finite planar substrates using perfectly matched layers, *IEEE Trans. Antennas and Propag.*, **49**(2), 185–195, 2001.

50. K. Jiang and W. P. Huang, Finite-difference-based mode-matching method for 3-D waveguide structures under semivectorial approximation, *J. Lightw. Technol.*, **23**(12), 4239–4248, 2005.

51. Y. Shani, R. Alferness, T. Koch, U. Koren, M. Oron, B. I. Miller, and M. G. Young, Polarization rotation in asymmetric periodic loaded rib wave-guides, *Appl. Phys. Lett.*, **59**(11), 1278–1280, 1991.

52. C. Vassallo, 1993–1995 optical mode solvers, *Opt Quantum Electron.*, **29**(2), 95–114, 1997.

53. Y. P. Chiou, Y. C. Chiang, and H. C. Chang, Improved three-point formulas considering the interface conditions in the finite-difference analysis of step-index optical devices, *J. Lightw. Tech*, **18**(2), 243–251, 2000.

54. M. S. Stern, Semivectorial polarized finite-difference method for optical wave-guides with arbitrary index profiles, *IEE Proc. J. Optoelectron.*, **135**(1), 56–63, 1988.

55. C. L. Xu, W. P. Huang, M. S. Stern, and S. K. Chaudhuri, Full-vectorial mode calculations by finite-difference method, *IEE Proc. J. Optoelectron.*, **141**(5), 281–286, 1994.

56. E. Anemogiannis, E. N. Glytsis, and T. K. Gaylord, Determination of guided and leaky modes in lossless and lossy planar multilayer optical waveguides: Reflection pole method and wavevector density method, *J. Lightw. Technol.*, **17**(5), 929–941, May 1999.

57. J. Chilwell and I. Hodgkinson, Thin-films field-transfer matrix-theory of planar multilayer waveguides and reflection from prism-loaded waveguides, *J. Opt. So Am A*, **1**(7), 742–753, 1984.

58. J. Petracek and K. Singh, Determination of leaky modes in planar multilayer waveguides, *IEEE Photon. Tech. Lett.*, **14**(6), 810–812, 2002.

59. C. Vassallo, Improvement of finite-difference methods for step-index optical wave-guides, *IEE Proc J. Optoelectron.*, **139**(2), 137–142, 1992.

60. Y. C. Chiang, Y. P. Chiou, and H. C. Chang, Improved full-vectorial finite-difference mode solver for optical waveguides with step-index profiles, *J. Lightw. Technol.*, **20**(8), 1609–1618, 2002.

61. W. P. Huang, C. L. Xu, W. Lui, and K. Yokoyama, The perfectly matched layer (PML) boundary condition for the beam propagation method, *IEEE Photon. Tech. Lett.*, **8**(5), 649–651, May 1996.

62. M. Reed, P. Sewell, T. M. Benson, and P. C. Kendall, Efficient propagation algorithm for 3D optical waveguides, *IEE Proc. Optoelectron.*, **145**(1), 53–58, 1998.

63. E. Silberstein, P. Lalanne, J. P. Hugonin, and Q. Cao, Use of grating theories in integrated optics, *J. Opt. Soc. Am. A*, **18**(11), 2865–2875, 2001.

64. T. Ando, T. Murata, H. Nakayama, J. Yamauchi, and H. Nakano, Analysis and measurement of polarization conversion in a periodically loaded dielectric waveguide, *IEEE Photon. Tech. Lett.*, **14**(9), 1288–1290, 2002.

65. K. Kawano, T. Kitoh, M. Kohtoku, T. Takeshita, and Y. Hasumi, 3-D semivectorial analysis to calculate facet reflectivities of semiconductor optical waveguides based on the bi-directional method of line BPM (MoL-BPM), *IEEE Photon. Tech. Lett.*, **10**(1), 108–110, 1998.

66. V. R. Almeida, Q. F. Xu, C. A. Barrios, and M. Lipson, Guiding and confining light in void nanostructure, *Opt. Lett.*, **29**(11), 1209–1211, 2004.

67. N. N. Feng, J. Michel, and L. C. Kimerling, Optical field concentration in low-index waveguides, *IEEE Quantum Electron.*, **42**(9–10), 885–890, 2006.

68. M. Lipson, Guiding, modulating, and emitting light on silicon: Challenges and opportunities, *J. Lightw. Technol.*, **23**(12), 4222–4238, 2005.

69. Q. F. Xu, V. R. Almeida, R. R. Panepucci, and M. Lipson, Experimental demonstration of guiding and confining light in nanometer-size low-refractive-index material, *Opt. Lett.*, **29**(14), 1626–1628, 2004.

70. C. Vassallo, *Optical Waveguide Concepts*. Amsterdam: Elsevier, 1991.

71. T. Baehr-Jones, M. Hochberg, G. X. Wang, R. Lawson, Y. Liao, P. A. Sullivan, L. Dalton, A. K. Y. Jen, and A. Scherer, Optical modulation and detection in slotted Silicon waveguides, *Opt. Express*, 13(14), 5216–5226, 2005.

72. T. Fujisawa and M. Koshiba, Polarization-independent optical directional coupler based on slot waveguides, *Opt. Lett.*, **31**(1), 56–58, 2006.

73. F. Dell'Olio and V. M. N. Passaro, Optical sensing by optimized silicon slot waveguides, *Opt. Express*, **15**(8), 4977–4993, 2007.

74. T. Fujisawa and M. Koshiba, Theoretical investigation of ultrasmall polarization-insensitive 1 × 2 multimode interference waveguides based on sandwiched structures, *IEEE Photon. Tech. Lett.*, **18**(9–12), 1246–1248, 2006.

75. H. Zhang, J. W. Mu, and W. P. Huang, Improved bidirectional beam-propagation method by a fourth-order finite-difference scheme, *J. Lightw. Technol.*, **25**(9), 2807–2813, 2007.

76. W. L. Barnes, A. Dereux, and T. W. Ebbesen, Surface plasmon subwavelength optics, *Nature*, **424**(6950), 824–830, 2003.

77. L. Chen, J. Shakya, and M. Lipson, Subwavelength confinement in an integrated metal slot waveguide on silicon, *Opt. Lett.*, **31**(14), 2133–2135, 2006.

78. G. Gay, O. Alloschery, B. V. De Lesegno, C. O'Dwyer, J. Weiner, and H. J. Lezec, The optical response of nanostructured surfaces and the composite diffracted evanescent wave model, *Nat. Phys.*, **2**(4), 262–267, 2006.

79. K. G. Lee and Q. H. Park, Coupling of surface plasmon polaritons and light in metallic nanoslits, *Phys. Rev. Lett.*, **95**(10), 103902, 2005.

80. S. A. Maier, Plasmonics: The promise of highly integrated optical devices, *IEEE J. Sel. Top. Quantum Electron.*, **12**(6), 1671–1677, 2006.

81. E. Ozbay, Plasmonics: Merging photonics and electronics at nanoscale dimensions, *Science*, **311**(5758), 189–193, 2006.

82. H. Raether, *Surface Plasmons*. New York: Springer-Verlag, 1988.

83. S. A. Darmanyan and A. V. Zayats, Light tunneling via resonant surface plasmon polariton states and the enhanced transmission of periodically nanostructured metal films: An analytical study, *Phys. Rev. B*, **67**(3), 035424-1-7, 2003.

84. M. Kretschmann and A. A. Maradudin, Band structures of two-dimensional surface-plasmon polaritonic crystals, *Phys. Rev. B*, **66**(24), 245408-1-7, 2002.

85. J. M. Steele, C. E. Moran, A. Lee, C. M. Aguirre, and N. J. Halas, Metallo-dielectric gratings with subwavelength slots: Optical properties, *Phys. Rev. B*, **68**(20), 205103-1-7, 2003.

86. A. Boltasseva, T. Nikolajsen, K. Leosson, K. Kjaer, M. S. Larsen, and S. I. Bozhevolnyi, Integrated optical components utilizing long-range surface plasmon polaritons, *J. Lightw. Technol.*, **23**(1), 413–422, 2005.

87. Z. H. Han, E. Forsberg, and S. L. He, Surface plasmon Bragg gratings formed in metal-insulator-metal waveguides, *IEEE Photon. Tech. Lett.*, **19**(2–4), 91–93, 2007.

88. A. Hosseini and Y. Massoud, A low-loss metal-insulator-metal plasmonic Bragg reflector, *Opt. Express*, **14**(23), 11318–11323, 2006.

89. S. Jette-Charbonneau, R. Charbonneau, N. Lahoud, G. A. Mattiussi, and P. Berini, Bragg gratings based on long-range surface plasmon-polariton waveguides: Comparison of theory and experiment, *IEEE J. Quantum Electron.*, **41**(12), 1480–1491, 2005.

90. T. Sondergaard, S. I. Bozhevolnyi, and A. Boltasseva, Theoretical analysis of ridge gratings for long-range surface plasmon polaritons, *Phys. Rev. B*, **73**(4), 045320-1-7, 2006.

91. V. M. Fitio and Y. V. Bobitski, HIgh transmission of system dielectric grating thin film dielectric grating, in *Laser Fiber Opt. Netw. Model.*, Yalta, Crimea, 2005.

92. J. Yamauchi, T. Yamazaki, K. Sumida, and H. Nakano, TE/TM wave splitters using surface plasmon polaritons, in *Integrated Photonics and Nanophotonics Research and Applications*, 2007.

93. V. Bhatia and A. M. Vengsarkar, Optical fiber long-period grating sensors, *Opt. Lett.*, **21**(9), 692–694, May 1 1996.

94. H. J. Patrick, G. M. Williams, A. D. Kersey, J. R. Pedrazzani, and A. M. Vengsarkar, Hybrid fiber Bragg grating/long period fiber grating sensor for strain/temperature discrimination, *IEEE Photon. Tech. Lett.*, **8**(9), 1223–1225, 1996.

95. A. M. Vengsarkar, P. J. Lemaire, J. B. Judkins, V. Bhatia, T. Erdogan, and J. E. Sipe, Long-period fiber gratings as band-rejection filters, *J. Lightw. Technol.*, **14**(1), 58–65, 1996.

96. O. Duhem, J. F. Henninot, M. Warenghem, and M. Douay, Demonstration of long-period-grating efficient couplings with an external medium of a refractive index higher than that of silica, *Appl. Opt.*, **37**(31), 7223–7228, 1998.

97. Y. Jeong, B. Lee, J. Nilsson, and D. J. Richardson, A quasi-mode interpretation of radiation modes in long-period fiber gratings, *IEEE J. Quantum Electron.*, **39**(9), 1135–1142, 2003.

98. Y. Koyamada, Numerical analysis of core-mode to radiation-mode coupling in long-period fiber gratings, *IEEE Photon. Tech. Lett.*, **13**(4), 308–310, 2001.

99. H. J. Patrick, A. D. Kersey, and F. Bucholtz, Analysis of the response of long period fiber gratings to external index of refraction, *J. Lightw. Technol.*, **16**(9), 1606–1612, 1998.

100. D. B. Stegall and T. Erdogan, Leaky cladding mode propagation in long-period fiber grating devices, *IEEE Photon Technol. Lett*, **11**(3), 343–345, 1999.

101. H. Yanagawa, T. Shimizu, S. Nakamura, and I. Ohyama, Index-and-dimensional taper and its application to photonic devices, *J. Lightw. Technol.*, **10**(5), 587–592, May 1992.

102. O. Mitomi, K. Kasaya, and H. Miyazawa, Design of a single-mode tapered wave-guide for low-loss chip-to-fiber coupling, *IEEE J. Quantum Electron.*, **30**(8), 1787–1793, 1994.

103. J. D. Love, W. M. Henry, W. J. Stewart, R. J. Black, S. Lacroix, and F. Gonthier, Tapered single-mode fibers and devices. 1. Adiabaticity criteria, *IEE Proc. J. Optoelectron.*, **138**(5), 343–354, 1991.

104. J. Haes, J. Willems, and R. Bates, Design of adiabatic tapers for high contrast step index waveguides, in *Proc. SPIE*. 2212, 1994, pp. 685–693.

105. D. Marcuse, Radiation losses of tapered dielectric slab waveguides, *Bell Sys. Tech. J.*, **49**, 273–290, 1970.

106. A. R. Nelson, Coupling optical waveguides by tapers, *Appl. Opt.*, **14**(12), 3012–3015, 1975.

107. J. A. Fleck, J. R. Morris, and M. D. Feit, Time-dependent propagation of high-energy laser-beams through atmosphere, *Appl. Phys.*, **10**(2), 129–160, 1976.

108. T. Nakamura and N. Suzuki, Spot-size converted laser diodes based on mode interference, *IEEE Photon. Tech. Lett.*, **12**(2), 143–145, 2000.

109. C. T. Lee, M. L. Wu, L. G. Sheu, P. L. Fan, and J. M. Hsu, Design and analysis of completely adiabatic tapered waveguides by conformal mapping, *J. Lightw. Technol.*, **15**(2), 403–410, 1997.

110. P. G. Suchoski and V. Ramaswamy, Exact numerical technique for the analysis of step discontinuities and tapers in optical dielectric wave-guides, *J. Opt. Soc. Am. A*, **3**(2), 194–203, 1986.

111. P. G. Suchoski and R. V. Ramaswamy, Design of single-mode step-tapered waveguide sections, *IEEE J. Quantum Electron.*, **23**(2), 205–211, 1987.

112. A. W. Snyder, Coupling of modes in a tapered dielectric cylinder, *IEEE Trans. Microw. Theor. Tech.*, **18**(7), 383–392, 1970.

113. A. F. Milton and W. K. Burns, Mode-coupling in tapered optical-waveguide structures and electro-optic switches, *IEEE Trans. Circuits Syst.*, **26**(12), 1020–1028, 1979.

114. A. F. Milton and W. K. Burns, Mode-coupling in optical-waveguide horns, *IEEE J. Quantum Electron.*, **13**(10), 828–835, 1977.

Chapter 3

Template-Guided Self-Assembly of Discrete Optoplasmonic Molecules and Extended Optoplasmonic Arrays

Yan Hong,[a] Wonmi Ahn,[a] Svetlana Boriskina,[b] Xin Zhao,[a] and Björn M. Reinhard[a]

[a]*Department of Chemistry and The Photonics Center, Boston University, Boston, Massachusetts 02215, USA*
[b]*Department of Mechanical Engineering, Massachusetts Institute of Technology, Cambridge, Massachusetts 02139, USA*

bmr@bu.edu

The integration of metallic and dielectric building blocks into optoplasmonic structures creates electromagnetic systems in which plasmonic and photonic modes can interact in the near-, intermediate- and far-field. The electromagnetic coupling between multiple building blocks in hybrid structures provides a multitude of opportunities for controlling electromagnetic fields in both spatial and frequency domain as well as for engineering the phase landscape and the local density of optical states (LDOS). Control over any of these properties requires, however, rational fabrication approaches that can integrate the different building blocks into hybrid structures with defined morphology. Template-

Integrated Nanophotonic Resonators: Fundamentals, Devices, and Applications
Edited by Ya Sha Yi

ISBN 978-981-4613-78-1 (Hardcover), 978-981-4613-79-8 (eBook)
www.panstanford.com

guided self-assembly strategies have been demonstrated to facilitate the assembly of a wide range of discrete optoplasmonic structures as well as optoplasmonic arrays or metasurfaces. We focus here on two fundamental strategies to control the photonic environment around plasmonic nanoantennas. In discrete optoplasmonic atoms or molecules optical microcavities (OMs) serve as whispering gallery mode (WGM) resonators that provide a discrete photonic mode spectrum. In extended hetero-nanoparticle arrays, on the other hand, in-plane scattered light can induce coherent interactions between the nanoparticles that result in geometry-dependent resonances. The electromagnetic field of these photonic modes interacts with the localized surface plasmon resonances (LSPRs) located at the metal nanoparticles. We characterize the fundamental working principles of both optoplasmonic approaches and review template-guided self-assembly methods that facilitate their fabrication.

3.1 Introduction

The advancements of nanofabrication methods in the last 10–20 years have facilitated plasmonic nanoantennas[1–3] for a broad range of applications in the visible range of the electromagnetic spectrum. Due to the superb light focusing properties of metal nanoparticle clusters, plasmonic gap antennas allow the creation of electromagnetic hot-spots in which the incident E-field is enhanced by many orders of magnitude.[4–8] The hot-spots are of high interest for enhancing excitation rates or for boosting non-linear optical effects.[9–12] At the same time, the density of states associated with plasmon resonances in the nanostructures provides means to boost radiative rates,[9,13] and the combination of quantum emitter and plasmonic antenna is, thus, commonly referred to as superemitters.[14] Due to Ohmic losses at optical frequencies in metals, plasmon resonances have short (femtoseconds to tens of femtosecond) lifetimes.[15] Consequently, gold and silver nanoparticles have relatively broad spectral features. High non-radiative losses ultimately limit the performance of plasmonic nanostructures in some applications. Different strategies are currently pursued to overcome this limitation.[16–19] One promising approach seeks to integrate plasmonic nanostructures into a defined photonic environment where photonic and plasmonic modes can synergistically interact.[20–25] The coupling of localized surface plasmon resonances (LSPRs) with in-plane diffracted Rayleigh resonances to form surface lattice resonances (SLRs) in extended nanoparticle arrays is one example for this approach.[26–29] In an array of metal nanoparticles, the individual nanoparticles experiences the incident E-field plus the E-field re-radiated by all of the other nanoparticles in the array. The effective polarizability, α_{eff} of the individual nanoparticles is then given as[18]

$$\alpha_{\text{eff}} = \frac{1}{\frac{1}{\alpha} - S},$$

where α refers to the single particle polarizability, and the retarded dipole sum, S, captures the effect of the array. A negative imaginary part of S is equivalent to an effective reduction of the radiative damping and results in a sharper spectral width for the hybrid mode.[18]

An alternative strategy for overcoming the limitations of conventional plasmonic nanocircuitry in a more compact footprint is based on the combination of plasmonic nanoantennas with dielectric microcavity resonators into structurally defined, discrete hybrid optoplasmonic structures.[21,22,30] Although WGM resonators have larger mode volumes and generate significantly lower E-field enhancements than metallic nanoantennas, their eigenmodes have low losses and, therefore, higher Q factors than metallic nanostructures.[31] In discrete optoplasmonic devices, both of these components are combined and the design strategy is to position plasmonic antennas at defined locations in the evanescent field of the dielectric resonators. The goal is to facilitate the formation of photonic-plasmonic hybrid-modes in which part of the mode volume is localized around metallic nanoparticles in electromagnetic hot-spots.[32] A representative example for this approach is given in Fig. 3.1. The hybrid structure contains electromagnetic hot spots located outside of the dielectric microsphere around the metal nanoparticles where the E-field is available for interactions with the ambient medium.

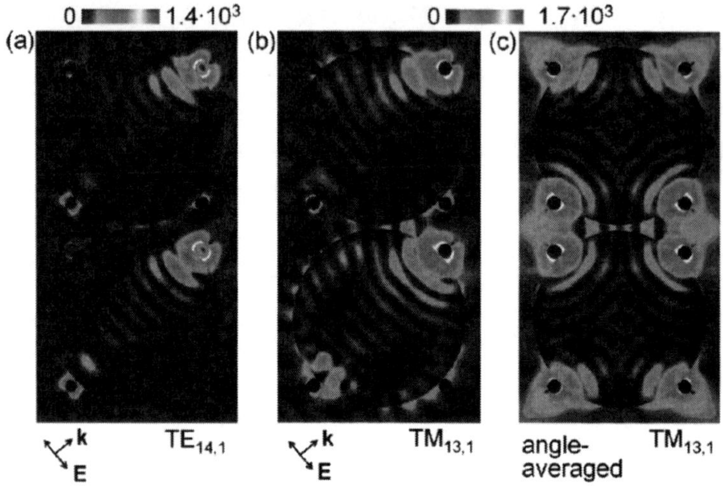

Figure 3.1 E-field intensity distributions in the optoplasmonic dimer at the wavelength of the hybrid $TE_{14,1}$ (a = 580.9 nm) and $TM_{13,1}$ (b = 590 nm) mode. The dimer is composed of two 2.048 μm microspheres, each encircled by four 148 nm diameter Au NPs separated from the microsphere surface by 1 nm gaps. The insets show the direction and polarization of the incident plane wave. (c) Near-field intensity distribution at the $TM_{13,1}$ mode wavelength averaged over the in-plane angle of incidence. Reprinted with permission from Ahn et al. *ACS Nano*, **6**, 951 (2012). Copyright 2012 American Chemical Society.

We note here in passing that discrete optoplasmonic structures do not only provide unique opportunities for enhancing E-fields but also for engineering LDOS in the vicinity of fluorescent superemitters. The effect of the enhanced LDOS on the emission efficiency of superemitters has been shown to differ depending on the superemitter's emission efficiency in the absence of the microsphere resonator.[33,34] For emitters with emission wavelengths in resonance with the antenna plasmon, two cases can be distinguished. For emitters with low emission rates, the radiative rate can be improved by enhanced LDOS. Intriguingly, the emission rate decreases, however, with increasing LDOS for bright emitters that operate close to the unitary limit.

A second class of optoplasmonic materials, the optoplasmonic nanoparticle array, is inspired by the formation of SLRs in conventional metal nanoparticle arrays. In optoplasmonic arrays metal nanoparticles are replaced as sole building block through a combination of high refractive index dielectric nanoparticles and clusters of strongly coupled metal nanoparticles.[24] Noble metal nanoparticle clusters are chosen as plasmonic component as they provide much higher *E*-field intensities than individual nanoparticles. Furthermore, plasmon coupling shifts the plasmon resonance wavelength into the red,[35,36] where the slope of the real part of the gold dielectric function as function of wavelength is steeper.[37] This behavior is beneficial for the design of sensitive colorimetric sensors.[38]

Conventional plasmonic nanoparticle cluster arrays (NCAs)[36,39,40] have been investigated in detail and have been shown to provide new opportunities for engineering a multiscale *E*-field enhancement through electromagnetic coupling in the near-, intermediate-, and far-field region. The availability of two building blocks in optoplasmonic arrays provides new opportunities for (i) controlling the light field distribution in the plane of the array and for (ii) efficiently squeezing light into electromagnetic hot-spots. In optoplasmonic arrays low-loss dielectric nanoparticles can be used to form grating-like structures that generate in-plane diffracted resonances.[24] In arrays with appropriate morphology, the photonic resonances can create enhanced *E*-field intensity at the interstitial spaces between the dielectric nanoparticles. If clusters of electromagnetically strongly coupled noble metal nanoparticles are located at these locations, the in-plane diffracted light is further localized by these efficient plasmonic antennas to achieve a cascaded *E*-field intensity enhancement.

Discrete optoplasmonic structures and optoplasmonic nanoparticle arrays have in common that they contain plasmonic antennas in a complex photonic environments to enhance or modulate the antenna characteristics. In the first case, this environment is defined by the WGM resonator, and in the second case it is generated by a two-dimensional nanoparticle array. The fabrication of both optoplasmonic materials requires the integration of building blocks with different chemical composition and, especially in the case of discrete optoplasmonic structures, of different sizes in precise geometric arrangements. In this chapter we will introduce template-guided self-assembly approaches to address these challenges and characterize selected examples of optoplasmonic materials obtained with them.

3.2 Results and Discussion

3.2.1 Discrete Optoplasmonic Atoms and Molecules

Discrete optoplasmonic devices that contain a single dielectric microsphere resonator and one or more plasmonic antennas are referred to as optoplasmonic atoms in the following.[21] We chose this nomenclature to indicate similarity with the more established concept of the "photonic atom"[41] as well as to imply that these structures represent the simplest possible unit of optoplasmonic materials of this kind. More complex optoplasmonic molecules can be obtained by integrating multiple optoplasmonic atoms into networks. Hybrid structures containing more than one OM as WGM resonator are referred to as optoplasmonic molecules.[21]

The particular challenge for the experimental realization of optoplasmonic atoms and molecules as the one outlined in Fig. 3.1 is the requirement to position a plasmonic antenna in close vicinity to a microsphere resonator. While this can be accomplished—in principle—by directly attaching noble metal nanoparticles to a dielectric resonator,[42–44] this approach is very limited in scope as it does not provide control over the location and, in the case of anisotropic nanoparticles, orientation of the noble metal nanoparticles. Geometric control over the positioning of nanoparticles is crucial for a successful integration of the optoplasmonic structures into an on-chip platform. Furthermore, it is difficult to position more complex antenna structures, such as gap antennas, on the OM by nanoparticle attachment, and the separation between the antenna and the dielectric resonator can also not be varied in a controlled fashion. Considering these fundamental limitations, we decided to develop new rational fabrication strategies for positioning top-down fabricated plasmonic nanoantennas at defined positions in the equatorial plane of WGM resonators. Due to the large differences in size between the plasmonic antennas (tens of nanometers) and the dielectric microspheres (microns), the proposed structures require the formation of plasmonic antennas microns above the substrate surface, which is a true three-dimensional fabrication task. One particular advantage of our approach as outlined in the following is that it facilitates the generation of networks of optoplasmonic atoms that contain the plasmonic components in one defined plane where they can interact synergistically.

3.2.1.1 Fabrication of optoplasmonic structures comprising OMs and nanoantennas

The template-guided self-assembly approach for positioning plasmonic antennas in the equatorial plane of WGM resonators is outlined in Fig. 3.2.[21,22] The strategy of this approach is to generate noble metal nanoparticle tipped pillars that define binding cavities for microspheres in a first step. In a subsequent step, these binding cavities are then filled with dielectric microspheres. Finally, opticalmicrospheres (OMs) are inserted into the generated binding sites by dispersing a solution of dielectric microspheres across the surface (convective self-assembly). The height (h) of the nanopillars (including the radius of the gold nanoparticles) is chosen to correspond to the radius of the microspheres of interest so thatthe metal nanoparticles are positioned in the equatorial plane of

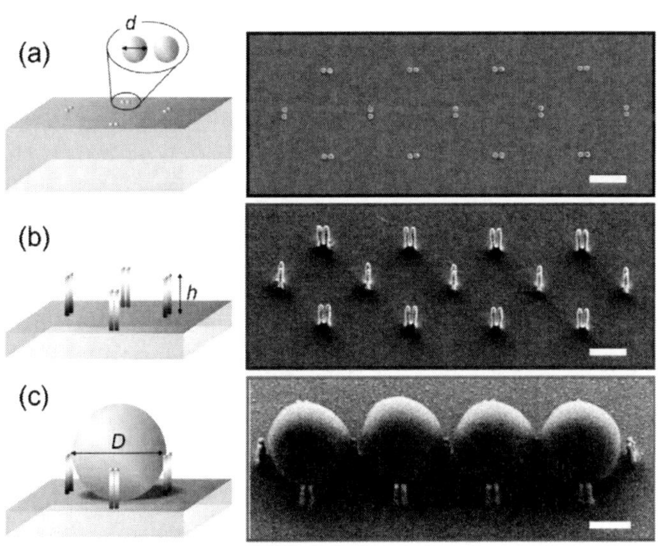

Figure 3.2 (a)–(c) Schematic outline of the individual steps (left) in the template guided self-assembly of optoplasmonic atoms and molecules and SEM images (right) of the corresponding hybrid structures. Reprinted with permission from Ahn et al. *ACS Nano*, **7**, 4470 (2013). Copyright 2013 American Chemical Society.

the WGM resonator. The separation between the nanopillars can be varied, for instance, to change the average separation between adjacent microspheres and, thus, to modulate the coupling between the OMs. In the following, we provide a detailed description of the fabrication of optoplasmonic structures containing microspheres of ~2 μm diameter. With some modifications, the same approach can also be applied to generate optoplasmonic structures containing microspheres with larger diameters.

Regular arrays of nanoholes of 150 nm diameter are patterned in 200 nm thick PMMA layers spin-coated on a quartz (or silicon) substrate through electron beam lithography (EBL) using a Zeiss SUPRA 40VP scanning electron microscope equipped with a Raith Beam Blanker and nanometer pattern generation system. A Ti/Au/Cr layer with a thickness of 10/80/35 nm, respectively, is then deposited on the EBL-patterned surface by electron beam evaporation with deposition rates of 0.5/1.0/0.5 Å/s for each metal to create optoplasmonic dimers that consist of Au nanoparticles. For photonic dimer controls that lack Au nanoparticles on top of nanopillars, only a Cr layer is deposited on the substrate surface. A Ti layer improves adhesion of the Au films to a quartz substrate, and a Cr layer protects the Au films from anisotropic dry etch gases. After the PMMA layer is lifted off by immersion in acetone for 1 min followed by 10, 20, and 30 s of sonication each in a fresh acetone bath, substrates are ready for reactive ion etching (RIE, Plasma-Therm. model 790). We commonly use a mixture of CHF_3/O_2 (50/5 sccm, standard cubic centimeters per minute) at a pressure of 200 mTorr and a power of 150 W for a total etching time of 18–20 min. The anisotropic etching process removes the SiO_2 only in exposed areas, which are not protected by Cr layers, resulting in the formation of nanopillar arrays with Cr layer tips. The height of the nanopillars can be controlled through the RIE process parameters, including the etching time, power, pressure, and ratio between the two mixed gases. A RIE chamber is purged with O_2 for 15 min and prerun with CHF_3/O_2 for 30 min prior to each RIE process on samples to ensure reproducible etching results. The Cr layer is subsequently removed by swirling in Cr etchant solution at 40°C for 15 s, after which the substrate is rinsed with copious amounts of distilled water. The substrate can then be heated in a rapid thermal annealer (Molecular Process Technology Corp.) to 800°C (3 min) to create spherical gold nanoparticles on top of the nanopillars. The regular arrays of micropillar cavities with defined widths and depths are then used as templates for the assembly of optoplasmonic structures. The created cavities are filled with PS microspheres using a convective self-assembly strategy. An aqueous suspension of PS microspheres (2.048 μm in diameter; 1%; 50 μL) is sandwiched between the patterned microcavity substrate and a blank quartz substrate separated by a 380 μm gap. Upon evaporation of the water, the meniscus of the microsphere-containing solution is dragged across the substrate surface by capillary forces. These forces have a component pointing perpendicular to the patterned surface,[45] which enables an efficient trapping of microspheres in the cavities formed by the fabricated pillars. The insertion of microspheres into the gold nanoparticle functionalized pillars completes the assembly of the optoplasmonic structures.

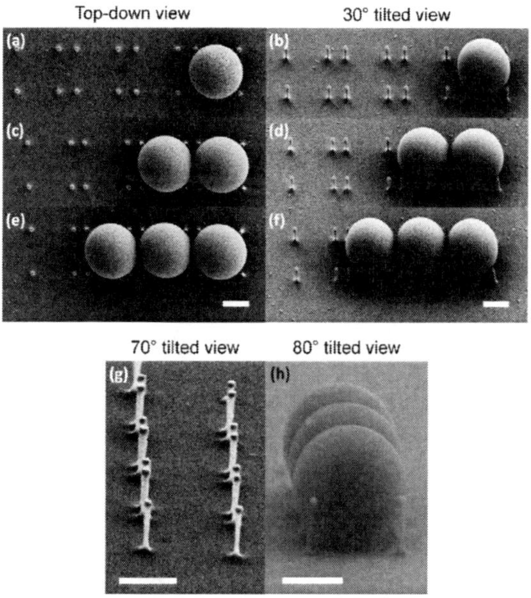

Figure 3.3 (a), (c), (e) Top and (b), (d), (f) side view (30° tilted angle) of optoplasmonic structures with 1, 2 or 3 OMs (diameter: 2 μm). Side views of pillar scaffolds (g) before and (h) after integration of OMs. The gold nanoparticles are clearly discernible at the tip of the pillars. Reprinted with permission from Ahn et al. *ACS Nano*, **6**, 951 (2012). Copyright 2012 American Chemical Society.

Figure 3.3 shows scanning electron microscope (SEM) images of optoplasmonic structures that contain one (a, b), two (c, d), or three (e, f) 2.048 μm diameter microspheres. The edge-to-edge distance (d_p) between neighboring pillars in the optoplasmonic molecules (c–f) was chosen as d_p = 330 nm to enable strong electromagnetic coupling between the individual constituent optoplasmonic atoms. At d_p = 330 nm the embedded microspheres are nearly touching, ensuring an efficient coupling of the WGM modes in the optoplasmonic molecules. The height of the pillars (h) in the structures was determined to be ~870 nm from SEM images acquired at a tilt angle of θ = 80° using the formula $h = h_m/\sin(\theta)$, where h_m is the measured length of pillars in the tilted SEM image. In the magnified view (θ = 70°) of a representative nanopillar scaffold in Fig. 3.3g the spherical Au nanoparticles on top of the created nanopillars are clearly visible. The diameter of the Au nanoparticles in these optoplasmonic structures is 148 nm. The side view in Fig. 3.3h confirms that the spherical Au nanoparticles are located in the equatorial plane of the PS microspheres inserted into the binding cavities.

3.2.1.2 Optical responses of discrete optoplasmonic structures in the near- and far field

Figure 3.4a shows the scattering image as well as the spectrally resolved scattering image of an optoplasmonic dimer obtained with unpolarized white-light. The imaging plane lies in the equatorial plane of the optoplasmonic dimer. The top panel in Fig. 3.4b shows representative normalized scattering spectra of an optoplasmonic molecule (red) that consists of two OMs, Au nanoparticles, and the SiO_2 nanopillars, and of a control with an overall identical structure but without Au nanoparticles on the nanopillars. The bottom panel in Fig. 3.4b contains the corresponding GMT-simulated scattering spectra of the optoplasmonic dimer (red) and the control (blue) in an ambient medium of refractive index n = 1. The experimental scattering spectra of the optoplasmonic dimer are broadened when compared with the control spectra. This broadening becomes more obvious in Fig. 3.4c, where the Full width at half maximum (FWHM) of the fitted peaks is plotted. The spectral broadening is more pronounced in the experiments than in the simulations, but the latter reproduce the general trend well. The TE modes are broader than the TM modes in the experimental spectra, although the simulations predict the opposite behavior. The detected TE modes have orbits that are pointing perpendicular to the substrate. Consequently, the TE modes experience more losses at the substrate than the detected TM modes whose orbits lie in a plane parallel to the substrate.[46] Substrate effects were not considered in the performed electromagnetic simulations.

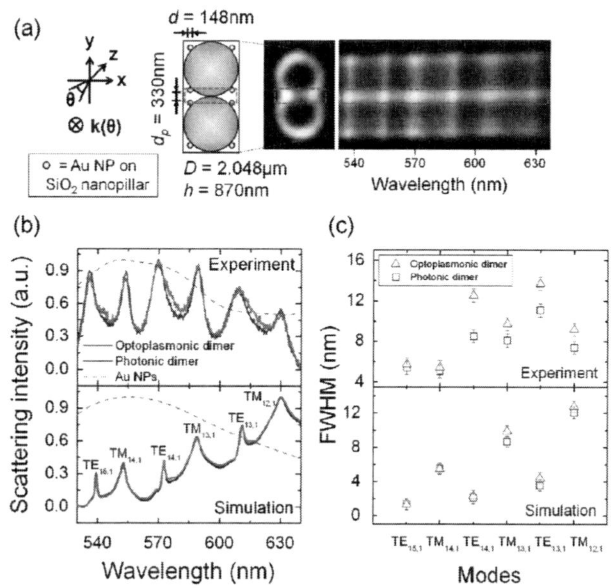

Figure 3.4 Characterization of optoplasmonic dimer through scattering darkfield microscopy/spectroscopy. (a) Geometry of the optoplasmonic dimer, scattering image and spectrally resolved scattering image. (b) Experimental (top) and simulated (bottom) scattering spectra of optoplasmonic dimer (red) and photonic control (blue) without gold nanoparticle tips. (c) FWHM of the individual modes shown in (b). Reprinted with permission from Ahn et al. *ACS Nano*, **6**, 951 (2012). Copyright 2012 American Chemical Society.

The spectral broadening of the TE and TM modes observed in the optoplasmonic dimer in the experimental and, albeit to a lesser degree, also in the simulated spectra is a direct consequence of photonic-plasmonic mode coupling. One consequence of this coupling is the redistribution of electromagnetic energy from the OMs into the nanoparticless. Dissipative losses in the Au nanoparticles shorten the lifetime of the hybridized modes and account for the observed broadening of the spectral features.

In the optoplasmonic dimer every OM is located in the effective field of four Au nanoparticles. Although this significant association of the OMs with Au nanoparticles leads to a measurable peak broadening, the induced broadening is relatively small, which reaffirms the advantage of our experimental strategy to place Au nanoparticles into the evanescent field of OMs in a controlled fashion.

Figure 3.4b also contains the scattering spectra of the Au nanoparticles used in this work as dashed gray lines. The nanoparticles show a broad resonance that peaks at approximately 555 nm. The scattering peaks of the optoplasmonic dimer are significantly sharper, which has tangible effects for a broad range of applications.

To characterize the *E*-field for microsphere-coupled nanoparticles, we performed generalized multiple particle Mie theory (GMT) simulations and analyzed two different nanoparticle diameters in Fig. 3.5. Our simulations of a nanoparticle in the vicinity of an OM indicate that the most efficient coupling between photonic and plasmonic components is achieved if the light is incident along the OM–nanoparticle axis. In this case the *E*-field intensity on the nanoparticle surface can be enhanced by two orders of magnitude (Fig. 3.5a). A comparison of the *E*-field intensity spectrum of the optoplasmonic structure with the spectra of the separate OM ("bead") and Au nanoparticle ("NP") confirms a gain in *E*-field intensity in the hybrid system for both nanoparticle diameters. This synergistic interaction justifies our interest in these unique electromagnetic materials.

Interestingly, the coupling of the nanoparticle LSPR with the WGM of the OM yields high resonant *E*-field intensities at the WGM resonances in the long wavelength tail of the LSPR spectrum. Close to the LSPR, however, strong electromagnetic coupling results in significant perturbations of the WGM peak shape. Figure 3.5b,c shows *E*-field intensity maps for two different modes. The maps show an *E*-field intensity concentration around the nanoparticle and, thus, outside of the OM.

The OM in the vicinity of a plasmonic antenna does not only enhance the peak *E*-field intensity and the spatial *E*-field intensity distribution but is also anticipated to modulate the emission properties of superemitters. Interestingly, we found that if superemitters (Cy-3 functionalized nanoparticle dimer) are located in the vicinity of an OM, the emission is highly directional into the OM, which efficiently traps and recirculates the photons.[22]

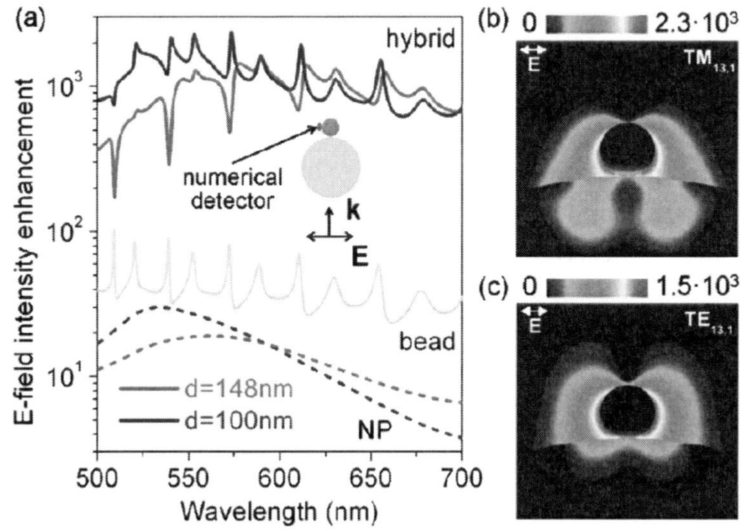

Figure 3.5 Near-field characterization of an optoplasmonic atom comprising an Au nanoparticle separated from a OM by a 1 nm gap. The diameter of the polystyrene microsphere is 2.048 μm and the diameter of the Au nanoparticles is 148 or 100 nm, as indicated. (a) *E*-field intensity enhancement in the gap between the Au nanoparticle and the OM. The location of the numerical director for evaluating the *E*-field intensity is indicated by a red dot in the inset. (b) and (c) Maps of the spatial *E*-field intensity enhancement around the Au nanoparticle at λ = 589.7 nm and λ = 614.4 nm. Reprinted with permission from Ahn et al. *ACS Nano*, **6**, 951 (2012). Copyright 2012 American Chemical Society.

An alternative approach to create a photonic environment around a plasmonic antenna is the integration of the antenna into a regular array of nanoparticles in which the individual elements are diffractively coupled.[27,47] Optoplasmonic arrays are unique in the sense that they combine high refractive index dielectric nanoparticles and noble metal nanoparticles (or their clusters) into arrays, which can use the dielectric component to focus the light onto nanoparticle clusters.[24] The nanoparticle clusters can then squeeze the light into deeply sub-diffraction limit hot-spots. The challenge of this approach is to position nanoparticles of different composition at pre-defined locations. A second complication derives from the need to minimize the interparticle separations in the metal clusters to maximize the near-field coupling. As we will show in the following, template-guided self-assembly is a versatile fabrication strategy capable of addressing both of these challenges.

3.2.1.3 Fabrication of optoplasmonic arrays through template-guided self-assembly

In the first step of the assembly procedure nanoparticle assembly sites are created through a lithographically fabrication technique (e.g. EBL) for the subsequent assembly of colloidal NPs onto the lithographically defined assembly sites.[24] While EBL makes it possible to define the separation, Λ, between individual assembly sites on the tens to hundreds of nanometer scale, the separation within the self-assembled clusters is determined by the assembly conditions as well as the surface of the assembled nanoparticles, and is typically on the few nanometer length scale or below.[48] This versatility in realizing nanoparticle separations over such a broad range, make the method a useful tool for realizing multiscale nanoparticle arrays.

Diameter (D) and shape of the assembly sites represent control parameters for guiding nanoparticles to specific assembly sites. In case of the optoplasmonic arrays this control parameters is utilized to generate an array that contains TiO_2 nanoparticles and clusters of 60 nm Au nanoparticles on separate lattice sites.[24] The assembly strategy of optoplasmonic arrays is schematically outlined in Fig. 3.6. First, two different binding sites are templated in the electron beam resist. Then the larger TiO_2 nanoparticles are immobilized on the binding sites with larger diameter. Subsequently, smaller gold nanoparticles are bound to the vacant binding sites with smaller diameter. For details regarding the assembly process, please refer to ref. [24].

Figure 3.7 shows SEM images of different arrays obtained from 60 nm Au nanoparticles and 250 nm diameter TiO_2 nanoparticles with binding sites containing diameters of 140 and 270 nm, respectively.[24] The assembly yielded regular arrays containing TiO_2 NPs located on the vertices of regular squares around a central Au nanoparticle cluster. The average number of Au nanoparticles in the clusters was ~6.

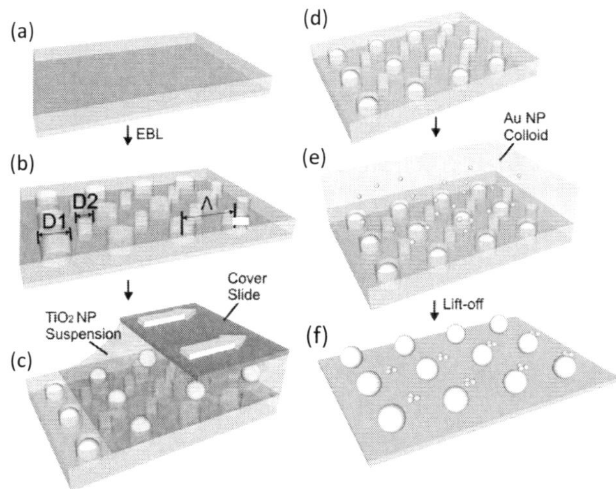

Figure 3.6 Scheme of template-guided self-assembly approach of optoplasmonic arrays. (a) An array with two different binding sites D1 and D2 is created in a PMMA film. (b) The separation between two identical binding sites is defined by the grating period, Λ. (c) Dispersing a suspension of TiO_2 nanoparticles that only bind to the larger binding site D1 creates (d) a dielectric nanoparticle array. (e) Smaller Au nanoparticles are then targeted to the vacant binding sites, (f) generating the optoplasmonic material after lift-off. Reprinted with permission from Hong et al. *Adv. Funct. Mater.*, **24**, 739 (2014). Copyright 2013 Wiley-VCH GmbH & Co. KgaA.

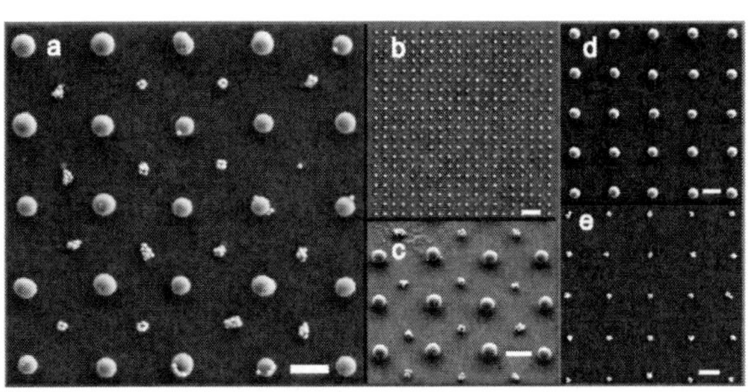

Figure 3.7 SEM images of optoplasmonic arrays. Scale bars are 500 nm in (a), (c)–(e), and 2 μm in (b). Reprinted with permission from Hong et al. *Adv. Funct. Mater.*, **24**, 739 (2014). Copyright 2013 Wiley-VCH GmbH & Co. KgaA.

3.2.1.4 Morphology-dependent *E*-field enhancement in optoplasmonic arrays

To quantify the synergistic gain that results from electromagnetic interactions between metallic and dielectric building blocks, we simulated the peak *E*-field intensity enhancement spectra for (i) a TiO_2 nanoparticle only array, (ii) a Au nanoparticle cluster only array, and (iii) the optoplasmonic hybrid array that contains both Au nanoparticle clusters and TiO_2 nanoparticles. In the hybrid array, Au nanoparticle clusters were positioned in the center of a square unit cell formed by four TiO_2 nanoparticles (Fig. 3.7). For simplicity we modeled the Au nanoparticle cluster as a trimer. The results of our multiple sphere T-matrix simulations[49] are summarized in Fig. 3.8. The *E*-field enhancement for the all-dielectric TiO_2 NP array in Fig. 3.8a shows discrete resonances, whose spectral positions red-shift with increasing grating period, Λ. As expected, the Au NCAs (Fig. 3.8b) show much higher *E*-field enhancements than the all-dielectric nanoparticle array. The near-field spectra of the metal arrays are dominated by the cluster plasmon resonance with a simulated peak wavelength of ~720 nm. The near-field intensity spectrum for an individual Au NP trimer is included for comparison in Fig. 3.8a.

Optoplasmonic arrays (Fig. 3.8c) achieve a further boost of the near-field intensities beyond the level of Au NCAs when the resonance wavelength of a photonic array mode overlaps with the cluster plasmon resonance. Under these conditions photonic-plasmonic mode coupling facilitates a cascaded enhancement of the *E*-field intensity provided by the individual clusters. According to our simulations, optoplasmonic arrays with $\Lambda \approx 800$ nm fulfill this requirement and, indeed, the overlap of array and nanoparticle cluster resonance yields an additional enhancement of the *E*-field intensity for this grating period.

To test the Λ-dependence of the *E*-field intensity enhancement experimentally, we set out to measure the surface enhanced Raman scattering (SERS) spectra of the small test molecule para-mercaptoaniline (pMA) chemisorbed onto optoplasmonic arrays with Λ = 700–1100 nm. As the SERS signal intensity of a molecule scale as the product of the *E*-field intensities at the pump and Raman scattering wavelengths,[50] the SERS signal intensity can provide valuable information of the E-field intensity at the probe and emission wavelengths. Figure 3.9 shows the background corrected SERS spectra in the spectral range 980–1140 cm^{-1} that contains the C-S stretch mode at 1077 cm^{-1}. The SERS signal intensity—even after correction of the different filling fractions (see inset)—shows a clear Λ-dependence and peaks at Λ = 900 nm. The Λ-dependence of the E-field intensity at the plasmon resonance wavelength is consistent with our simulations. The discrepancy in the precise value of Λ that maximizes the E-field intensity between simulations and experiment can be understood in terms of necessary simplifications in the computational model. Overall, both electromagnetic simulations and SERS spectroscopy confirm the synergistic interaction between dielectric and metallic components in the optoplasmonic array.

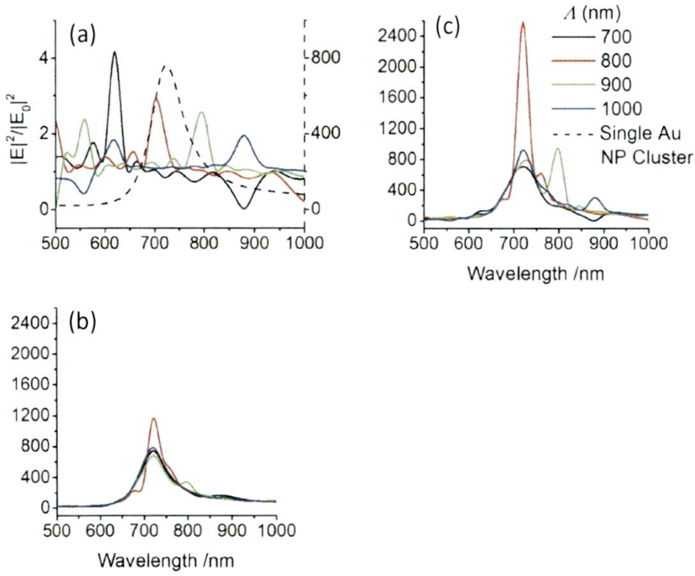

Figure 3.8 Simulated E-field intensity enhancement spectra at the hottest spot in (a) an array of 20 × 20 TiO_2 nanoparticles (solid) and a single trimer of Au nanoparticles (dashed), (b) array of 20 × 20 Au nanoparticle trimers, (c) combined optoplasmonic array. The incident k-vector is normal to the array plane in all models. Reprinted with permission from Hong et al. *Adv. Funct. Mater.*, **24**, 739 (2014). Copyright 2013 Wiley-VCH GmbH & Co. KgaA.

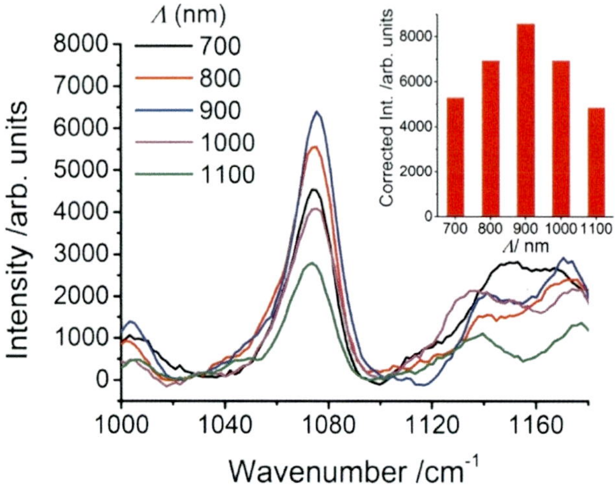

Figure 3.9 SERS spectra of para-mercaptoaniline (pMA) obtained on optoplasmonic arrays with Λ = 700–1100 nm. The inset shows filling fraction-corrected peak SERS signal intensities as function of Λ. Reprinted with permission from Hong et al. *Adv. Funct. Mater.*, **24**, 739 (2014). Copyright 2013 Wiley-VCH GmbH & Co. KgaA.

3.3 Conclusion

Template-guided self-assembly is a versatile fabrication strategy for the fabrication of novel electromagnetic materials from different building blocks. We have taken advantage of the method's ability to combine OM and nanogap antennas into discrete optoplasmonic atoms and molecules with defined geometry. Furthermore, we utilized template-guided assembly strategies to integrate high refractive index dielectric nanoparticles and clusters of electromagnetically strongly coupled noble metal nanoparticles into hetero-nanoparticle arrays. The ability to position metallic and dielectric nano- and micro-structures with high lateral resolution through template-guided self-assembly strategies provides rational control over the interplay of photonic and plasmonic modes in these new optoplasmonic materials. The combination of mutually synergistic properties of metallic and dielectric materials in optoplasmonic structures generates new functionalities for sensing, light harvesting, information processing and many other application areas.

Acknowledgments

This work was supported by the U.S. Department of Energy, Office of Basic Energy Sciences, Division of Materials Science and Engineering under Award DOE DE-SC0010679.

References

1. P. Muhlschlegel, H. J. Eisler, O. J. F. Martin, B. Hecht, D. W. Pohl (2005). Resonant optical antennas, *Science*, **308**, 1607–1609.
2. L. Novotny. (2007). Effective wavelength scaling for optical antennas, *Phys. Rev. Lett.*, **98**, 266802.
3. L. Novotny and N. van. Hulst (2011). Antennas for light, *Nat. Photon.*, **5**, 83–90.
4. N. J. Halas, S. Lal, W. S. Chang, S. Link, and P. Nordlander (2011). Plasmons in strongly coupled metallic nanostructures, *Chem. Rev.*, **111**, 3913–3961.
5. P. Bharadwaj, B. Deutsch, L. Novotny (2009). Optical antennas, *Adv. Opt. Photon.*, **1**, 438–483.
6. J. J. Li, A. Salandrino, N. Engheta (2007). Shaping light beams in the nanometer scale: A Yagi-Uda nanoantenna in the optical domain, *Phys. Rev. B*, **76**, 245403.
7. O. L. Muskens, V. Giannini, J. A. Sanchez-Gil, J. G. Rivas (2007). Optical scattering resonances of single and coupled dimer plasmonic nanoantennas, *Opt. Expr.*, **15**, 17736–17746.
8. J. A. Schuller, et al. (2010). Plasmonics for extreme light concentration and manipulation, *Nat. Mater.*, **9**, 193-192-194.
9. O. L. Muskens, V. Giannini, J. A. Sanchez-Gil, J. Gomez Rivas (2007). strong enhancement of the radiative decay rate of emitters by single plasmonic nanoantennas, *Nano Lett.*, **7**, 2871–2875.
10. J. Zhang, Y. Fu, M. H. Chowdhury, J. R. Lakowicz (2007). Metal-enhanced single-molecule fluorescence on silver particle monomer and dimer: Coupling effect between metal particles, *Nano Lett.*, **7**, 2101–2107.
11. K. R. Li, M. I. Stockman, D. J. Bergman (2003). Self-similar chain of metal nanospheres as an efficient nanolens, *Phys. Rev. Lett.*, **91**, 227402.
12. J. P. Camden, et al. (2008). Probing the structure of single-molecule surface-enhanced raman scattering hot spots, *J. Am. Chem. Soc.*, **130**, 12616–12617.

13. V. Giannini, A. I. Fernandez-Diminguez, S. C. Heck, S. A. Maier (2011). Plasmonic nanoantennas: fundamentals and their use in controlling the radiative properties of nanoemitters, *Chem. Rev.*, **111**, 3888–3912.
14. J. N. Farahani, D. W. Pohl, H. J. Eisler, B. Hecht (2005). Single quantum dot coupled to a scanning optical antenna: A tunable superemitter, *Phys. Rev. Lett.*, **95**, 017402.
15. S. Link and M. A. El-Sayed (1999). Spectral properties and relaxation dynamics of surface plasmon electronic oscillations in gold and silver nanodots and nanorods, *J. Phys. Chem. B*, **103**, 8410–8426.
16. J. A. Fan, et al. (2010). Self-assembled plasmonic nanoparticle clusters, *Science*, **328**, 1135–1138.
17. M. A. Noginov, et al. (2009). Demonstration of spaser-based nanolaser, *Nature*, **460**, 1110–1112.
18. S. L. Zou, G. C. Schatz (2004). Narrow plasmonic/photonic extinction and scattering line shapes for one and two dimensional silver nanoparticle arrays, *J. Chem. Phys.*, **121**, 12606–12612.
19. D. Brunazzo, E. Descrovi, J. M. Olivier (2009). Narrowband optical interactions in a plasmonic nanoparticle chain coupled to a metallic film, *Opt. Lett.*, **34**, 1405–1407.
20. O. Benson (2011). Assembly of hybrid photonic architectures from nanophotonic constituents, *Nature*, **480**, 193–199.
21. W. Ahn, S. V. Boriskina, Y. Hong, B. M. Reinhard (2012). Photonic-plasmonic mode coupling in on-chip integrated optoplasmonic molecules, *ACS Nano*, **6**, 951–960.
22. W. Ahn, Y. Hong, S. V. Boriskina, B. M. Reinhard (2013). Demonstration of efficient on-chip photon transfer in self-assembled optoplasmonic networks, *ACS Nano*, **7**, 4470–4478.
23. Y. Hong, M. Pourmand, S. V. Boriskina, B. M. Reinhard (2013). Enhanced light focusing in self-assembled optoplasmonic clusters with subwavelength dimensions, *Adv. Mater.*, **25**, 115–119.
24. Y. Hong, Y. Qiu, T. Chen, and B. M. Reinhard (2014). Rational assembly of optoplasmonic hetero-nanoparticle arrays with tunable photonic-plasmonic resonances, *Adv. Funct. Mater.*, **24**, 739–746.
25. I. Mukherjee, G. Hajisalem, R. E. Gordon (2011). One-step integration of metal nanoparticle in photonic crystal nanobeam cavity, *Opt. Expr.*, **19**, 22462–22469.
26. G. Vecchi, V. Giannini, J. Gomez Rivas (2009). Shaping the fluorescent emission by lattice resonances in plasmonic crystals of nanoantennas, *Phys. Rev. Lett.*, **102**, 146807.

27. G. Vecchi, V. Giannini, J. Gomez Rivas (2009). Surface modes in plasmonic crystals induced by diffractive coupling of nanoantennas, *Phys. Rev. B*, **80**, 201401.

28. B. Auguie and W. L. Barnes (2008). Collective resonances in gold nanoparticle arrays, *Phys. Rev. Lett.*, **101**, 143902.

29. A. I. Vakevainen, et al. (2013). Plasmonic surface lattice resonances at the strong coupling regime, *Nano Lett.*, **14**, 1721–1727.

30. S. Zou and G. C. Schatz (2006). Combining micron-size glass spheres with silver nanoparticles to produce extraordinary field enhancements for surface-enhanced raman scattering applications, *Isr. J. Chem.*, **46**, 293–297.

31. D. W. Vernooy, A. Furusawa, N. P. Georgiadis, V. S. Ilchenko, H. J. Kimble (1998). Cavity QED with high-Q whispering gallery modes, *Phys. Rev. A*, **57**, R2293.

32. W. Ahn, S. V. Boriskina, Y. Hong, B. M. Reinhard (2012). Electromagnetic field enhancement and spectrum shaping in plasmonically integrated optical vortices, *Nano Lett.*, **12**, 219–227.

33. M. Frimmer and A. F. Koenderink (2012). Superemitters in hybrid photonic systems: A simple lumping rule for the local density of optical states and its breakdown at the unitary limit, *Phys. Rev. B.*, **86**, 235428.

34. M. Frimmer and A. F. Koenderink (2013). Spontaneous emission control in a tunable hybrid photonic system, *Phys. Rev. Lett.*, **110**, 217405.

35. B. Yan, S. V. Boriskina, B. M. Reinhard (2011). Optimizing gold nanoparticle cluster configurations ($n \leq 7$) for array applications, *J. Phys. Chem. C*, **115**, 4578–4583.

36. B. Yan, et al. (2009). Engineered SERS substrates with multiscale signal enhancement: Nanoparticle cluster arrays, *ACS Nano*, **3**, 1190–1202.

37. P. B. Johnson, R. W. Christy (1972). *Phys. Rev. B*, **6**, 4370.

38. M. Konig, et al. (2014). Unveiling the correlation between sub-10-nm alkanethiol molecule monolayer sensitivity and plasmonic near-field enhancement in coupled metal nanoparticle oligomers, *ACS Nano*, in press.

39. B. Yan, S. V. Boriskina, B. M. Reinhard (2011). Design and implementation of noble metal nanoparticle cluster arrays for plasmon enhanced biosensing, *J. Phys. Chem. C*, **115**, 24437–24453.

40. L. Yang, et al. (2010). Engineering nanoparticle cluster arrays for bacterial biosensing: The role of the building block in multiscale SERS substrates, *Adv. Funct. Mater.*, **20**, 2619–2628.

41. S. Arnold (2001). Microspheres, photonic atoms and the physics of nothing, *Am. Sci.*, **89**, 414.

42. S. Shopova, C. Blackledge, A. Rosenberger (2008). Enhanced evanescent coupling to whispering-gallery modes due to gold nanorods grown on the microresonator surface, *Appl. Phys. B: Laser Opt.*, **93**, 183–187.

43. D. Melnikau, D. Savateeva, A. Chuvilin, R. Hillenbrand, Y. P. Rakovich (2011). Whispering gallery mode resonators with J-aggregates, *Opt. Expr.*, **19**, 22280.

44. M. A. Santiago-Cordoba, S. V. Boriskina, F. Vollmer, M. C. Demirel, (2011). Nanoparticle-based protein detection by optical shift of a resonant microcavity, *Appl. Phys. Lett.*, **99**, 073701.

45. Y. N. Xia, Y. D. Yin, Y. Lu, J. McLellan (2003). Template-assisted self-assembly of spherical colloids into complex and controllable structures, *Adv. Funct. Mater.*, **13**, 907–918.

46. T. Mitsui, Y. Wakayama, T. Onodera, Y. Takaya, H. Oikawa (2008). Observation of light propagation across 90° corner in chains of microspheres on a patterned substrate, *Opt. Lett.*, **33**, 1189–1191.

47. F. J. G. de Abajo (2007). Colloquium: Light scattering by particle and hole arrays, *Rev. Mod. Phys.*, **79**, 1267–1290.

48. T. Chen, M. Pourmand, A. Feizpour, B. Cushman, B. M. Reinhard (2013) Tailoring plasmon coupling in self-assembled one-dimensional Au nanoparticle chains through simultaneous control of size and gap separation, *J. Phys. Chem. Lett.*, **4**, 2147–2152.

49. D. W. Mackowski, M. I. Mishchenko,(1996). Calculation of the T matrix and the scattering matrix for ensembles of spheres, *J. Opt. Soc. Am. A*, **13**, 2266–2278.

50. E. C. Le Ru and P. G. Etchegoin (2006). Rigorous justification of the |E|4 enhancement factor in surface enhanced Raman spectroscopy, *Chem. Phys. Lett.*, **423**, 63–66.

Chapter 4

Nanophotonic Resonators for Enhancement of Absorption and Transmission Cross Sections of Subwavelength Plasmonic Devices

Georgios Veronis,[a,b] Changjun Min,[b,c] Yin Huang,[a,b,d] and Liu Yang[a,b]

[a]*School of Electrical Engineering and Computer Science,*
Louisiana State University, Baton Rouge, Louisiana 70803, USA
[b]*Center for Computation and Technology, Louisiana State University,*
Baton Rouge, Louisiana 70803, USA
[c]*Current address: Institute of Modern Optics,*
Nankai University, Tianjin 300071, China
[d]*Current address: School of Physics and Electronics,*
Central South University, Changsha 410083, China

gveronis@lsu.edu

4.1 Introduction

In this chapter, we show that nanophotonic resonators can greatly enhance the absorption and transmission cross sections of subwavelength plasmonic devices.

We first introduce a compact submicron structure consisting of multiple optical microcavities at both the entrance and exit

Integrated Nanophotonic Resonators: Fundamentals, Devices, and Applications
Edited by Ya Sha Yi

ISBN 978-981-4613-78-1 (Hardcover), 978-981-4613-79-8 (eBook)
www.panstanford.com

sides of a subwavelength plasmonic slit filled with an absorbing material. We show that such microcavity structures at the entrance side of the slit can greatly enhance the coupling of the incident light into the slit, by improving the impedance matching between the incident plane wave and the slit mode. In addition, the microcavity structures can also increase the reflectivities at both sides of the slit, and therefore the resonant field enhancement. Thus, such structures can greatly enhance the absorption cross section of the slit. An optimized submicron structure consisting of two microcavities at each of the entrance and exit sides of the slit leads to ~9.3 times absorption enhancement at the opticalcommunication wavelength compared to an optimized slit without microcavities.

We also introduce compact wavelength-scale slit-based structures for coupling free space light into metal–dielectric–metal (MDM) subwavelength plasmonic waveguides. We first show that for a single-slit structure the coupling efficiency is limited by a trade-off between the light power coupled into the slit, and the transmission of the slit-MDM waveguide junction. We next consider a two-section slit structure, and show that for such a structure the upper slit section enhances the coupling of the incident light into the lower slit section. The optimized two-section slit structure results in ~2.3 times enhancement of the coupling into the MDM plasmonic waveguide compared to the optimized single-slit structure. We finally consider a symmetric double-slit structure, and show that for such a structure the surface plasmons excited at the metal–air interfaces are partially coupled into the slits. Thus, the coupling of the incident light into the slits is greatly enhanced, and the optimized double-slit structure results in ~3.3 times coupling enhancement compared to the optimized single-slit structure. In all cases, the coupler response is broadband.

4.2 Nanophotonic Resonators for Enhancement of Optical Absorption in Subwavelength Slits

Following the observation of extraordinary optical transmission through arrays of subwavelength apertures in metallic films (Ebbesen, 1998), there has been enormous interest in the properties of plasmonic structures consisting of subwavelength apertures (Thio, 2001; Lezec, 2002; Garcia-Vidal, 2003; Barnes, 2003; Skigin, 2005; Genet, 2007; Garcia-Vidal, 2010). Resonant subwavelength plasmonic apertures can efficiently concentrate light into deep subwavelength regions, and therefore significantly enhance the optical transmission through the apertures (Lezec, 2002; Garcia-Vidal, 2003; Barnes, 2003; Skigin, 2005; Genet, 2007; Garcia-Vidal, 2010), or the absorption in the apertures (Yu, 2006; White, 2009). In addition, grating structures, consisting of periodic arrays of grooves patterned on the metal film surrounding the aperture, are commonly used to enhance the coupling of incident light into the aperture through the excitation of surface plasmons (Lezec, 2002; Genet, 2007; Yu, 2006; Martin-Moreno, 2003; Lopez-Tejeira, 2007; Ishi, 2005; Cai, 2009; Gbur, 2005; Fujikata, 2004; Shinada, 2003; Guo, 2007). For efficient surface plasmon excitation, however, the period of the grating has to be equal to the surface plasmon wavelength, and several grating periods are required. Thus, such structures need to be several microns long to operate at optical frequencies.

In this section, we show that a compact submicron structure consisting of multiple optical microcavities at both the entrance and exit sides of a subwavelength plasmonic slit filled with an absorbing material can greatly enhance the absorption cross section of the slit (Min, 2011). Our reference structure is a single optimized subwavelength slit in a metal film deposited on a substrate. We show that such microcavity structures at the entrance side of the slit can greatly enhance the coupling of the incident light into the slit, by improving the impedance matching between the incident plane wave and the slit mode. In addition, the microcavity structures can increase the reflectivities at both sides of the slit, and therefore the resonant field enhancement in the slit. An optimized submicron structure consisting of two microcavities at each of the entrance and exit sides of the slit leads to ~9.3 times absorption enhancement at the optical communication wavelength compared to the optimized reference slit without microcavities. We also show that while the microcavity-enhanced structures are optimized at a single wavelength, the operation wavelength range for high absorption is broad.

In previous studies, it was demonstrated that the use of a single microcavity at the entrance and exit sides of a subwavelength slit can enhance the transmission cross section of the slit (Verslegers, 2010; Min, 2008).

The remainder of this section is organized as follows. In Subsection 4.2.1, we define the absorption cross section and absorption enhancement factor of the slit, and employ single-mode scattering matrix theory to account for their behavior. The results obtained for the reference structure of a slit without microcavities, as well as for the microcavity-enhanced structures are presented in Subsections 4.2.2–4.2.5.

4.2.1 Absorption Cross Section and Absorption Enhancement Factor

We consider a structure consisting of a single slit in a silver film with N microcavities at the entrance side, and M microcavities at the exit side of the slit deposited on a silica substrate (Fig. 4.1a).

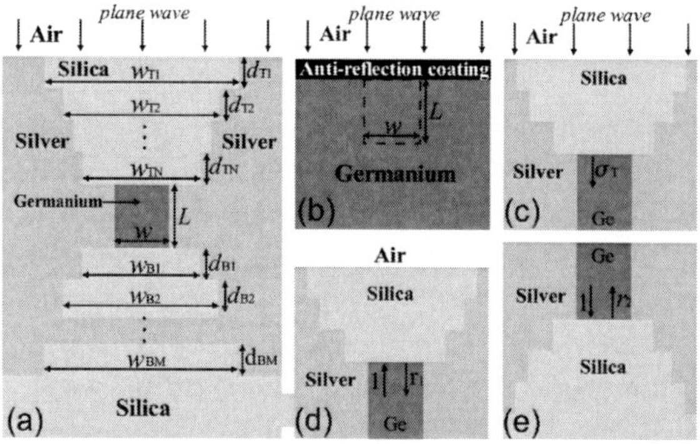

Figure 4.1 (a) Schematic of a structure consisting of a slit in a silver film with N microcavities at the entrance side, and M microcavities at the exit side of the slit deposited on a silica substrate. The slit is filled with germanium, while the microcavities are filled with silica. (b) Schematic of a bulk germanium photodetector with an anti-reflection coating. (c) Schematic defining the transmission cross section σ_T of a silver–germanium–silver waveguide through the structure above the entrance side of the slit of Fig. 4.1a for a normally incident plane wave from air. (d) Schematic defining the reflection coefficient r_1 of the fundamental TM mode of a silver–germanium–silver waveguide at the interface of such a waveguide with the structure above the entrance side of the slit of Fig. 4.1a. (e) Schematic defining the reflection coefficient r_2 of the fundamental TM mode of a silver-germanium-silver waveguide at the interface of such a waveguide with the structure below the exit side of the slit of Fig. 4.1a.

The slit is filled with germanium, which is one of the most promising materials for near-infrared photodetectors in integrated optical circuits (Cao, 2010), while the microcavities are filled with silica. We consider compact structures in which all microcavity dimensions are limited to less than 1 μm.

We use a two-dimensional finite-difference frequency-domain (FDFD) method (Veronis, 2007b) to numerically calculate the absorption in the material filling the slit. This method allows us to directly use experimental data for the frequency-dependent dielectric constant of metals such as silver (Palik, 1985), including both the real and imaginary parts, with no approximation. We use perfectly matched layer (PML) absorbing boundary conditions at all boundaries of the simulation domain (Jin, 2002). We also use the total-field-scattered-field formulation to simulate the response of the structure to a normally incident plane wave input (Taflove, 2005).

For comparison of different configurations, we define the absorption cross section σ_{A} of the slit as the total light power absorbed by the material (germanium) in the slit, normalized by the incident plane wave power flux (Yu, 2006). In two dimensions, the cross section is in the unit of length. We also calculate the absorption cross section for the same volume of germanium in a uniform thick slab with an anti-reflection coating, which is a typical configuration for conventional photodetectors (Yu, 2006) (Fig. 4.1b). The ratio between these two absorption cross sections defines the absorption enhancement factor η, which is a measure of the enhancement of the light absorption for a unit volume of absorbing material. We employ single-mode scattering matrix theory to account for the absorption cross section σ_{A} of the slit in the structure of Fig. 4.1a (Kocabas, 2008). We define the transmission cross section σ_{T} of a silver–germanium–silver MDM waveguide of width w (in the unit of length in two dimensions) as the transmitted power into the waveguide from the structure above the entrance side of the slit of Fig. 4.1a, normalized by the incident plane wave power flux (Fig. 4.1c). We also define $r_1(r_2)$ as the complex magnetic field reflection coefficient for the fundamental propagating TM mode in a silver–germanium–silver MDM waveguide of width w at the interface of such a waveguide with the structure above the entrance side (below the exit side) of the slit of Fig. 4.1a (Figs. 4.1d,e). We use FDFD to numerically extract σ_T, r_1, and r_2 (Kocabas, 2008). The absorption cross section σ_A of the slit can then be calculated using scattering matrix theory as

$$\sigma_A = \frac{f_{Ge}\sigma_T(1-|\exp(-\gamma L)|^2)|(1+r_2\exp(-\gamma L)|^2}{|1-r_1 r_2 \exp(-2\gamma L)|^2}, \tag{4.1}$$

where γ is the complex wave vector of the fundamental propagating TM mode in a silver–germanium–silver MDM waveguide of width w, L is the length of the slit, and f_{Ge} is the ratio of the power absorbed in germanium to the total power absorbed in the slit, which also includes the power absorbed in the metal. Based on Eq. (4.1), we observe that for fixed slit dimensions, γ, L, and f_{Ge} are fixed, so that the absorption cross section σ_A of the slit is solely determined by σ_T, r_1, and r_2. These three parameters, in turn, can be tuned by adjusting the geometrical dimensions of the microcavities at the entrance and exit sides of the slit. In addition, the absorption cross section for the same volume of germanium in a conventional bulk photodetector with anti-reflection coating (Fig. 4.1b) is

$$\tau_{A,bulk} = \omega(1-|\exp(-\gamma_0 L|^2), \tag{4.2}$$

where γ_0 is the complex wave vector of a plane wave propagating in germanium. Thus, the absorption enhancement factor for the slit is

$$\eta \equiv \frac{\sigma_A}{\sigma_{A,bulk}} = \frac{f_{Ge}\sigma_T(1-|\exp(-\sigma L)|^2)(1+r_2\ |\exp(-\gamma L)|^2)}{\omega(1-|\exp(-\gamma_0 L)|^2)|1-r_1 r_2\exp(-2\gamma L)|^2} = \eta_1\eta_2\eta_3,$$

where $\eta_1 \equiv \frac{f_{Ge}(1-|\exp(-\gamma L)|^2)}{1-|\exp(-\gamma_0 L)|^2}$, $\eta_2 \equiv \frac{\sigma_T}{\omega}$ and $\eta_3 \equiv \frac{1+|r_2\exp(-\gamma L)|^2}{|1-r_1 r_2 r\exp(-2\gamma L)|^2}$. (4.3)

We note that η_1 is fixed when the slit dimensions w and L are fixed, and does not depend on the microcavity structures above and below the slit. As an example, for a silver–germanium–silver slit with w = 50 nm, L = 122 nm at λ_0 = 1.55 μm we find $\eta_1 \sim 1.42$. In addition, η_2 is the transmission cross section enhancement factor of the MDM waveguide with respect to its geometrical cross section, associated with the microcavities above the entrance side of the slit. Finally, η^3 is the resonance enhancement factor, associated with the slit resonance. We note that η^3 is a function of the reflection coefficients r_1 and r_2 at both sides of the slit, and therefore depends on both the structure above and the structure below the slit. We also observe that the resonance enhancement factor η^3 is maximized for $|r_1|, |r_2| < 1$, and when the slit resonance condition $\arg(r_1) + \arg(r_2) - 2\text{Im}(\gamma)L = -2m\pi$ is satisfied, where m is an integer. We note, however, that there is a tradeoff between the resonance enhancement factor η_3, and the transmission cross section enhancement factor η_2. As a result, in the optimized structures we have $|r_1| < 1$ in all cases.

4.2.2 Single-Slit Structure

We first consider our reference structure consisting of a single subwavelength slit in a metal film deposited on a substrate (Fig. 4.2a). In Fig. 4.2b, we show the absorption cross section σ_A, and the absorption enhancement factor η for the structure of Fig. 4.2a as a function of the slit length L calculated using FDFD. We observe that as the slit length L increases, both the absorption cross section and the absorption enhancement factor exhibit peaks, corresponding to the Fabry–Pérot resonances in the slit. In Fig. 4.2b we also show σ_A and η calculated using scattering matrix theory (Eqs. (4.1–4.3)). We observe that there is excellent agreement between the scattering matrix theory results and the exact results obtained using FDFD. Similarly, excellent agreement between the results of these two methods is observed for all the structures considered in this section (Table 4.1). The maximum absorption enhancement factor η of ~43.9 with respect to a conventional photodetector is obtained at the first peak (L = 122 nm), due to the strong electromagnetic field enhancement associated with the Fabry–Pérot resonance in the slit (White, 2009). For such a structure, the transmission cross section enhancement factor is η_2 ~ 0.96 (Table 4.1). In other words, the transmission cross section of a silver–germanium–silver MDM waveguide with w = 50 nm at λ_0 = 1.55 μm is approximately equal to its geometrical cross section. In addition, the resonance enhancement factor is η_3 ~ 32.3 (Table 4.1). Such large resonance enhancement is due to the strong reflectivities r_1, r_2 at both sides of the slit (Table 4.1), associated with the strong impedance mismatch between the fundamental MDM mode in the slit and propagating plane waves in air and in the silica substrate (Yu, 2006).

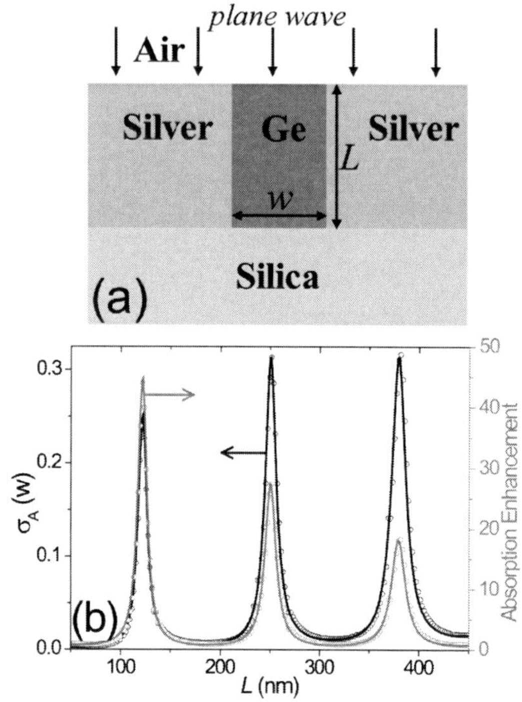

Figure 4.2 (a) Schematic of a structure consisting of a single slit in a silver film deposited on a silica substrate. The slit is filled with germanium. (b) Absorption cross section σ_A in units of w (black line and circles), and absorption enhancement factor η (red line and circles) for the structure of Fig. 4.2a as a function of slit length L calculated using FDFD (circles) and scattering matrix theory (solid line). Results are shown for w = 50 nm and λ_0 = 1.55 μm.

Table 4.1 Absorption enhancement factor η, transmission cross section enhancement factor η_2, resonance enhancement factor η_3, and reflection coefficients r_1, r_2

(*N*, *M*)	**(0,0)**	**(1,0)**	**(1,1)**	**(2,2)**
η	43.9	75.7	133.6	410.6
η_2	0.96	2.98	1.87	2.96
η_3	32.3	17.9	50.4	97.9

(N, M)	(0,0)	(1,0)	(1,1)	(2,2)
r_1	$0.940e^{2.91i}$	$0.416e^{2.91i}$	$0.972e^{2.86i}$	$0.998e^{2.91i}$
$\sigma_A(w)$ S-matrix	0.250	0.416	0.744	2.299
FDFD	0.246	0.423	0.747	2.295

Note: Also shown is the absorption cross section σ_A in units of w, calculated using scattering matrix theory and FDFD. Results are shown for the optimized structures of Fig. 4.1a with (N, M) = (0, 0), (1, 0), (1, 1), and (2, 2).

4.2.3 Structure with a Single Microcavity at the Entrance of the Slit

We next consider a structure consisting of a slit in a metal film with a single microcavity at the entrance side of the slit (Fig. 4.3a). In Fig. 4.3b, we show the absorption cross section σ_A for the structure of Fig. 4.3a as a function of the width w_{T1} and length d_{T1} of the microcavity. For the range of parameters shown, we observe two absorption peaks associated with different resonant modes of the microcavity. When the microcavity dimensions approach a resonance, the light power collected by the microcavity increases, and the coupling to the slit is enhanced. However, we found that the maximum absorption cross section in the slit is not obtained when the microcavity is on resonance. This is because the on-resonance field pattern in the microcavity is a standing wave, which does not correspond to optimum coupling to the slit (Min, 2008). The maximum absorption cross section of $\sigma_A \sim 0.423w$ is obtained for such a structure at w_{T1} = 380 nm and d_{T1} = 200 nm, and the corresponding absorption enhancement factor is $\eta \sim 75.7$ (Table 4.1). We observe that for such a structure the transmission cross section enhancement factor is $\eta^2 \sim 2.98$ (Table 4.1), which is ~3.1 times larger compared to a slit without the microcavity. In other words, the microcavity can greatly enhance the coupling of the incident light into the slit mode, by improving the impedance matching between the incident plane wave and the slit mode. On the other hand, the resonance enhancement factor for the optimized structure of Fig. 4.3a is $\eta_3 \sim 17.9$ (Table 4.1), which is ~1.8 times smaller than the one of a slit without a microcavity. This is because the reflectivity $|r_1|^2$ at the interface between the silver–germanium–silver slit and the silver–silica–silver microcavity is smaller than the one at the interface between the silver–germanium–silver slit and air (Table 4.1). Thus, overall the use of an optimized single microcavity at the entrance side of the slit (Fig. 4.3a) results in 3.1/1.8 $\simeq$ 1.7 times larger absorption cross section compared to the optimized reference slit without a microcavity (Fig. 4.2a). In Fig. 4.3c, we show the magnetic field profile for the structure of Fig. 4.3a with dimensions optimized for maximum absorption cross section σ_A. We observe that the field in the microcavity is weaker than the field in the slit. The maximum magnetic field amplitude enhancement in the slit with respect to the incident plane wave is ~18 (Fig. 4.3c).

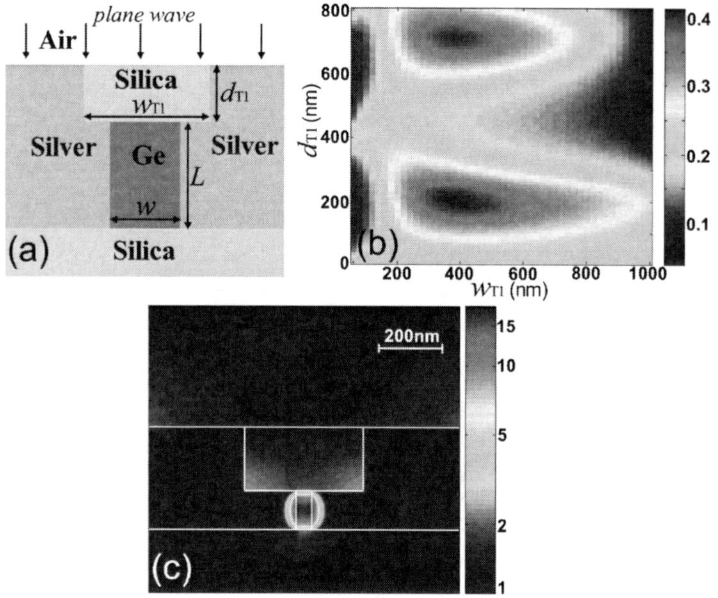

Figure 4.3 (a) Schematic of a structure consisting of a slit in a silver film deposited on a silica substrate with a single microcavity at the entrance side of the slit. The slit is filled with germanium, while the microcavity is filled with silica. (b) Absorption cross section σ_A in units of w for the structure of Fig. 4.3a as a function of width w_{T1} and length d_{T1} of the microcavity calculated using FDFD. Results are shown for w = 50 nm, L = 122 nm, and λ_0 = 1.55 μm. (c) Profile of the magnetic field amplitude enhancement with respect to the field amplitude of the incident plane wave for the structure of Fig. 4.3a for w_{T1} = 380 nm and d_{T1} = 200 nm. All other parameters are as shown in Fig. 4.3b.

4.2.4 Structure with a Single Microcavity at the Entrance and Exit of the Slit

We next consider a structure with a single microcavity at each of the entrance and exit sides of the slit (Fig. 4.1a with $N = M = 1$). We use a genetic global optimization algorithm in combination with FDFD (Krishnakumar, 1989; Veronis, 2007a) to optimize the width and length of both microcavities in the structure for maximum absorption cross section σ_A. All microcavity dimensions are limited to less than 1 μm. The maximum absorption cross section for such a structure is found to be $\sigma_A \sim 0.747w$, and the corresponding absorption enhancement factor is $\eta \sim 133.6$ (Table 4.1). Similar to the optimized ($N = 1$, $M = 0$) structure (Fig. 4.3a), the microcavity at the entrance side of the slit results in larger (by a factor of $\sim$1.95) transmission cross section enhancement factor η_2, and smaller reflectivity $|r_1|^2$ compared to a slit without microcavities (Table 4.1). However, the presence of a microcavity at the exit side of the slit for the ($N = 1$, $M = 1$) structure results in larger reflectivity $|r_2|^2$ compared to a slit without a microcavity (Table 4.1). This is because such a cavity can be tuned to either resonantly enhance or resonantly suppress the reflectivity. In the optimized ($N = 1$, $M = 1$) structure, resonant enhancement of the reflectivity $|r_2|^2$ is achieved by proper choice of the microcavity length d_{B1}. The increase of the reflectivity $|r_2|^2$ at the exit side of the slit for the optimized ($N = 1$, $M = 1$) structure overcompensates the decrease of the reflectivity $|r_1|^2$ at the entrance side, so that its resonance enhancement factor $\eta_3 \sim 50.4$ (Table 4.1) is $\sim$1.5 times larger than the one of a slit without a microcavity. Thus, the use of a single microcavity at both the entrance and exit sides of the slit enables increasing both the transmission cross section enhancement factor η^2, and the resonance enhancement factor η_3. Overall, such a structure, when optimized, results in $1.95 \times 1.5 \simeq 3$ times larger absorption cross section compared to the optimized reference slit without microcavities (Fig. 4.2a).

4.2.5 Structure with Multiple Microcavities at the Entrance and Exit of the Slit

To further enhance the absorption cross section of the slit, we consider structures with multiple microcavities at both the entrance and exit sides of the slit (Fig. 4.1a). More specifically, we use the genetic optimization algorithm to optimize the widths and lengths of all microcavities in a (N = 2, M = 2) structure. As before, the dimensions of the structures at both the entrance and the exit sides of the slit are limited to less than 1 μm. The maximum absorption cross section for such a structure is found to be $\sigma_A \sim 2.295w$, and the corresponding absorption enhancement factor is $\eta \sim 410.6$ (Table 4.1). We observe that the use of multiple microcavities at the entrance side of the slit (N = 2, M = 2) increases the transmission cross section enhancement factor η_2 compared to the single microcavity (N = 1, M = 1) structure (Table 4.1). This is because multiple-section structures can improve the impedance matching and therefore the coupling between optical modes (Veronis, 2007a). In addition, the use of multiple microcavities at the entrance and exit sides of the slit (N = 2, M = 2) also increases the corresponding reflectivities $|r_1|^2$ and $|r_2|^2$ compared to the single microcavity (N = 1, M = 1) structure (Table 4.1). Thus, a large resonance enhancement factor $\eta_3 \sim 97.9$ is obtained (Table 4.1). This is because multiple-section structures can be more finely tuned than single-section structures, and can therefore provide larger resonantly enhanced reflectivity. This is analogous to the multilayer Bragg reflectors, which can provide larger reflectivity compared to single layer structures. Overall, the optimized (N = 2, M = 2) structure results in ~3.1 times larger transmission cross section enhancement factor η_2, ~3 times larger resonance enhancement factor η_3, and therefore ~9.3 times larger absorption cross section compared to the optimized reference slit without microcavities (Fig. 4.2a).

In Figs. 4.4a,b, we show the magnetic field profile for the (N = 1, M = 1) and (N = 2, M = 2) cases, respectively, with dimensions optimized for maximum absorption cross section σ_A. We find that as the number of microcavities increases, the field in the microcavities at the entrance side is stronger, because more incident light is collected in the microcavities. The maximum magnetic field amplitude enhancement in the slit with respect to the incident plane wave is ~23 and ~40 for the (N = 1, M = 1) and (N = 2, M = 2) cases, respectively.

The microcavity-enhanced structures were optimized at a single wavelength of λ_0 = 1.55 μm. In Fig. 4.5, we show the absorption cross section σ_A as a function of incident wavelength for the optimized structures of Fig. 4.2a ($N = M = 0$), Fig. 4.4a ($N = M = 1$), and Fig. 4.4b ($N = M = 2$). We observe that the operation wavelength range for high absorption is broad. For example, the full width at half maximum (FWHM) of the absorption peak in the ($N = M = 2$) case is ~50 nm. This is because in all cases the enhanced absorption is not associated with any strong resonances. In other words, the quality factors Q of the microcavities are low.

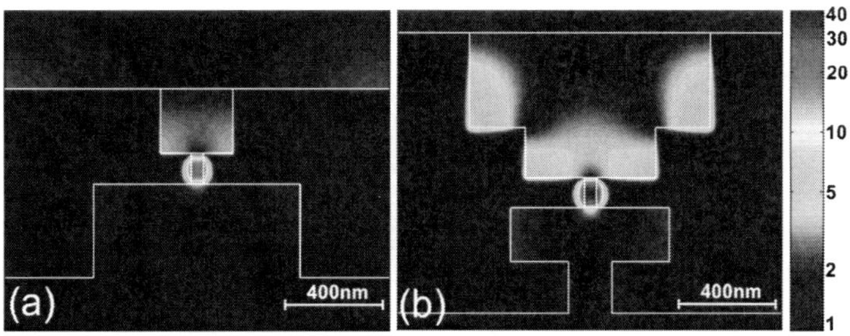

Figure 4.4 (a) Profile of the magnetic field amplitude enhancement with respect to the field amplitude of the incident plane wave for the optimized structure of Fig. 4.1a with $N = M = 1$. Results are shown for (w_{T1}, d_{T1}, w_{B1}, d_{B1}) = (300, 260, 860, 380) nm. All other parameters are as shown in Fig. 4.3b. (b) Profile of the magnetic field amplitude enhancement with respect to the field amplitude of the incident plane wave for the optimized structure of Fig. 4.1a with $N = M = 2$. Results are shown for (w_{T1}, d_{T1}, w_{T2}, d_{T2}, w_{B1}, d_{B1}, w_{B2}, d_{B2}) = (1000, 390, 540, 200, 660, 220, 180, 210) nm. All other parameters are as shown in Fig. 4.3b.

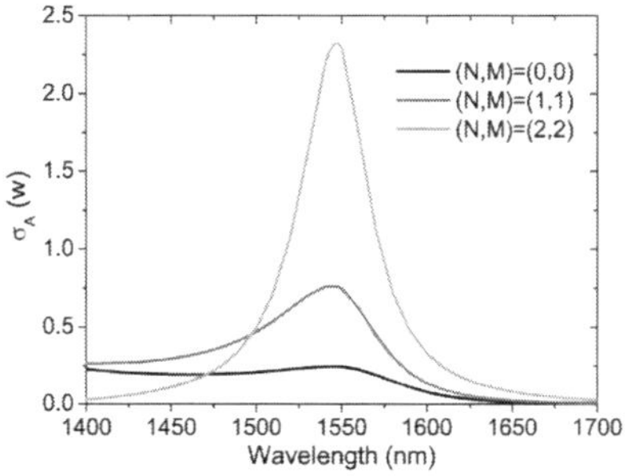

Figure 4.5 Absorption cross section σ_A in units of w as a function of wavelength for the optimized structures of Fig. 4.1a with $N = M = 0$ (black line), $N = M = 1$ (red line), and $N = M = 2$ (green line). All other parameters for the $N = M = 0$, $N = M = 1$, and $N = M = 2$ cases are as shown in Figs. 4.3b, 4.4a,b, respectively.

4.3 Nanophotonic Resonators for Enhanced Coupling to Metal–Dielectric–Metal Plasmonic Waveguides

Plasmonic waveguides have shown the potential to guide subwavelength optical modes, the so-called surface plasmon polaritons, at metal–dielectric interfaces. Several different nanoscale plasmonic waveguiding structures have been proposed, such as metallic nanowires, metallic nanoparticle arrays, V-shaped grooves, and MDM waveguides (Krenn, 2002; Maier, 2003; Bozhevolnyi, 2006; Zia, 2004; Veronis, 2005; Hosseini, 2007; Matsuzaki, 2008; Lin, 2008). Among these, MDM plasmonic waveguides, which are the optical analogue of microwave two-conductor transmission lines (Pozar, 1998), are of particular interest because they support modes with deep subwavelength scale over a very wide range of frequencies extending from DC to visible (Economou, 1969). Thus, MDM waveguides could provide an interface between conventional optics and subwavelength electronic and optoelectronic devices.

For applications involving MDM plasmonic waveguides, it is essential to develop compact structures to couple light efficiently into such waveguides (Veronis, 2007a). Several different couplers between MDM and dielectric waveguides have been investigated both theoretically and experimentally (Veronis, 2007a; Feigenbaum, 2007; Wahsheh, 2009; Yang, 2010; Tian, 2009; Delacour, 2010). In addition, structures for coupling free space radiation into MDM waveguides have also been investigated. In particular, Preiner et al. (Preiner, 2008) investigated subwavelength diffraction gratings as coupling structures into MDM waveguide modes. However, in diffraction grating structures, several grating periods are required for efficient waveguide mode excitation, so that such structures need to be several microns long when designed to operate at frequencies around the optical communication wavelength (λ_0 = 1.55 μm). In addition, in several experimental investigations of MDM waveguides and devices, a single slit was used to couple light from free space into MDM plasmonic waveguides (Dionne, 2006); Lezec, 2007; Bozhevolnyi, 2009; Neutens, 2009; Diest, 2009). While single-slit coupling structures are more compact, slit-based coupler designs have not been investigated in detail.

In this section, we investigate compact wavelength-scale slit-based structures for coupling free space light into MDM plasmonic waveguides (Huang, 2012). We show that for a single-slit structure the coupling efficiency is limited by a trade-off between the light power coupled into the slit, and the transmission of the slit-MDM waveguide junction. We next consider a two-section slit structure, and show that for such a structure the upper slit section enhances the coupling of the incident light into the lower slit section, by improving the impedance matching between the incident plane wave and the lower slit mode. The optimized two-section slit structure results in ~2.3 times enhancement of the coupling into the MDM plasmonic waveguide compared to the optimized single-slit structure. We then consider a symmetric double-slit structure.

We show that for such a structure, the surface plasmons excited at the metal–air interfaces are partially coupled into the slits, and thus the coupling of the incident light into the slits is greatly enhanced. The optimized double-slit structure results in ~3.3 times coupling enhancement compared to the optimized single-slit structure. Finally, we show that while all incoupling structures are optimized at a single wavelength, the operation wavelength range for high coupling efficiency is broad.

The remainder of this section is organized as follows. In subsection 4.3.1, we first define the transmission cross section of the MDM plasmonic waveguide for a given coupling structure, and briefly describe the simulation method used for the analysis of the couplers. The results obtained for the single-slit, two-section slit, and double-slit coupling structures are presented in Subsections 4.3.2, 4.3.3, and 4.3.4, respectively.

4.3.1 Transmission Cross Section

We consider a silver–silica–silver MDM plasmonic waveguide in which the upper metal layer has a finite thickness (Fig. 4.6a). The minimum thickness of this metal layer is chosen to be 150 nm. For such a thickness, the field profile and wave vector of the fundamental TM mode supported by such a waveguide at optical frequencies are essentially identical to the ones of a MDM plasmonic waveguide with semi-infinite metal layers. We consider compact wavelength-scale structures for incoupling a normally incident plane wave from free space into the fundamental mode of the silver–silica–silver MDM plasmonic waveguide. In all cases, the total width of the incoupling structure is limited to less than 1.1 μm, which approximately corresponds to one wavelength in silica ($\lambda_s = \lambda_0/n_s$, where $n_s = 1.44$), when operating at the optical communication wavelength ($\lambda_0 = 1.55$ μm).

Owing to the symmetry of all coupling structures considered in this section, the same amount of power couples into the left and right propagating silver–silica–silver MDM waveguide modes. In other words, half of the total incoupled power couples into each of the left and right propagating MDM waveguide modes. For comparison of different incoupling configurations, we define the transmission cross section σ_T of the silver–silica–silver MDM waveguide as the total light power coupled into the right propagating fundamental TM mode of the waveguide, normalized by the incident plane wave power flux density (Veronis, 2007a). In two dimensions, the transmission cross section is in the unit of length.

As shown in Section 4.2, we use the FDFD method to numerically calculate the transmission in the MDM plasmonic waveguide. PML absorbing boundary conditions are used at all boundaries of the simulation domain. We also use the total-field-scattered-field formulation to simulate the response of the structure to a normally incident plane wave input.

4.3.2 Single-Slit Coupler

We first consider a structure consisting of a single slit for incoupling a normally incident plane wave from free space into the fundamental mode of the silver–silica–silver MDM plasmonic waveguide with dielectric core thickness w. The slit extends half way into the dielectric core of the MDM waveguide (Fig. 4.6a). In Fig. 4.6b, we show the transmission cross section σ_T of the silver–silica–silver MDM waveguide in units of w for the single-slit structure of Fig. 4.6a as a function of the width d and length h of the slit. For the range of parameters shown, we observe one transmission peak. The maximum cross section of $\sigma_T \sim 4.67\ w$ is obtained for such an incoupling structure at $d = 250$ nm and $h = 205$ nm (Fig. 4.6b).

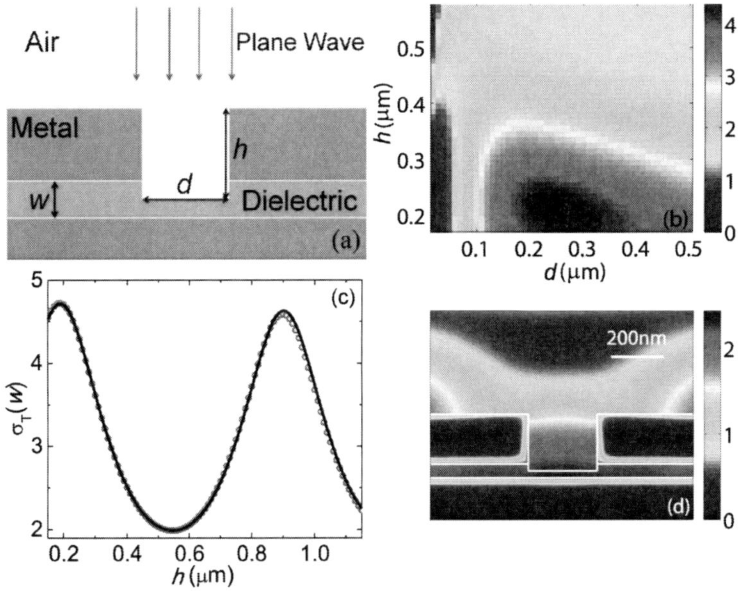

Figure 4.6 (a) Schematic of a structure consisting of a single slit for incoupling a normally incident plane wave from free space into the fundamental mode of a MDM plasmonic waveguide. (b) Transmission cross section σ_T of the MDM plasmonic waveguide in units of w for the structure of Fig. 4.6a as a function of the slit width d and length h calculated using FDFD. Results are shown for a silver–silica–silver structure with w = 50 nm at λ_0 = 1.55 μm. (c) Transmission cross section σ_T for the structure of Fig. 4.6a as a function of the slit length h calculated using FDFD (red circles) and scattering matrix theory (black solid line). Results are shown for d = 220 nm. All other parameters are as shown in Fig. 4.6b. (d) Profile of the magnetic field amplitude for the structure of Fig. 4.6a for d = 250 nm and h = 205 nm, normalized with respect to the field amplitude of the incident plane wave. All other parameters are as shown in Fig. 4.6b.

Both the silver–silica–silver MDM waveguide and the silver–air–silver slit have subwavelength widths, so that only the fundamental TM mode is propagating in them. Thus, we can use single-mode scattering matrix theory to account for the behavior of the system (Kocabas, 2008). We use FDFD to numerically extract the transmission cross section σ_{T1} of a silver–air–silver MDM waveguide with air core thickness d (Fig. 4.7a). We also use FDFD to extract the complex magnetic field reflection coefficient r_1 and transmission coefficient t_1 of the fundamental mode of a silver–air–silver MDM waveguide at the T-shaped junction with a silver–silica–silver MDM waveguide (Fig. 4.7b), as well as the reflection coefficient r_2 at the interface between the silver–air–silver MDM waveguide and air (Fig. 4.7c).

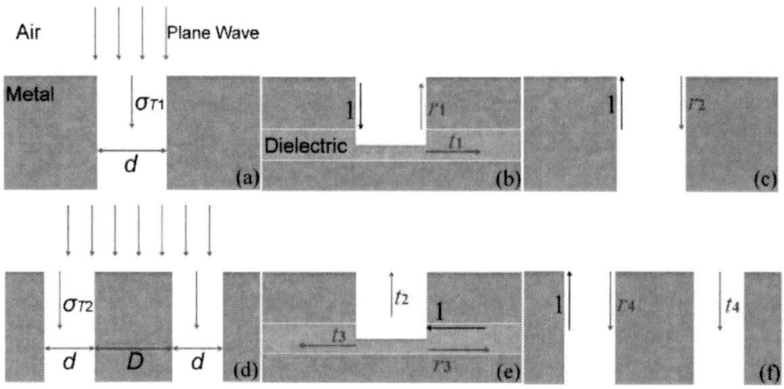

Figure 4.7 (a) Schematic defining the transmission cross section σ_{T1} of a semi-infinite MDM waveguide when a plane wave is normally incident on it. (b) Schematic defining the reflection coefficient r_1, and transmission coefficient t_1 when the fundamental TM mode of a metal–air–metal waveguide is incident at the junction with a metal–dielectric–metal waveguide. (c) Schematic defining the reflection coefficient r_2 of the fundamental TM mode of a MDM waveguide at the waveguide/air interface. (d) Schematic defining the transmission cross section σ_{T2} of two semi-infinite MDM waveguides when a plane wave is normally incident on them. (e) Schematic defining the reflection coefficient r_3, and transmission coefficients t_2, t_3 when the fundamental TM mode of a metal–dielectric–metal waveguide is incident at the junction with a metal–air–metal waveguide. (f) Schematic of a structure consisting of two semi-infinite MDM waveguides defining the reflection coefficient r_4 of the fundamental TM mode of one of the MDM waveguides at the waveguide/air interface, and the transmission coefficient t_4 into the other MDM waveguide.

The transmission cross section σ_T of the silver–silica–silver MDM waveguide for the single-slit structure of Fig. 4.6a can then be calculated using scattering matrix theory as (Kocabas, 2008):

$$\sigma_{\mathrm{T}} = \sigma_{\mathrm{T}}, \eta_{\mathrm{res}}, T_{\mathrm{splitter}}, \tag{4.4}$$

where $T_{\mathrm{splitter}} = |t_1|^2$ is the power transmission coefficient of the T-shaped junction of Fig. 4.7b, $\eta_{\mathrm{res1}} = \left|\dfrac{\exp(-\gamma h)}{1 - r_1 r_2 \exp(-2\gamma h)}\right|^2$ resonance enhancement factor associated with the silver–air–silver slit resonance, and $\gamma_1 = \alpha_1 + i\beta_1$ is the complex wave vector of the fundamental propagating TM mode in a silver–air–silver MDM waveguide with air core thickness d. We note that η_{res1} is a function of the reflection coefficients r_1 and r_2 at both sides of the silver–air–silver slit. We also observe that the resonance enhancement factor η_{res1} exhibits a maximum when the slit Fabry–Pérot resonance condition $\arg(r_1) + \arg(r_2) - 2\beta_1 h = -2m\pi$ is satisfied, where m is an integer. Thus, for a given silver–air–silver slit width d, the transmission cross section σ_{T} of the silver–silica–silver MDM waveguide is maximized for slit lengths h which satisfy the above Fabry–Pérot resonance condition.

In Fig. 4.6c, we show the transmission cross section σ_{T} of the silver–silica–silver MDM waveguide for the single-slit structure of Fig. 4.6a as a function of the slit length h calculated using FDFD. We observe that as the slit length h increases, the transmission cross section σ_{T} exhibits peaks, corresponding to the Fabry–Pérot resonances in the slit. The maximum transmission cross section σ_{T} is obtained at the first peak associated with the first Fabry–Pérot resonance in the slit. In Fig. 4.6c, we also show σ_{T} calculated using scattering matrix theory (Eq. (4.4)). We observe that there is excellent agreement between the scattering matrix theory results and the exact results obtained using FDFD.

For the optimized single-slit structure (d = 250 nm, h = 205 nm), the transmission cross section σ_{T1} of the corresponding silver–air–silver MDM waveguide with air core thickness d = 250 nm (Fig. 4.7a) is ~7.71w = 385.5 nm (Table 4.2). In other words, the silver–air–silver subwavelength MDM waveguide collects light from an area significantly larger than its geometric cross-sectional area (Veronis, 2007a). In addition, for the optimized single-slit structure the power transmission coefficient of the T-shaped junction is $T_{\mathrm{splitter}} \sim 0.37$, and the resonance enhancement factor is $\eta_{\mathrm{res1}} \sim 1.64$ (Table 4.2). Thus, ~2 × 37 = 74% of the incident power at the junction is transmitted to the left and right propagating modes of the silver–silica–silver MDM waveguide.

Table 4.2 Transmission cross sections $\sigma_{T1/2}$ and σ_T in units of w, power transmission coefficient of the T-shaped junction $T_{splitter}$, and resonance enhancement factors $\eta_{res1/2}$ calculated using scattering matrix theory

	Single slit	Two-section slit	Double-silt
$\sigma_{T1/2}(w)$	7.71	12.33	18.49
$T_{splitter}$	0.37	0.28	0.41
$\eta_{res1/2}$	1.64	3.11	2.02
$\sigma_T(w)$	4.67	10.75	15.29

Note: Results are shown for the optimized single-slit, two-section-slit, and double-slit structures of Figs. 4.6a, 4.9a, and 4.10a, respectively.

In Fig. 4.8, we show the transmission cross section σ_{T1} of a silver–air–silver MDM waveguide (Fig. 4.7a) as a function of the waveguide air core thickness d. We observe that, as expected, σ_{T1} increases monotonically as the thickness d increases. In other words, the light power collected by the waveguide increases as the air core thickness of the waveguide increases. On the other hand, the properties of the T-shaped junction (Fig. 4.7b) can be described using the concept of characteristic impedance and transmission line theory (Veronis, 2005; Pozar, 1998; Ramo, 1994). Based on transmission line theory, the structure is equivalent to the junction of three transmission lines. The load connected to the input transmission line at the junction consists of the series combination of the two output transmission lines. The characteristic impedances of the input and output transmission lines are $Z_1 = \frac{\gamma_1}{j\omega\varepsilon_0}$ and $Z_2 = \frac{\gamma_2}{j\omega\varepsilon}w$, respectively, where $\gamma_2 = \alpha_2 + i\beta_2$ is the complex wave vector of the fundamental propagating TM mode in a silver–silica–silver MDM waveguide with dielectric core thickness w, and $\simeq$ is the dielectric permittivity of silica (Veronis, 2005; Min, 2009). Thus, the equivalent load impedance is $Z_L = 2Z_2$, and the maximum transmission coefficient $T_{splitter}$ is obtained when the impedance matching condition $Z_1 = Z_L = 2Z_2$ is satisfied. The transmission coefficient $T_{splitter}$ of the T-shaped junction (Fig. 4.7b) therefore does not increase monotonically with d. As a result, the coupling efficiency of the single-slit structure is limited by a trade-off between the power incident at the slit-MDM waveguide junction, and the transmission coefficient T_{splitter} of the T-shaped junction. More specifically, the width of the optimized single slit is d = 250 nm, as mentioned above. If the slit width d decreased, the impedance matching between the silver–air–silver MDM input waveguide and the two silver–silica–silver MDM output waveguides would improve, and T_{splitter} therefore would increase. However, if d decreased, the transmission cross section σ_{T1} of the silver–air–silver MDM waveguide would decrease (Fig. 4.8). In addition, the reflectivity $|r_1|^2$ at the bottom side of the slit, and therefore the resonance enhancement factor η_{res1} would also decrease. Thus, the power incident at the junction between the slit and the silver–silica–silver MDM waveguide would decrease.

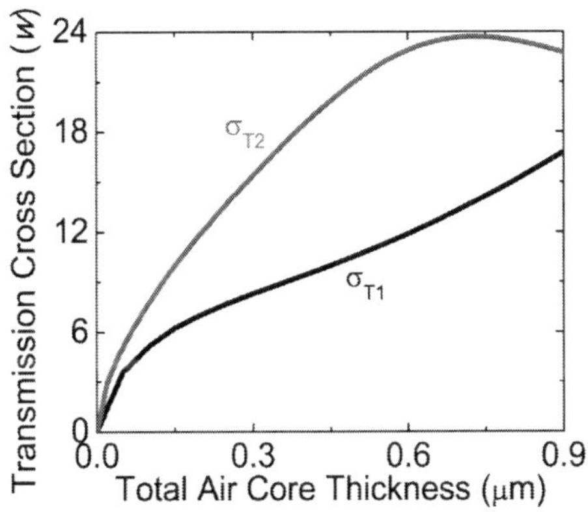

Figure 4.8 Transmission cross sections (in units of w = 50 nm) of a single silver–air–silver MDM waveguide σ_{T1} (Fig. 4.7a), and of a double silver–air–silver MDM waveguide σ_{T2} (Fig. 4.7d), as a function of their total air core thickness (d for the single and $2d$ for the double waveguide). The total width of the double waveguide is $2d + D$ = 1.1 μm.

In Fig. 4.6d, we show the magnetic field profile for the structure of Fig. 4.6a when the slit dimensions are optimized for maximum transmission cross section σ_{T}. We observe that since the transmission cross section of the silver–silica–silver MDM waveguide $\sigma_{\text{T}} \sim 4.67w$ is larger than its geometrical cross section w, the field in the MDM waveguide is enhanced with respect to the incident plane wave field. We find that the maximum magnetic field amplitude enhancement in the silver–silica–silver waveguide with respect to the incident plane wave is ~2.4 (Fig. 4.6d).

4.3.3 Two-Section Slit Coupler

To enhance the transmission cross section σ_T of the silver–silica–silver MDM plasmonic waveguide, we next consider a structure consisting of a two-section slit for incoupling light into the waveguide (Fig. 4.9a). The lengths h_1, h_2 and widths d_1, d_2 of these slit sections are optimized using a genetic global optimization algorithm in combination with FDFD to maximize the transmission cross section σ_T of the silver–silica–silver MDM waveguide. As before, the width of the incoupling structure is limited to less than 1.1 μm. Using this approach, the maximum transmission cross section of the silver–silica–silver MDM waveguide for such a two-section slit structure is found to be $\sigma_T \sim 10.75w$ (Table 4.2) for $d_1 = 410$ nm, $d_2 = 1100$ nm, $h_1 = 230$ nm, and $h_2 = 540$ nm.

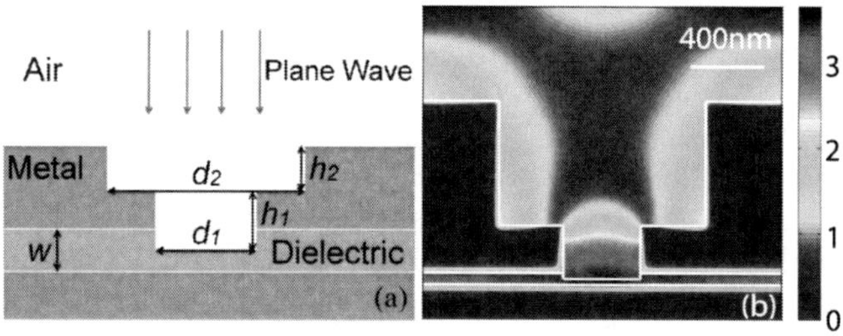

Figure 4.9 (a) Schematic of a structure consisting of a two-section slit for incoupling a normally incident plane wave from free space into the fundamental mode of a MDM plasmonic waveguide. (b) Profile of the magnetic field amplitude for the optimized structure of Fig. 4.9a with $d_1 = 410$ nm, $d_2 = 1100$ nm, $h_1 = 230$ nm, and $h_2 = 540$ nm, normalized with respect to the field amplitude of the incident plane wave. All other parameters are as shown in Fig. 4.6b.

We observe that for such a structure the transmission cross section of the corresponding silver–air–silver MDM waveguide (with air core thickness d_1) is $\sigma_{T1} \sim 12.33w$ (Table 4.2), which is ~1.6 times larger compared to the optimized single-slit coupler. In other words, the upper slit section can enhance the coupling of the incident light into the lower slit section, by improving the impedance matching between the incident plane wave and the lower slit mode (Min, 2011). In addition, the resonance enhancement factor of the optimized two-section slit structure is $\eta_{res1} \sim 3.11$ (Table 4.2), which is ~1.9 times larger compared to the optimized single-slit coupler. We found that the increase in the resonance enhancement factor η_{res1} of this two-section slit structure is due to larger reflectivities $|r_1|^2$ and $|r_2|^2$ at both sides of the lower slit section compared to the optimized single-slit coupler. On the other hand, the transmission coefficient of the T-shaped junction for the optimized two-section slit structure of Fig. 4.9a is $T_{splitter} \sim 0.28$ (Table 4.2), which is ~1.3 times smaller than the one of the optimized single-slit structure. This is due to larger mismatch between the characteristic impedance of the input waveguide Z_1 and the load impedance $Z_L = 2Z_2$ at the T-shaped junction. Thus, overall the use of an optimized two-section slit coupler (Fig. 4.9a) results in $1.6 \times 1.9/1.3 \simeq 2.3$ times larger transmission cross section σ_T of the silver–silica–silver MDM waveguide compared to the single-slit coupler case (Fig. 4.6a). In Fig. 4.9b, we show the magnetic field profile for the structure of Fig. 4.9a with dimensions optimized for maximum transmission cross section σ_T of the silver–silica–silver MDM waveguide. The field in the narrower lower slit section is stronger than the field in the upper slit section. The maximum magnetic field amplitude enhancement in the silver–silica–silver MDM waveguide with respect to the incident plane wave is ~3.6 (Fig. 4.9b).

4.3.4 Double-Slit Coupler

To further enhance the transmission cross section σ_T of the silver–silica–silver MDM plasmonic waveguide, we consider a symmetric double-slit structure for incoupling light into the waveguide (Fig. 4.10a). As before, the total width $2d + D$ of the incoupling structure is limited to less than 1.1 μm. For such a double-slit coupling structure, we found that if $2d + D \leq 1.1$ μm, the maximum transmission cross section σ_T is obtained when $2d + D = 1.1$ μm. In the following, we therefore set $2d + D = 1.1$ μm. In Fig. 4.10b, we show the transmission cross section σ_T of the silver–silica–silver MDM waveguide in units of w for the structure of Fig. 4.10a as a function of the width d and length h of the slits. For the range of parameters shown, we observe one transmission peak in the silver–silica–silver MDM waveguide. The maximum transmission cross section of $\sigma_T \sim 15.29w$ is obtained for such an incoupling structure at $d = 200$ nm ($D = 700$ nm) and $h = 250$ nm. We also note that for $d \sim 400$ nm ($D \sim 300$ nm) the transmission into the silver–silica–silver MDM waveguide is almost zero (Fig. 4.10b). We found that this is because for a slit distance of $D \sim 300$ nm the incident light strongly couples into the silver–silica–silver waveguide resonator between the slits. In addition, there is almost no light coupled into the left and right propagating modes of the silver–silica–silver MDM waveguide, due to destructive interference between the wave directly coupled through the slit, and the wave coupled through the silver–silica–silver waveguide resonator.

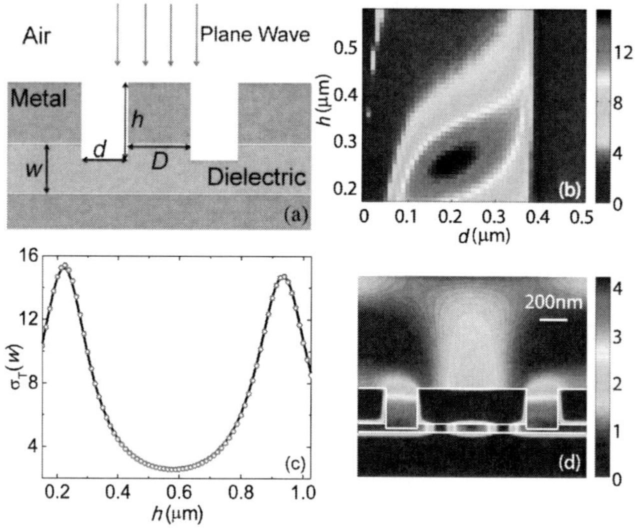

Figure 4.10 (a) Schematic of a double-slit structure for incoupling a normally incident plane wave from free space into the fundamental mode of a MDM plasmonic waveguide. (b) Transmission cross section σ_T of the MDM plasmonic waveguide in units of w for the structure of Fig. 4.10a as a function of the slit width d and length h calculated using FDFD. The total width of the incoupling structure is $2d + D = 1.1$ μm. All other parameters are as shown in Fig. 4.6b. (c) Transmission cross section σ_T for the structure of Fig. 4.10a as a function of the slit length h calculated using FDFD (red circles) and scattering matrix theory (black solid line). Results are shown for $d = 220$ nm. All other parameters are as shown in Fig. 4.10b. (d) Profile of the magnetic field amplitude for the structure of Fig. 4.10a for $d = 200$ nm and $h = 250$ nm, normalized with respect to the field amplitude of the incident plane wave. All other parameters are as shown in Fig. 4.10b.

We use again single-mode scattering matrix theory to account for the behavior of the system. We use FDFD to numerically extract the transmission cross section σ_{T2} of a double silver–air–silver MDM waveguide as shown in Fig. 4.7d. We also use FDFD to extract the complex magnetic field reflection coefficient r_3 and transmission coefficients t_2, t_3 of the fundamental mode of a silver–silica–silver MDM waveguide at the *T*-shaped junction with a silver–air–silver MDM waveguide (Fig. 4.7e). Note that $t_1 = t_2$ due to reciprocity (Pozar, 1998). Finally, we also extract the reflection coefficient r_4 at the interface between the silver–air–silver MDM waveguide and air, and the transmission coefficient t_4 into the other MDM waveguide, for the double MDM waveguide structure (Fig. 4.7f). The transmission cross section σ_T of the silver–silica–silver MDM plasmonic waveguide for the double-slit coupling structure of Fig. 4.10a is then calculated using scattering matrix theory as

$$\sigma_{\mathrm{T}} = \sigma_{\mathrm{T}}, \eta_{\mathrm{res}}, \mathrm{T}_{\mathrm{splitter}}, \tag{4.5}$$

where, as before, $T_{\mathrm{splitter}} = |t_1|^2 = |t_2|^2$ is the power transmission coefficient of the *T*-shaped junction, $\eta_{\mathrm{res2}} = \left|\frac{\exp(-\gamma_1 h)(1+t_3 A)}{1-(r_1+t_1 t_2 A)(r_4+t_4)\exp(-2\gamma_1 h)}\right|^2$ is the resonance enhancement factor associated with the complex resonator formed by the two silver–air–silver slits and the silver–silica–silver MDM waveguide resonator of length D between them, and $A = \frac{\exp(-\gamma_2 D) + r_3 \exp(-2\gamma_2 D)}{1 - r_3^2 \exp(-2\gamma_2 D)}$. Thus, we observe that the resonant enhancement factor η_{res_2} for such a complex resonator is similar to that of a Fabry–Pérot resonator with effective reflectivities $r_{\mathrm{eff1}} = r_1 + t_1 t_2 A$ and $r_{\mathrm{eff2}} = r_4 + t_4$.

In Fig. 4.10c, we show the transmission cross section σ_T for the structure of Fig. 4.10a as a function of the slit length h calculated using FDFD. We observe that as the slit length h increases, the transmission cross section σ_T exhibits peaks, associated with the resonances of the double-slit structure. The maximum transmission cross section σ_T is obtained at the first peak associated with the first resonant length of the slits. In Fig. 4.10c, we also show σ_T calculated using scattering matrix theory

(Eq. (4.5)). We observe that there is excellent agreement between the scattering matrix theory results and the exact results obtained using FDFD.

We found that for the optimized double-slit structure the transmission cross section of the corresponding double silver–air–silver MDM waveguide (Fig. 4.7d) is $\sigma_{T2} \sim 18.49w$ (Table 4.2), which is ~2.4 times larger compared to the transmission cross section $\sigma_{T1} \sim 7.71w$ of the single silver–air–silver MDM waveguide corresponding to the optimized single-slit coupler (Fig. 4.7a). In Fig. 4.8 we show the transmission cross sections of a single silver–air–silver MDM waveguide σ_{T1} (Fig. 4.7a), and of a double silver–air–silver MDM waveguide σ_{T2} (Fig. 4.7d) as a function of their total air core thickness (d for the single and $2d$ for the double waveguide). We observe that a double silver–air–silver MDM waveguide collects more light than a single silver–air–silver MDM waveguide with the same total air core thickness. This is because when a plane wave is incident on a semi-infinite MDM waveguide, surface plasmon waves are excited at the air–metal interfaces. In the double MDM waveguide structure (Fig. 4.7d), the power of these surface plasmon waves is partially coupled into the MDM waveguides, thus increasing the total light power collected by the structure. In addition, the resonance enhancement factor of the optimized double-slit structure $\eta_{res2} \sim 2.02$ (Table 4.2) is slightly larger than the resonance enhancement factor of the optimized single-slit coupler ($\eta_{res} \sim 1.64$). Overall, the use of an optimized double-slit coupler (Fig. 4.10a) results in ~3.3 times larger transmission cross section σ_T of the silver–silica–silver MDM waveguide compared to the optimized single-slit coupler case (Fig. 4.6a). In Fig. 4.10d, we show the magnetic field profile for the structure of Fig. 4.10a with dimensions optimized for maximum transmission cross section. The maximum magnetic field amplitude enhancement in the silver–silica–silver waveguide with respect to the incident plane wave is ~4.2.

The incoupling structures were all optimized at a single wavelength of λ_0 = 1.55 μm. In Fig. 4.11, we show the transmission cross section σ_T of the silver–silica–silver MDM plasmonic waveguide as a function of frequency for the optimized structures of Fig. 4.6d (single slit), Fig. 4.9b (two-section slit), and Fig. 4.10d (double slit). We observe that the operation frequency range for high transmission is broad. This is because in all cases the enhanced transmission is not associated with any strong resonances. In other words, the quality factors Q of the slit coupling structures are low. In Fig. 4.11, we also show the transmission cross section σ_T for the double-slit structure, if the metal in the MDM waveguide is lossless ($\varepsilon_{metal} = \text{Re}(\varepsilon_{metal})$), neglecting the imaginary part of the dielectric permittivity ($\text{Im}(\varepsilon_{metal})$). We observe that material losses in the metal do not significantly affect the transmission efficiency of the incoupling structures. This is because the dimensions of the incoupling structures are much smaller than the propagation lengths of the fundamental TM modes in the silver–silica–silver and the silver–air–silver waveguides. We found that neither the coupling of the incident light into the silver–air–silver slits nor the coupling between the slits and the silver–silica–silver MDM plasmonic waveguide are significantly affected by material losses in the metal.

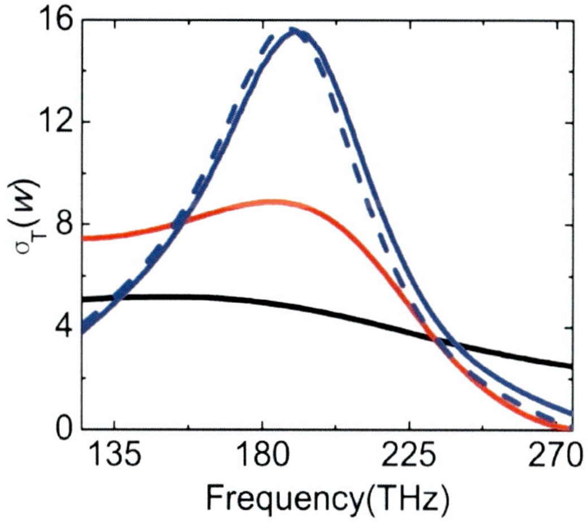

Figure 4.11 Transmission cross section σ_T spectra in units of w for the three optimized incoupling structures in Figs. 4.6a (single slit), 4.9(a) (two-section slit), and 4.10(a) (double slit). Results are shown for the structure of Fig. 4.6a with d = 250 nm, h = 205 nm (black line), for the structure of Fig. 4.9a with d_1 = 410 nm, d_2 = 1100 nm, h_1 = 230 nm, and h_2 = 540 nm (red line), and for the structure of Fig. 4.10a with d = 200 nm, h = 250 nm (blue line). Also shown are the transmission cross section σ_T spectra for the double-slit structure (Fig. 4.10a), if the metal in the MDM waveguide is lossless (blue dashed line). All other parameters are as shown in Fig. 4.6b.

4.4 Conclusions

In this chapter, we first investigated compact submicron structures consisting of multiple optical microcavities at both the entrance and exit sides of a subwavelength plasmonic slit filled with an absorbing material, with the goal to increase the absorption in the slit. Our reference structure consisted of a single subwavelength slit in a metal film deposited on a substrate. For such a structure, the maximum absorption enhancement factor with respect to a conventional photodetector is $\eta \sim 43.9$, and the absorption enhancement is due to the large resonant field enhancement in the slit, associated with the strong reflectivities at both sides of the slit.

To further enhance the absorption in the slit, we first considered a structure with a single microcavity at the entrance side of the slit. We found that the microcavity greatly enhances the coupling of the incident light into the slit, by improving the impedance matching between the incident plane wave and the slit mode. On the other hand, the microcavity reduces the reflectivity at the entrance side of the slit, and therefore the resonance enhancement factor. Overall, the use of an optimized single microcavity at the entrance side of the slit resulted in an absorption enhancement factor of $\eta \sim 75.7$, which is $\sim$1.7 times larger compared to the slit without a microcavity. We then considered a structure with a single microcavity at each of the entrance and exit sides of the slit. We found that the microcavity at the exit side of the slit results in larger reflectivity, and therefore larger resonant field enhancement. Overall, the use of an optimized single microcavity at both the entrance and exit sides of the slit resulted in an absorption enhancement factor of $\eta \sim 133.6$, which is $\sim$3 times larger compared to the slit without a microcavity.

We finally considered structures with multiple microcavities at both the entrance and exit sides of the slit. We found that the use of multiple microcavities at the entrance side of the slit further enhances the coupling of the incident light into the slit through improved impedance matching. In addition, the use of multiple microcavities at the entrance and exit sides of the slit also increases the reflectivities at both sides of the slit, and therefore the resonant field enhancement. Overall, the use of two optimized microcavities at both the entrance and exit sides of the slit resulted in an absorption enhancement factor of $\eta \sim 410.6$, which is ~9.3 times larger compared to the slit without a microcavity. We also found that while the microcavity-enhanced structures were optimized at a single wavelength, the operation wavelength range for high absorption is broad.

We note that for the fabrication of each microcavity of the structure, one can make first a silica ridge using lithography and etching processes. This step can then be followed by metal deposition and lift-off processes to form the metal parts of the microcavity (Yu, 2006).

In this chapter, we also investigated compact slit-based structures for coupling free space light into silver–silica–silver MDM plasmonic waveguides. In all cases, the total width of the incoupling structure was limited to less than 1.1 μm, which approximately corresponds to one wavelength in silica $\lambda_s = \lambda_0/n_s$, when operating at $\lambda_0 = 1.55$ μm. We first considered a coupling structure consisting of a single slit extending half way into the dielectric core of the MDM waveguide. We found that the coupling efficiency of such a single-slit structure is limited by a trade-off between the light power coupled into the slit, and the transmission of the slit-MDM waveguide T-shaped junction.

To enhance the coupling into the silver–silica–silver MDM plasmonic waveguide, we next considered a two-section slit structure. We found that for such a structure the upper slit section enhances the coupling of the incident light into the lower slit section, by improving the impedance matching between the incident plane wave and the lower slit mode. In addition, the use of the optimized two-section slit structure increases the reflectivities at both sides of the lower slit section, and therefore the resonance enhancement factor. On the other hand, the transmission of the T-shaped junction for the optimized two-section slit structure is smaller than the one of the optimized single-slit structure. Overall, the use of an optimized two-section slit coupler resulted in ~2.3 times enhancement of the coupling into the MDM plasmonic waveguide compared to the optimized single-slit coupler.

To further enhance the coupling into the silver–silica–silver MDM plasmonic waveguide, we considered a symmetric double-slit structure. We found that such a structure greatly enhances the coupling of the incident light into the slits. This is because the incident light excites surface plasmons at the air–metal interfaces.

In the case of a double-slit structure, these plasmons are partially coupled into the slits, thus increasing the total light power collected by the structure. In addition, the resonance enhancement factor of the optimized double-slit coupler is slightly larger than the resonance enhancement factor of the optimized single-slit coupler. Overall, the use of an optimized double-slit coupler resulted in ~3.3 times enhancement of the coupling into the MDM plasmonic waveguide compared to the optimized single-slit coupler. We also found that while the incoupling structures were all optimized at a single wavelength, the operation wavelength range for high coupling efficiency is broad.

As final remarks, for wavelength-scale slit-based structures, the double-slit structure results in optimal coupling performance. We verified that if three or more slits are used in a wavelength-scale coupler, the performance is always worse due to destructive interference between the waves coupled through the slits. Moreover, if a reflector is introduced in one of the two silver–silica–silver MDM output waveguides, then all the incoupled power will couple into the other silver–silica–silver MDM output waveguide. In addition, the proposed slit-based structures can also be used to couple light from a MDM plasmonic waveguide into free space. We found that when the single-slit structure is used to outcouple light, the radiation pattern of the structure is approximately isotropic (Verslegers, 2010). On the other hand, we found that the two-section slit and double-slit structures introduce anisotropy in the radiation pattern, with stronger radiation in the normal direction (Verslegers, 2010). Finally, we note that there are some analogies between the proposed coupling structures and the slot antennas used in the microwave frequency range (Balanis, 2005).

References

Balanis, C. A. (2005). *Antenna Theory: Analysis and Design*, 3rd ed. (Wiley).

Barnes, W. L., Dereux, A., and Ebbesen, T. W. (2003). Surface plasmon subwavelength optics, *Nature*, **424**, 824–830.

Bozhevolnyi, S. I. (2009). *Plasmonic Nanoguides and Circuits* (World Scientific).

Bozhevolnyi, S. I., Volkov, V. S., Devaux, E., Laluet, J. Y., and Ebbesen, T. W. (2006). Channel plasmon subwavelength waveguide components including interferometers and ring resonators, *Nature*, **440**, 508–511.

Cao, L., Park, J. S., Fan, P., Clemens, B., and Brongersma, M. L. (2010). Resonant germanium nanoantenna photodetectors, *Nano Lett.*, **10**, 1229–1233.

Cai, W., White, J. S., and Brongersma, M. L. (2009). Compact, high-speed and power-efficient electrooptic plasmonic modulators, *Nano Lett.*, **9**, 4403–4411.

Delacour, C., Blaize, S., Grosse, P., Fedeli, J. M., Bruyant, A., Salas-Montiel, R., Lerondel, G., and Chelnokov, A. (2010). Efficient directional coupling between silicon and copper plasmonic nanoslot waveguides: Toward metal-oxide-silicon nanophotonics, *Nano Lett.*, **10**, 2922–2926.

Diest, K., Dionne, J. A., Spain, M., and Atwater, H. A. (2009). Tunable color filters based on metal-insulator-metal resonators, *Nano Lett.*, **9**, 2579–2583.

Dionne, J. A., Lezec, H. J., and Atwater, H. A. (2006). Highly confined photon transport in subwavelength metallic slot waveguides, *Nano Lett.*, **6**, 1928–1932.

Ebbesen, T. W., Lezec, H. J., Ghaemi, H. F., Thio, T., and Wolff, P. A. (1998). Extraordinary optical transmission through sub-wavelength hole arrays, *Nature*, **391**, 667–669.

Economou, E. N. (1969). Surface plasmons in thin films, *Phys. Rev.*, **182**, 539–554.

Feigenbaum, E., and Orenstein, M. (2007). Modeling of complementary void plasmon waveguiding, *J. Lightwave Technol.*, **25**, 2547–2562.

Fujikata, J., Ishi, T., Yokota, H., Kato, K., Yanagisawa, M., Nakada, M., Ishihara, K., Ohashi, K., Thio, T., and Linke, R. A. (2004). Surface plasmon enhancement effect and its application to near-field optical recording, *Trans. Magn. Soc. Jpn.*, **4**, 255–259.

Garca-Vidal, F. J., Martn-Moreno, L., Ebbesen, T. W., and Kuipers, L. (2010). Light passing through subwavelength apertures, *Rev. Mod. Phys.*, **82**, 729–787.

Garcia-Vidal, F. J., Lezec, H. J., Ebbesen, T. W., and Martin-Moreno, L. (2003). Multiple paths to enhance optical transmission through a single subwavelength slit, *Phys. Rev. Lett.*, **90**, 213901.

Gbur, G., Schouten, H. F., and Visser, T. D. (2005). Achieving superresolution in near-field optical data readout systems using surface plasmons, *Appl. Phys. Lett.*, **87**, 191109.

Genet, C., and Ebbesen, T. W. (2007). Light in tiny holes, *Nature*, **445**, 39–46.

Guo, B., Song, G., and Chen, L. (2007). Plasmonic very-small-aperture lasers, *Appl. Phys. Lett.*, **91**, 021103.

Hosseini, A., and Massoud, Y. (2007). Nanoscale surface plasmon based resonator using rectangular geometry, *Appl. Phys. Lett.*, **90**, 181102.

Huang, Y., Min, C., and Veronis, G. (2012). Compact slit-based couplers for metal-dielectric-metal plasmonic waveguides, *Opt. Express*, **20**, 22233–22244.

Ishi, T., Fujikata, J., Makita, K., Baba, T., and Ohashi, K. (2005). Si nano-photodiode with a surface plasmon antenna, *Jpn J. Appl. Phys.*, **44**, 364–366.

Jin, J. (2002). *The Finite Element Method in Electromagnetics* (Wiley).

Kocabas, S. E., Veronis, G., Miller, D. A. B., and Fan, S. (2008). Transmission line and equivalent circuit models for plasmonic waveguide components, *IEEE J. Sel. Top. Quantum Electron.*, **14**, 1462–1472.

Krenn, J. R., Lamprecht, B., Ditlbacher, H., Schider, G., Salerno, M., Leitner, A., and Aussenegg, F. R. (2002). Non-diffraction-limited light transport by gold nanowires, *Europhys. Lett.*, **60**, 663–669.

Krishnakumar, K. (1989). Micro-genetic algorithms for stationary and non-stationary function optimization, *Proc. SPIE*, **1196**, 289–296.

Lezec, H. J., Degiron, A., Devaux, E., Linke, R. A., Martin-Moreno, L., Garcia-Vidal, F. J., and Ebbesen, T. W. (2002). Beaming light from a subwavelength aperture, *Science*, **297**, 820–822.

Lezec, H. J., Dionne, J. A., and Atwater, H. A. (2007). Negative refraction at visible frequencies, *Science*, **316**, 430–432.

Lin, X. S., and Huang, X. G. (2008). Tooth-shaped plasmonic waveguide filters with nanometeric sizes, *Opt. Lett.*, **33**, 2874–2876.

Lpez-Tejeira, F., Rodrigo, S. G., Martin-Moreno, L., Garca-Vidal, F. J., Devaux, E., Ebbesen, T. W., Krenn, J. R., Radko, I. P., Bozhevolnyi, S. I., Gonzalez, M. U., Weeber, J.-C., and Dereux, A. (2007). Efficient unidirectional nanoslit couplers for surface plasmons, *Nat. Phys.*, **3**, 324–328.

Maier, S. A., Kik, P. G., Atwater, H. A., Meltzer, S., Harel, E., Koel, B. E., and Requicha, A. A. G. (2003). Local detection of electromagnetic energy transport below the diffraction limit in metal nanoparticle plasmon waveguides, *Nat. Mater.*, **2**, 229–232.

Martin-Moreno, L., Garcia-Vidal, F. J., Lezec, H. J., Degiron, A., and Ebbesen, T. W. (2003). Theory of highly directional emission from a single subwavelength aperture surrounded by surface corrugations, *Phys. Rev. Lett.*, **90**, 167401.

Matsuzaki, Y., Okamoto, T., Haraguchi, M., Fukui, M., and Nakagaki, M. (2008). Characteristics of gap plasmon waveguide with stub structures, *Opt. Express,* **16**, 16314–16325.

Min, C., and Veronis, G. (2009). Absorption switches in metal-dielectric-metal plasmonic waveguides, *Opt. Express,* **17**, 10757–10766.

Min, C., Yang, L., and Veronis, G. (2011). Microcavity enhanced optical absorption in subwavelength slits, *Opt. Express,* **19**, 26850–26858.

Min, Q., and Gordon, R. (2008). Surface plasmon microcavity for resonant transmission through a slit in a gold film, *Opt. Express,* **16**, 9708–9713.

Neutens, P., Dorpe, P. V., Vlaminck, I. D., Lagae, L., and Borghs, G. (2009). Electrical detection of confined gap plasmons in metal-insulator-metal waveguides, *Nat. Photon.,* **3**, 283–286.

Palik, E. D. (1985). *Handbook of Optical Constants of Solids* (Academic).

Pozar, D. M. (1998). *Microwave Engineering* (Wiley).

Preiner, M. J., Shimizu, K. T., White, J. S., and Melosh, N. A. (2008). Efficient optical coupling into metal-insulator-metal plasmon modes with subwavelength diffraction gratings, *Appl. Phys. Lett.,* **92**, 113109.

Ramo, S., Whinnery, J. R., and Van Duzer, T. (1994). *Fields and Waves in Communication Electronics* (Wiley).

Skigin, D. C., and Depine, R. A. (2005). Transmission resonances of metallic compound gratings with subwavelength slits, *Phys. Rev. Lett.,* **95**, 217402.

Shinada, S., Hashizume, J., and Koyama, F. (2003). Surface plasmon resonance on microaperture vertical-cavity surface-emitting laser with metal grating, *Appl. Phys. Lett.,* **83**, 836–838.

Taflove, A., and Hagness, S. C. (2005). *Computational Electrodynamics: The Finite-Difference Time-Domain Method,* 3rd ed. (Artech House).

Thio, T., Pellerin, K. M., Linke, R. A., Lezec, H. J., and Ebbesen, T. W. (2001). Enhanced light transmission through a single subwavelength aperture, *Opt. Lett.,* **26**, 1972–1974.

Tian, J., Yu, S., Yan, W., and Qiu, M. (2009). Broadband high-efficiency surface- plasmon-polariton coupler with silicon-metal interface, *Appl. Phys. Lett.,* **95**, 013504.

Veronis, G., and Fan, S. (2005). Bends and splitters in subwavelength metal-dielectric-metal plasmonic waveguides, *Appl. Phys. Lett.,* **87**, 131102.

Veronis, G., and Fan, S. (2007a). Theoretical investigation of compact couplers between dielectric slab waveguides and two-dimensional metal-dielectric-metal plasmonic waveguides, *Opt. Express,* **15**, 1211–1221.

Veronis, G., and Fan, S. (2007b). Overview of simulation techniques for plasmonic devices, in *Surface Plasmon Nanophotonics* (Springer).

Verslegers, L., Yu, Z., Catrysse, P. B., and Fan, S. (2010). Temporal coupled-mode theory for resonant apertures, *J. Opt. Soc. Am. B,* **27**, 1947–1956.

Wahsheh, R. A., Lu, Z. L., and Abushagur, M. A. G. (2009). Nanoplasmonic couplers and splitters, *Opt. Express,* **17**, 19033–19040.

White, J. S., Veronis, G., Yu, Z., Barnard, E. S., Chandran, A., Fan, S., and Brongersma, M. L. (2009). Extraordinary optical absorption through subwavelength slits, *Opt. Lett.*, **34**, 686–688.

Yang, R. X., Wahsheh, R. A., Lu, Z. L., and Abushagur, M. A. G. (2010). Efficient light coupling between dielectric slot waveguide and plasmonic slot waveguide, *Opt. Lett.,* **35**, 649–651.

Yu, Z., Veronis, G., Brongersma, M. L., and Fan, S. (2006). Design of mid-infrared photodetectors enhanced by surface plasmons on grating structures, *Appl. Phys. Lett.*, **89**, 151116.

Zia, R., Selker, M. D., Catrysse, P. B., and Brongersma, M. L. (2004). Geometries and materials for subwavelength surface plasmon modes, *J. Opt. Soc. Am. A,* **21**, 2442–2446.

Chapter 5

Photoluminescent Centers Interacting with Silicon-Based Photonic Devices

Xingjun Wang and Zhiping Zhou

State Key Laboratory of Advanced Optical Communication Systems and Networks,
School of Electronics Engineering and Computer Science,
Peking University, Beijing, 100871, China

xjwang@pku.edu.cn, zjzhou@pku.edu.cn

5.1 Silicon-Based Light Source Introduction

Silicon photonics has attracted increasing attention and research effort in recent years because of its potential for low-cost integration using existing complementary metal–oxide–semiconductor (CMOS) technology and the small footprints of silicon photonic devices.[1–4] Some important components such as high-speed modulators and detectors have already been realized, while a silicon-based or silicon-compatible light source has not been developed yet because silicon, as an indirect band gap semiconductor, does not emit light efficiently. Just like any other lasers, to develop a silicon laser, three key parts are needed: first, an active material that should be luminescent in the active region and be able to amplify light;

Integrated Nanophotonic Resonators: Fundamentals, Devices, and Applications
Edited by Ya Sha Yi

ISBN 978-981-4613-78-1 (Hardcover), 978-981-4613-79-8 (eBook)
www.panstanford.com

second, an optical cavity into which the active material should be placed to provide the positive optical feedback; and third, a suitable and efficient pumping scheme to achieve and sustain the laser action preferably by electrical injection for the advantage of integration.

For a silicon-based active material, the two major approaches were adopted to overcome the difficulty caused by the indirect bandgap of silicon in order to obtain optical gain, microstructuring the silicon materials or introducing light-emitting impurities into silicon-based materials. For the microstructured silicon, such as porous silicon and silicon nanocrystals,[5] the light emission mechanism is complicated, and these materials are subjected to strong light-scattering and absorption, thus limiting the optical gain. For silicon containing light-emitting impurities such as erbium, it is usually difficult to reach high doping concentrations because erbium has a very low solid solubility in silicon. The most severe limitation was an energy back-transfer from the excited erbium ion to the silicon lattice at room temperature.[6] Later it was discovered that silicon nanocrystals excite erbium ions efficiently, significantly increasing the light emission from erbium in silicon oxide or silicon nitrides.[7] The study of erbium-doped silicon nanocrystalline (nc-Si)-rich silicon oxide and nitride is currently a very active research field. Compared with the resonant absorption of a photon, erbium ions can have about two orders of magnitude higher excitation cross section with the sensitization of silicon nanocrystals, thus these material systems emit light efficiently. Waveguides and light-emitting devices based on these materials have been fabricated. However, the optical gain is limited by the high carrier absorption of silicon nanocrystals.

Instead of including erbium ions as dopants, another approach is to use erbium silicate materials. In erbium silicate, erbium is one of the three major elements, therefore reaching high concentrations due to the chemical stoichiometry without segregation.[8] Priolo's group fabricated erbium silicate by using radio frequency magnetron sputtering. Efficient room temperature photoluminescence (PL) at 1535 nm was obtained and upconversion was limited to high optical pump powers.[9] They found that oxygen annealing can remove defects efficiently and enhance the PL intensity strongly.[10] Electroluminescence (EL) of erbium silicate–riched silicon oxide was obtained at around 1530 nm with a high forward bias of about 19 V.[11] These investigations showed that erbium silicate has very good optical and electrical properties, and can be fabricated with compatible silicon fabrication technology, thus it is very promising for the application of silicon-based light sources. However, the phase of all reported materials could not be determined accurately through X-ray diffraction measurements. The influence of the mixture of different phases on the optical properties of the materials is not clear. Energy upconversion processes also need to be investigated in detail. Net gain is still not obtained from these materials. In order to improve the light emission properties of erbium silicate and get net gain, detailed materials and device studies have to be performed.

A resonant cavity for Er-doped laser requires a high-Q microresonator cavity because of the low gain of the active material. Microdisk and micro-toroidal lasers with Er-doped silicon oxide as the active material were demonstrated, and an optical fiber was adopted as the coupler for the pump and output light, which is not adaptable for chip-scale integration.[12] As an initial step, the fabrication of microdisks or microring cavities is a straightforward way to test lasing as long as the refractive index of the material is high enough. However, if the index is as low as silicon oxide, only slot-waveguides are available for high-Q resonators. Slot-waveguides are waveguides that sandwich a low index material between high index materials and can therefore confine light in the low-index layer. For Er-doped silicon nanocrystalline (nc-Si) material systems, multiple slot structures can separate the light distribution in nc-Si layers and Er-doped layers and thus avoid the light absorption of nc-Si. The high-index layers of the slot structure can adopt crystalline silicon or poly-silicon, both having good electrical conductivity and are suitable for current injection to realize electrical pumping. More recently, losses of below 7 dB/cm of silicon-based slot waveguides were reported.[13] Although still too high to obtain enough modal gain, additional reduction of loss can reach acceptable values. The Q-factors of slot-waveguide-based microrings can be higher than 80,000. Recently, a polysilicon slot waveguide for electrical injection was demonstrated experimentally.[14] Both high-index layers were made of electrically conductive silicon and the tunneling voltage can be as low as several volts. The whole structure has a propagation loss of 7 dB/cm and a *Q*-factor of 83,000. Different type of slot-waveguides were designed and demonstrated such as horizontal, vertical, and multiple slot waveguides as well as couplers from channel to slot waveguides. These advances in slot-waveguide research are important for erbium-based light sources and lasers.

The last key part for the realization of a silicon-based laser is a suitable and efficient pumping scheme. Light pumping is usually easy to implement, however, electrical pumping is required for the CMOS integration of the device. As mentioned before, slot waveguide structures are suitable for current injection. However, the loss is very large. In addition, the solutions are strongly dependent on the active materials. Thus, above three key parts must be carefully considered in order to obtain the laser.

In this chapter, Er, used as an optical dopant and utilizing its intra-4f transition at 1530 nm, will be first introduced as a viable means for on-chip light generation that has the advantage of being compatible with long-distance telecom as well. Second, Er-doped silicon-rich silicon oxide (SRSO) and nitride (SRSN), and Er silicates will be introduced as the promising host material for compact on-chip light sources. Finally, some basic photonic components using Er-doped SRSO and SRSN, and Er-silicates will be presented.

5.2 Er-Doped Silicon-Riched Silicon Oxide and Nitride Photonic Devices

During the past two decades, several experimental efforts have been proposed in order to develop an efficient and reliable light source by using Er^{3+} in Si.[15] The room temperature emission with an external quantum efficiency of 0.1% in a Er^{3+}-doped Si LED has been successfully reported.[16] However, the main problem associated with Er^{3+} in Si is the back transfer of energy from the Er^{3+} ions to the Si host, which causes a lowering of the diode emission efficiency,[17] since a resonant level of the Si bandgap couples with the Er^{3+} levels. In order to reduce this back-transfer process, the bandgap of the host Si should be enlarged so that the resonance between the defect level and the internal Er^{3+} levels is lost.[17] Silicon-rich silicon oxide dielectric was thus proposed as the host.[18] In the SRSO host, the silicon is nanocrystal (nc). The Si-nc are very efficient sensitizers of the Er^{3+} luminescence with typical transfer efficiency as high as 70%.[19–21]

Figure 5.1 shows the diagram of the excitation process of Er^{3+} ions via a Si-nc with the main related cross sections. The main internal energy levels of the Er^{3+} are shown in the right-hand side of figure. Excitation of Er^{3+} occurs via an energy transfer from photoexcited electron–hole pairs, which are excited in the Si-nc, the overall efficiency of light generation at 1.535 μm through direct absorption in the Si-nc is described by an effective Er^{3+} excitation cross section σ_{exc}. On the other hand, the direct absorption of the Er^{3+} ions, without the mediation of the Si-nc, and the emission from the Er ions are described by absorption σ_{abs} and emission σ_{em} cross sections, respectively. It is important to note the five orders of magnitude increase in σ_{exc}. This value is conserved also when electrical injection is used to excite the Si-nc.[22] A few groups have performed Er-doped silicon-rich silicon oxide waveguide amplifier experiments.[23–26] The most successful result was reported in.[26] A very low Si-nc concentration has been used, and an internal gain of 7 dB/cm has been deduced.[26] However, the large net gain amplifier cannot still be obtained. The most relevant problem is related to the fact that the coupling of Er with Si-nc is not complete. Indeed, most of the reported works show that only up to 5% of the Er ions are coupled to the Si-nc, while the other can be excited only through direct Er absorption.[27] This is the main problem limiting the achievement of a net overall gain in the waveguide amplifier.

Recently, several electrically pumped Er-SRSO LED based on a MOS structure are reported.[28,29] The injected carriers will first enter into oxide layers, and Si nanocrystal act as the sensitizers that transfer the energy to Er ions to achieve EL.[28] The main obstacle in this issue is to obtain a high emitting efficiency. In addition to the free carrier absorption (FCA) from carrier injection, another detrimental factor is that the solubility of Er in SRO layers is still not large enough. Due to this fact, Er ions tend to form Er_2O_3 clusters when the intensity of Er is relatively large. However, since the dielectric constant of Er_2O_3 is even higher than that of SiO_2, the relatively weak electrical field in Er_2O_3 clusters cannot accelerate electrons efficiently in order to excite the Er ions.[30] Therefore, further research is required to make the distribution of Er ions in SRO layers more uniform. Excellent progress has been achieved in this field. An EL of Si-nc sensitized Er-doping device with power efficiency of 10^{-2} and 20% of Er population inversion has been demonstrated by Jambois' group, as shown in Fig. 5.2.[29] Furthermore, how to decrease the threshold voltage and to realize laser operation in room temperature should also be a crucial topic in future investigations.

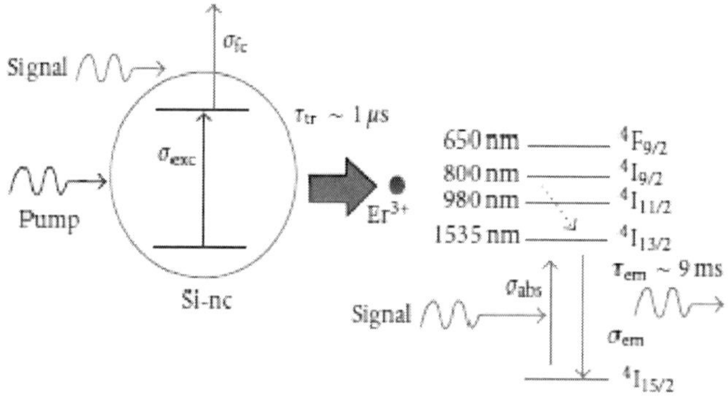

Figure 5.1 Diagram of the excitation process of Er^{3+} ions via an Si-nc with the main related cross sections.

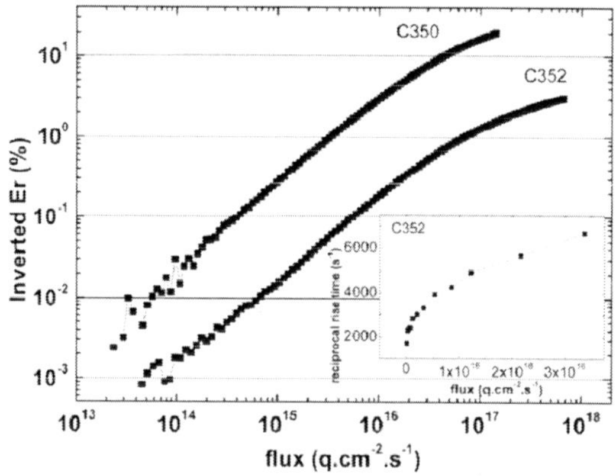

Figure 5.2 Estimation of the inverted Er fraction for both layers. Inset shows the non linear behavior of the reciprocal rise rate with flux.

However, the use of SRSO host material faces some fundamental difficulties. First, the solubility of Er in SiO_2 or SRSO host material is low. Because the Er light emission occurs via the atomic intra-4f transition, the amount of light that can be generated is limited strictly by the number of optically active Er^{3+} ions. The concentration of optically active Er that can be doped into silica without clustering is limited to 10^{20} cm^{-3}.[31] Given that the emission cross section of Er^{3+} is 10^{-21}–10^{-20} cm^2, the maximum gain of any SiO_2-based material was limited to a few dB/cm at most. Thus, obtaining a net gain from such a material will require a large device and/or highly advanced processing techniques to reduce any parasitic losses. Second, for a stable operation in integrated devices, all photonic devices need outer cladding layers. However, because the refractive index of SRSO is very close to that of SiO_2, it requires waveguides with a width ≥1 μm, a bending radius ≥100 μm, and an areal footprint $\geq 10^8$ μm^2 for a 3 dB gain.[32] These values do not present any problem for stand-alone telecom devices but are unacceptably large for electro-photonic integration. Indeed, the only reported case of an on-chip light source based on Er-doped silica is micro-disk resonator of 120 μm diameter with air-cladding.[33]

Silicon-rich silicon nitride recently was chosen for Er doping host materials. There are three reasons. First, nitrides are used widely in the Si industry, and thus are fully CMOS compatible. Second, the refractive indices are much higher than that of oxide and can be controlled by varying the silicon content, thereby allowing compact optical devices on a chip without requiring cladding layers that are several microns thick. Third, nitrogen has been reported to be even more effective than oxygen in optically activating Er in Si,[34] indicating that nitrides may offer an efficient environment of Er^{3+} luminescence. Finally, nitrides have a bandgap that is much smaller than that of oxide, which would allow easier injection of current for the possible development of electrically excited devices.[35] Figure 5.3 shows the anneal-temperature dependence of the Er^{3+} PL intensity of SRSN:Er, SRSO:Er, and SiO_2:Er, obtained under resonant optical excitation using a 980 nm laser. Note the decrease in the Er^{3+} intensity from SRSO:Er and SiO_2:Er thin films at anneal temperatures above 1000°C, indicating optical de-activation of Er as distinct from de-coupling of Er from nc-Si. However, the de-activation effect of Er in SRSN host materials is weak. In addition, are much more resistant to Er clustering, such that higher Er concentrations and annealing temperatures can be used.[36]

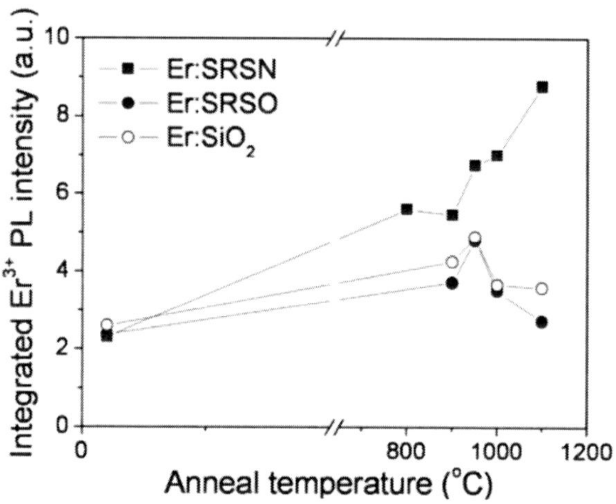

Figure 5.3 The anneal-temperature dependence of the Er^{3+} PL intensity of SRSN:Er, SRSO:Er, and SiO_2:Er under resonant optical excitation using a 980 nm laser.

5.3 Er-Silicates Photonic Devices

Er silicates (Er_2SiO_5, $Er_2Si_2O_7$) have been attracting considerable attentions as materials for new light sources and optical amplifiers in silicon photonics. Er silicates contain high density of Er^{3+} ions as one of the constituent elements of the crystals and show strong emissions around 1.5 μm due to the intra-4f transition, corresponding to the low-loss window of standard silica based optical fibers.[8–10] In 2004, Isshiki et al.[8] first fabricated Er silicate compounds (Er_2SiO_5) by a wet chemical method using $ErCl_3$ deposited on Si substrates. The fine photoluminescence (PL) structure due to Stark splitting with a full width at half-maximum as small as 7.0 nm (4 meV) at around 1.53 μm has been observed at room temperature. Various methods have been adopted to prepare this new structure, including metal organic molecular beam epitaxy (MOMBE),[37] magnetron sputtering,[9–10] reaction of

$ErCl_3$ and Si nanowires,[38] and laser ablation.[39] Due to the crystalline nature of Er silicates, they show almost no segregation of Er atoms and formation of defects even at such a high Er density of $\sim 10^{22}$ cm^{-3} in contrast to Er^{3+}-doped Si-based materials (10^{16}–10^{20} cm^{-3}). Therefore, Er silicates are expected as small size and high optical gain waveguide amplifiers and emitters for silicon photonics. However, such a high concentration results in upconversion due to near distances of Er ions that limits the Er luminescence. Therefore, characterizing and controlling Er ions distances in such Er silicates are necessary. To overcome this problem, ytterbium (Yb) cations and yttrium (Y) were added into the structure to dilute Er ions, which can substitute Er ions in the silicate lattice and prevent neighboring Er ions from causing upconversion.[40–43] Moreover, the Er concentrations of silicates can be easily continuously changed through Y- or Yb co-doping, which is necessary to reduce the deleterious effects for 1.53 μm gain and luminescent efficiencies, such as the concentration quenching and upconversions due to the neighboring Er ions. The Yb co-doped Er silicates have another advantage than the Y co-doping that is the sensitization effect of Yb for Er, which can enhance the excitation efficiency of Er. In this chapter, we summarize our group's recent work about Er/Yb/Y silicates. First, the Er/Yb/Y materials structure and optical properties were optimized. Then, we fabricated three kinds of waveguide structure, the strip-loaded, slot and hybrid Er/Yb/Y silicates waveguides, and the optical amplification was observed in these waveguide structures. In addition, 1.53 μm EL in ErYb silicates was also realized using hot carriers' impact excitations of Er ions. The current conduction behavior, EL emissions, and impact excitation cross sections were studied.

5.3.1 Materials Fabrication and Optical Properties Optimization

$Er_xY_{2-x}SiO_5$ and $Er_xYb_{2-x}SiO_5$ films for (x = 0 to 2) were fabricated by using a mixture of Er-O, Y-O(Yb-O) and Si-O (Er(Y or Yb):Si 2:1) sol solutions. The sol solutions were first spin-coated on Si (100) substrates, dried at 140°C for 30 min in air and baked at 600°C for 30 min in Ar. This procedure was repeated 5 times to obtain about 360 nm thick films. Finally, the coated surface of the samples were covered with a polished surface of another Si(100) wafer, and then sintered in the well heated furnace at 1200°C for 30 min at a heating rate of 100°C min^{-1} in Ar atmosphere.

Phase structure of the samples was examined using a Rigaku X-ray diffractometer (XRD) with Cu*K* radiation. Photoluminescence measurements were measured by two kinds of setup systems. First, the samples were performed by using a semiconductor laser diode with a 30 mW power at 654 nm. This wavelength corresponds to the direct excitation of Er^{3+} ions from $^4F_{9/2}$ to $^4I_{15/2}$. The PL emissions were detected by a liquid nitrogen–cooled Ge detector. Another system was performed using a 1 W semiconductor laser of 980 nm as the excitation source pumping. The detector is InGaAs. The fluorescent lifetime was detected by a time-gated photon counting method using an electrically cooled InGaAs photomultiplier.

Figure 5.4 shows the PL spectra of $Er_xY_{2-x}SiO_5$ (x = 0–2) films at the wavelength pump of 654 nm. It can be seen that the same PL shape with main peak of 1.528 μm, corresponding to the typical PL spectrum of Er_2SiO_5 phase,[8] was observed for $Er_xY_{2-x}SiO_5$ films at the different Er^{3+} concentrations. It indicated that local atomic structure of Er^{3+} ions is similar for all samples. However, PL intensity has a significant change at the different Er^{3+} concentrations. PL intensity first increased ~30 times with the decrease of Er^{3+} concentrations from 25 at.% (x = 2.0) to 1.25 at.% (x = 0.1), and then decreased slightly when the Er^{3+} concentrations were further decreased to 0.5 at.% (x = 0.04). The two factors, crystal structure, and radiative and nonradiative transition rates were considered to study this phenomenon.

Figure 5.5 shows the XRD patterns of $Er_xY_{2-x}SiO_5$ films. The strong peaks of (100), (200) and (300) at 10.3°, 20.6° and 31.2° from Er_2SiO_5 films (JCPDS 52-1809) were observed for the high Er^{3+} concentrations above 12.5 at.% (x = 1.0). The three strong peaks became blurred when Er^{3+} concentrations were decreased to below 6.25 at.% (x = 0.5). The highly ordered crystalline phase of $Er_xY_{2-x}SiO_5$ films for high Er^{3+} concentrations obtained here was due to the high ordered Er_2SiO_5 structure. The poor ordered orientation for low Er^{3+} concentrations should be the result of the poor ordered Y_2SiO_5 structure. However, the XRD peak positions are same for all samples ranging from pure Y_2SiO_5 to pure Er_2SiO_5, which agree well with the peak positions of standard Er_2SiO_5 phase and Y_2SiO_5 (JCPDS 52-1810) phase. It shows that the local atomic structure of Er^{3+} for all samples is identical, which results in the similar PL spectrum shape, as shown in Fig. 5.4.

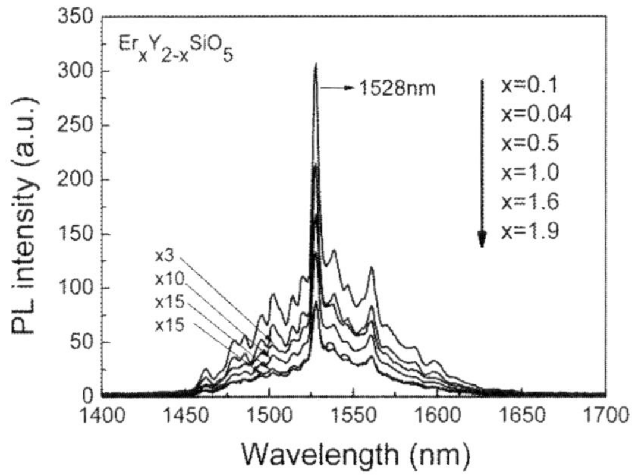

Figure 5.4 PL spectra of $Er_xY_{2-x}SiO_5$ films (x = 0–2) at the wavelength pump of 654 nm.

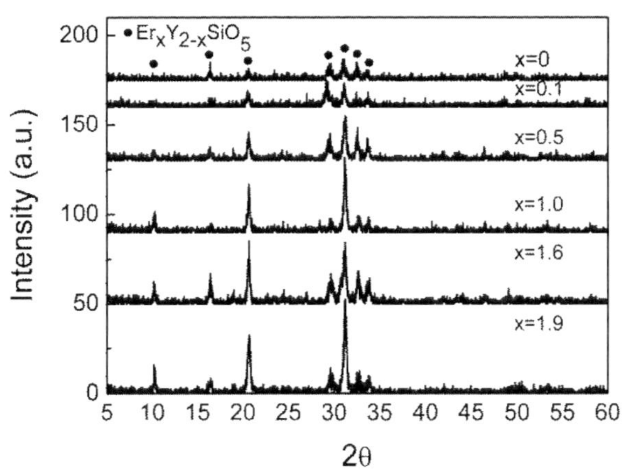

Figure 5.5 XRD patterns of $Er_xY_{2-x}SiO_5$ (x = 0–2) films.

Figure 5.6a shows the decay time of $Er_xY_{2-x}SiO_5$ films. For high Er^{3+} concentrations of 20 at.% and 23.75 at.% (x = 1.6 and 1.9), fast decay time (~20 μs) was observed, and slow decay time (~2 ms) appeared for low Er^{3+} concentration of 1.25 at.% ($x = 0.1$). Figure 5.6b shows the integrated PL intensity from 1400 to 1700 nm and decay time dependent on x value for the $Er_x Y_{2-x}SiO_5$ films. Decay time was increased about 100 times when Er^{3+} concentrations were decreased from 23.75 at.% to 1.25 at.%. The results may suggest that the incorporation of upconversion results in the fast decay observed in pure Er_2SiO_5. Y additions fully separate Er^{3+} ions and weaken upconversion of Er^{3+} for Er_2SiO_5, which lead to the decrease of nonradiative transition rate. It shows that (~30 times) increase of PL intensity can be explained partly due to increase of decay time.

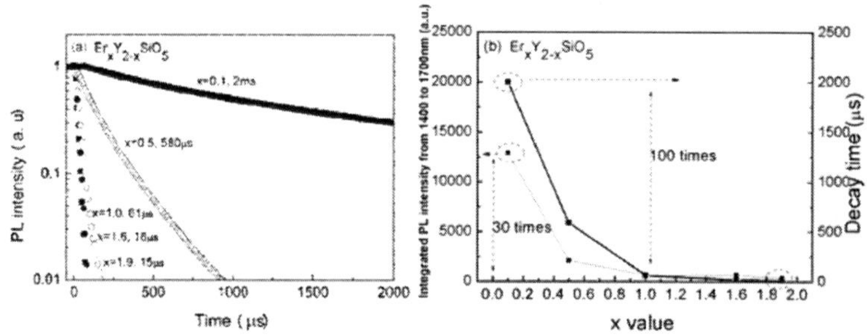

Figure 5.6 (a) Decay time of $Er_xY_{2-x}SiO_5$ films, (b) Integrated PL intensity from 1400 to 1700 nm and decay time dependent on x value for the $Er_xY_{2-x}SiO_5$ films.

Figure 5.7a shows the PL spectra of $Er_{2-x}Yb_xSiO_5$ (x = 0–2) films on Si(100) substrates at the wavelength pump of 654 nm. It can be seen that the typical PL spectrum with main peak at 1.528 μm of Er_2SiO_5 phase (JCPDS 52-1809) was observed without Yb additions (x = 0).[8] The PL spectra have no significant change with the increase of Yb concentration to 2.5 at.% (x = 0.2) compared with that of Er_2SiO_5 phase. However, the peak intensity at 1.528 μm becomes weak, and another two strong peaks at 1.535 and 1.545 μm appear when the Yb concentration was further increased to above 12.5% (x = 1.0). It shows that the local environment of Er ions has been changed compared with that of samples having low Yb concentrations. The 1.53 μm integrated PL intensity is similar for the different Yb concentrations at 654 nm pump wavelength. In order to study the effect of Yb on Er, the 980 nm wavelength laser was used as pump source. Figure 5.7b shows the 1.53 μm integrated PL intensity of $Er_{2-x}Yb_xSiO_5$ films on SiO_2/Si substrates and Si(100) substrates as a function of Yb concentration at 980 and 654 nm pump wavelengths. PL intensity by pumping at 980 nm has a significant increase than that by pumping at 654 nm. The above 10 times enhanced PL intensity for $Er_{2-x}Yb_xSiO_5$ film on Si substrate was obtained with the Yb concentration increased to 23.75 at.% (x = 1.9), and then decreased slightly when the Yb concentration was further increased to 24.5 at.% (x = 1.96). At the base of above 10 times enhancement, another 20 times enhanced PL intensity for the $Er_{2-x}Yb_xSiO_5$ (x = 1.9) film on SiO_2/Si substrate was observed compared with that on Si substrate by pumping at 980 nm.

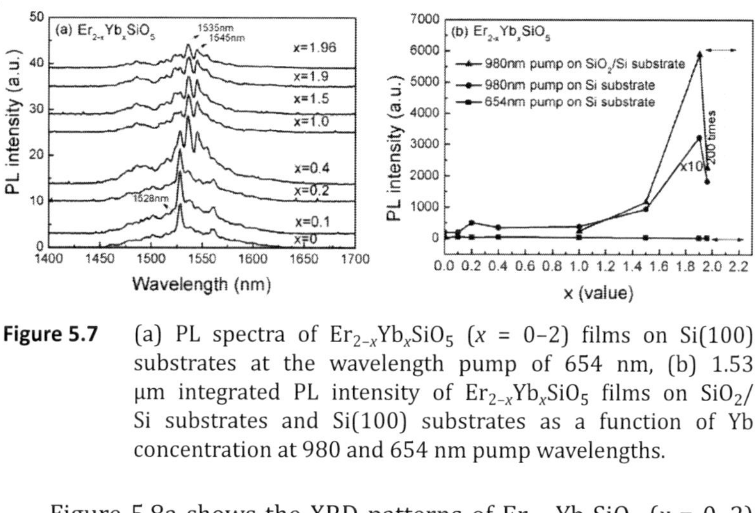

Figure 5.7 (a) PL spectra of $Er_{2-x}Yb_xSiO_5$ (x = 0–2) films on Si(100) substrates at the wavelength pump of 654 nm, (b) 1.53 μm integrated PL intensity of $Er_{2-x}Yb_xSiO_5$ films on SiO_2/Si substrates and Si(100) substrates as a function of Yb concentration at 980 and 654 nm pump wavelengths.

Figure 5.8a shows the XRD patterns of $Er_{2-x}Yb_xSiO_5$ (x = 0–2) films on Si(100) substrates. The strong peaks of (100), (200) and (300) at 10.3°, 20.6°, and 31.2° from Er_2SiO_5 film (JCPDS 52-1809) were observed without Yb additions (x = 0). With the increase of Yb concentration to 2.5 at.% (x = 0.2), the XRD pattern has no significant change, indicating the phase structures are similar for the samples with low Yb concentrations. However, the three strong peaks became blurred when the Yb concentration was increased to above 12.5 at.% (x = 1.0). Another series of peaks appear, and are completely consistent with another Er_2SiO_5 phase (JCPDS 40-0384). So it can be confirmed that one Er_2SiO_5

phase (JCPDS 52-1809) formed below the Yb concentration of 2.5 at.% (x = 0.2), and another Er_2SiO_5 phase with monoclinic structure (JCPDS 40-0384) formed above the Yb concentration of 12.5 at.% (x = 1.0). It also can explain the reason for PL shape change when the Yb concentration was increased to 12.5 at.% (x = 1.0). The two strong peaks at 1.535 μm and 1.545 μm for PL spectrum should come from another Er_2SiO_5 phase (JCPDS 40-0384). Figure 5.2b shows the XRD patterns of $Er_{2-x}Yb_xSiO_5$ (x = 0–2) films on SiO_2/Si substrates with the increase of Yb concentration from 12.5 at.% (x = 1.0) to 24.5 at.% (x = 1.96). The peak positions are consistent with that on Si substrate, showing the substrate has no evident effect on phase structure of $Er_{2-x}Yb_xSiO_5$ films.

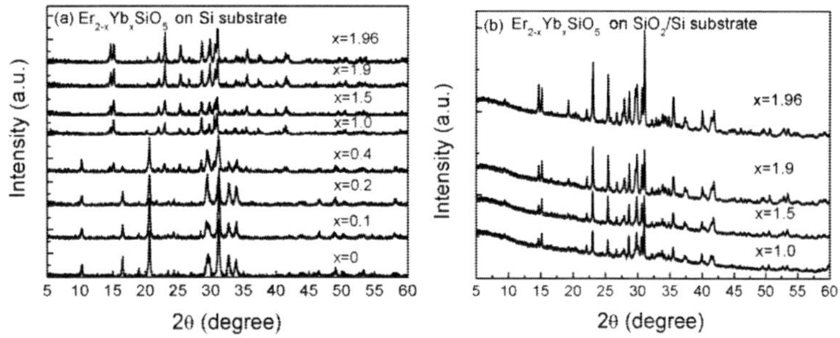

Figure 5.8 XRD patterns of $Er_{2-x}Yb_xSiO_5$ films (a) Si(100) substrates, (b) SiO_2/Si(100) substrates.

Figure 5.9 shows the decay time of $Er_{2-x}Yb_xSiO_5$ (x = 0–2) films on SiO_2/Si substrates and Si substrates. For pure Er_2SiO_5 phase on Si substrate, fast decay time of ~20 μs was observed. The slow decay time of ~0.7 ms, was observed for $Er_{2-x}Yb_xSiO_5$ films on SiO_2/Si substrate for high Yb concentration of 18.75 at.% (x = 1.5). With the further increase of Yb concentration to 23.75 at.% (x = 1.9) and 24.5 at.% (x = 1.96), decay time become more longer ~1.8 and ~3.5 ms, above 100 times of pure Er_2SiO_5. The smaller amount of Er and higher amount of O supplied from the SiO_2 underneath layer compared with Er_2SiO_5 may result in the longer decay time. The enhancement of decay time with increasing the composition of Yb additions can be explained by the decrease in the number of nonradiative decay channels, which may involved reduction of concentration quenching of Er ions at high Yb concentrations. Energy transfers among Er ions become very efficient when the distances of neighboring Er ions get to be smaller. The energy finally dissipates to quenching centers such as –OH when they meet with an excited Er ion.

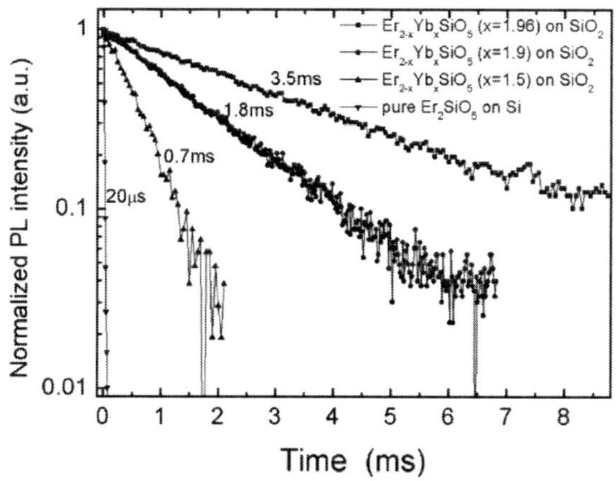

Figure 5.9 Decay time of $Er_{2-x}Yb_xSiO_5$ films on SiO_2/Si substrates and Si substrates.

Above 200 times enhanced PL intensity was observed for the $Er_{0.1}Yb_{1.9}SiO_5$ film on SiO_2/Si substrate by pumping at 980 nm compared with pure Er_2SiO_5 film on Si substrate at 654 nm. First, Yb ions have a larger optical cross section, and can transfer energy to Er ions efficiently for 980 nm pump wavelength.[44] The absorption curve of $Er_{0.1}Yb_{1.9}SiO_5$ film was measured, and the absorption for 980 nm wavelength is about 5 times than that for 654 nm wavelength. In addition, a proper molar ratio of Yb and Er was needed to gain the maximum PL intensity. Dong et al.[45] fabricated the Er-Yb-doped Al_2O_3 materials and found that ratio of Yb and Er [(10–20):1] can get maximum PL efficiency. The sensitization effect was also observed in a series of the Er-Yb co-doped phosphate glasses [$LiEr_xYb_yLa_{(1-x-y)}P_4O_{12}$] prepared using electronic furnace in a temperature of 1200–1300°C by Zhang et al.[46] The maximum PL intensity was observed when the molar ratio of Yb and Er is about 20:1. In addition, we considered decay time of the samples. Photoluminescence intensity and life time are related through rate equations governing the excitation and decay of the equations.[47]

$$I = \sigma\phi N\tau/\tau_{rad}, \quad (5.1)$$

where σ is the excitation cross section, ϕ is the incident photon flux, N is the optically active Er content, τ is the total lifetime and τ_{rad} is the radiative lifetime. Therefore, using fixed σ and τ_{rad} according to above Equation, PL intensity (I) can be explained by measuring τ for the two samples under the same excitation conditions. The lifetime τ for $Er_{0.1}Yb_{1.9}SiO_5$ film is about 1.8 ms, about 100 times than 20 μs for Er_2SiO_5, as shown in Fig. 5.9. So we can conclude that above two orders of magnitude enhanced PL for $Er_{0.1}Yb_{1.9}SiO_5$ film may be due to the higher radiative transition rate. However, although decay time becomes longer about 3.5 ms for $Er_{0.04}Yb_{1.96}SiO_5$ film with the Yb concentration further increased to 24.5 at.% (x = 1.96), PL intensity decreases. First, we can assume that all Er ions are optically active for $Er_{0.04}Yb_{1.96}SiO_5$ film due to such low Er concentration (Er ions is 0.5 at.%). It is suggested that σ and τ_{rad} for $Er_{0.1}Yb_{1.9}SiO_5$ and $Er_{0.04}Yb_{1.96}SiO_5$ are same. So $N_{(x = 1.9)} = N_{(x = 1.96)}I_{(x = 1.9)}\,\tau_{(x = 1.96)}/I_{(x = 1.96)}\,\tau_{(x = 1.9)}$ is approximately equal to 1.26 at% according to Equation, which is quite consistent with real Er concentration of 1.25 at% for $Er_{0.1}Yb_{1.9}SiO_5$ film. It can be concluded that all Er ions are optically active for $Er_{0.1}Yb_{1.9}SiO_5$ film. It indicated the $Er_{0.1}Yb_{1.9}SiO_5$ film is sought candidate material for compact waveguide amplifiers and emitters in silicon photonics integration.

5.3.2 Optical Waveguide Amplifier

Based on above optimized materials, we fabricated the three kinds of waveguides: the strip-loaded ErYb/Y silicates waveguide,[48] hybrid SiN_x-ErYb/Y silicates,[49] and slot ErYb/Y silicates waveguides.[50]

First, a mixed target was used in sputtering process, which is a mixture of Er_2O_3, Yb_2O_3, and SiO_2 with a molar ratio of 1:19:20. The composition of the deposited film is close to the molar ratio of the target, which was confirmed by Rutherford backscattering spectra (RBS) and X-ray photoelectron spectroscopy (XPS) measurements. The deposited film has a thickness of 350 nm on SiO_2 substrate and was annealed for 30 min at a temperature of 900°C to form a polycrystalline $Er_{0.1}Yb_{1.9}SiO_5$ film. Following the active layer deposition, a 450 nm-thick SiO_2 film was formed by plasma enhanced chemical vapor deposition (PECVD) method. Finally, 2.4 μm-wide SiO_2 strip-loaded waveguides were fabricated and cleaved to a length of 7.8 mm. To measure signal enhancement, a tunable laser with wavelength range of 1440–1640 nm was used as a signal light source. A polarization controller was employed to ensure the light to be transverse-electric (TE) mode propagating along the waveguide. Fiber-coupled continuous wave 1480 nm laser with a maximum power of 372 mW in the fiber output port was adopted as a pump light source. A multimode fiber was placed closely above the sample and normal to the surface to collect the emitted visible light, which was sent to an iHR550 triple-grating monochromator.

Figure 5.10 shows the scanning electron microscopy (SEM) image of Er/Yb silicate strip loaded waveguide. The Er/Yb silicate has a refractive index of about 1.63 using ellipsometer measurement. For this structure, the pumping light at 1480 nm and signal light at 1530 nm can be well confined in the active layer. The inset shows the calculated fundamental TE-mode profile of the signal light source. The calculated core-mode overlap is approximately 0.35.

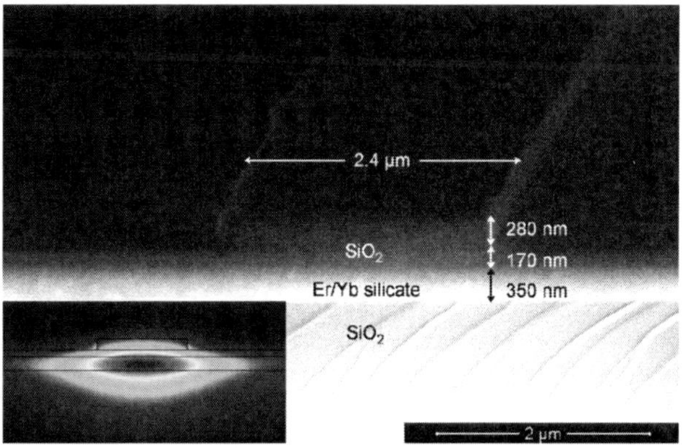

Figure 5.10 SEM micrograph profile of $Er_{0.1}Yb_{1.9}SiO_5$ strip loaded waveguide. Inset: Calculated fundamental TE-mode profile of fabricated waveguide.

Figure 5.11 shows the measured photoluminescence (black-solid line) and calculated absorption (red-dash line) spectra of $Er_{0.1}Yb_{1.9}SiO_5$ film. The PL spectrum with the broadband extending from 1400 to 1700 nm and centered at 1533 nm was given. The sublevels of the first excited state do not discriminate clearly, due to the low annealing temperature of 900°C. To ensure the feasibility of efficient 1480 nm pumping, an absorption spectrum was calculated using McCumber theory.[48] By comparing the PL and absorption spectra, it suggests that the absorption cross section is larger than the emission cross section in the wavelength range of 1460–1490 nm. This ensures the more pumping absorption of 1480 nm light source. The optical propagation and coupling loss of the $Er_{0.1}Yb_{1.9}SiO_5$ waveguide were measured at 1640 nm, where Er absorption loss is weak and can be neglected. For two straight waveguides with length of L_1 and L_2, without absorption loss, the insertion loss α_i only includes coupling loss (α_c) and propagation loss (α_p). Thus, α_{i1} and α_{i2} are written as $2x\alpha_c + \alpha_p x L_1$ and $2x\alpha_c + \alpha_p x L_2$, respectively. The propagation and coupling loss can be determined as $\alpha_p = (\alpha_{i2} - \alpha_{i1})/(L_2 - L_1)$ and $\alpha_c = (\alpha_{i1} - a_p x L_1) = (\alpha_{i2} - a_p x L_2)/2$. By averaging data from five 4.5 mm-long waveguides and five 7.8 mm-long waveguides, the coupling and propagation losses were determined to be 7.8 dB/facet and 8.1 dB/cm, respectively. It suggests that the large coupling loss comes from facet roughness, which may be significantly reduced through facet polishing. The propagation loss mainly comes from scattering loss induced by the roughness of the waveguide's surface and rough sidewall, as well as the nonuniform distribution of refractive index caused by the polycrystalline nature of the $Er_{0.1}Yb_{1.9}SiO_5$. To describe signal enhancement trend, different pumping powers were applied to measure the transmission spectrum for the fixed signal intensity.

Figure 5.12a shows the transmission spectra through a 7.8 mm-long waveguide at 1515–1580 nm wavelength range for 1480 nm pump power of 0 mW (pump off), 38, 131, 218, 299, and 372 mW, respectively. Under the pump-off condition, the rip is about 10.6 dB offset by intensity at 1580 nm where the absorption loss is about 1 dB. With increasing pump power, the rip depth decreases quickly, in accordance with anticipation of increasing signal enhancement due to absorption loss reduction resulting from the decrease of ground state population. Figure 5.12b shows the 1533 nm signal light enhancement as a function of pump power. Despite large coupling and propagation loss, a 5.5 dB signal enhancement was observed in the as-fabricated waveguides at the pump power of 372 mW. The curve shows that the signal enhancement keeps linearly increasing even at the maximum pump power. With higher pump power or pumping at forward and backward of the waveguide simultaneously, the signal enhancement can be further increased. To characterize the enhancement for different wavelength, an enhancement spectrum in the range of 1515–1580 nm for the fixed pump power of 372 mW was shown in the inset of Fig. 5.12b. As shown in Fig. 5.11, PL peak and absorption dip reasonably appear at the same wavelength of 1533 nm. For the wavelength with little absorption such as 1580 nm, the enhancement effect is relatively small. It indicated the Er/Yb silicate film is the sought candidate material for compact waveguide amplifiers and emitters in silicon photonics integration.

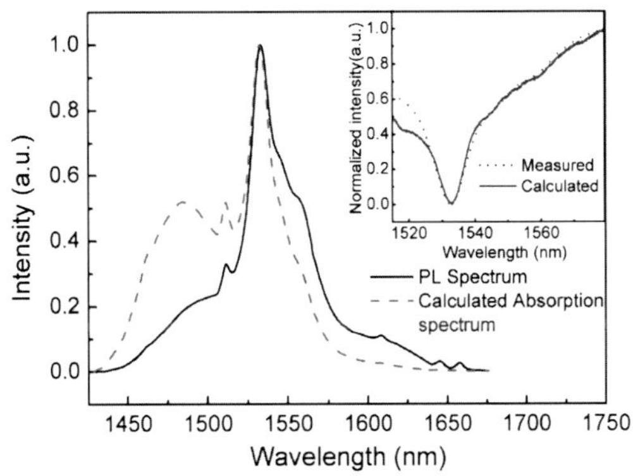

Figure 5.11 Photoluminescence spectrum of $Er_{0.1}Yb_{1.9}SiO_5$ film and calculated absorption spectrum using McCumber theory.

However, many questions still remain. For instance, the Er silicate was very difficult to etch, which makes it very hard to fabricate waveguide directly. Second, important optical parameters such as Er^{3+} optical absorption/emission cross sections and the cooperative upconversion coefficient (CUC) of Er/Yb silicate have not yet been reported, which limit the deep understand for this material. The most serious problem is that the propagation loss of the Er/Yb silicate waveguide is too high due to the scattering of the silicate crystal, which negates the advantage of optical integration in silicon photonics. The propagation loss in the Er/Yb silicate strip loaded waveguide was reported to be 8.1 dB/cm, which is too high for amplifier application.[48]

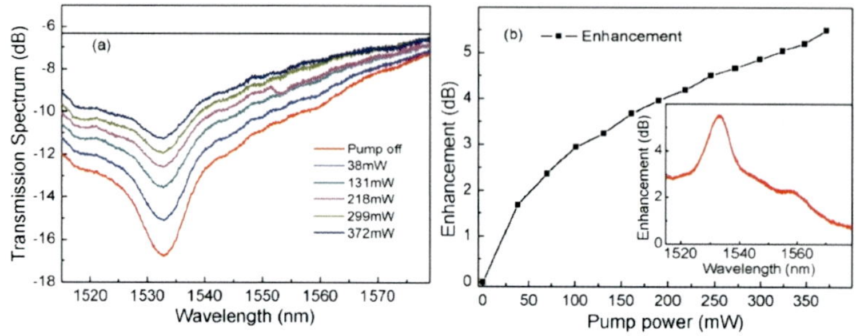

Figure 5.12 (a) Transmission spectra for signal light through a 7.8 mm-long waveguide at different pump power. (b) Measured enhancement of 1533 nm signal light as a function of pump power. Inset: enhancement spectrum for the fixed pump power of 372 mW.

In order to solve above the question, we fabricated a Si_3N_4 strip, then, deposited a 400 nm-thick $Er_{0.2}Yb_{1.8}SiO_5$ thin film above the strip.[49] In this way, the light can propagate through low loss Si_3N_4 strip, leading that propagation loss can be largely reduced. On the other hand, the Er silicate does not need to etch, and hard etch problem for this materials can be avoid. Pumped by 1480 nm laser, signal enhancement at 1532 nm was observed despite relatively high coupling loss. The CUC of Er silicate has been also calculated.

Figure 5.13a shows designed structure diagram of waveguide. The refractive index of the embedded rib Si_3N_4 is 1.9. Er/Yb silicate has a refractive index of about 1.67 at wavelength of 1530 nm using ellipsometer measurement. For this structure, the pumping light at 1480 nm and signal light at 1530 nm can be well confined in the interface of Si_3N_4 strip and silicate film. Figures 5.13b,c show the calculated fundamental TE-mode profile of fabricated waveguide and SEM image of the cleaved cross section of waveguide. The calculated confinement factor in the Si_3N_4 and silicate is approximately 0.48 and 0.33 separately. With half of the power transmitted in the Si_3N_4 strip, the propagation loss in the hybrid waveguide can be reduced dramatically.

The measurement setup is same with above strip loaded waveguide system. By averaging data from five 2.96 mm-long waveguides and five 5.9 mm-long waveguides, the coupling and propagation loss were determined to be 7.7 + 0.2 dB/facet and

3.2 + 0.3 dB/cm, respectively. It suggests that the large coupling loss comes from facet roughness, which may be significantly reduced through facet polishing. Compared with the previous reported propagation loss in Er/Yb silicate strip loaded waveguide,[48] this value has been reduced to a tolerable level of the waveguide, benefit from the embedded Si_3N_4 strip beneath the Er/Yb silicate film. The main reason is that the scattering loss induced by the roughness of the waveguide's surface and rough sidewall has been reduced due to low loss Si_3N_4 strip compared with the polycrystalline nature of the $Er_{0.2}Yb_{1.8}SiO_5$ silicate.

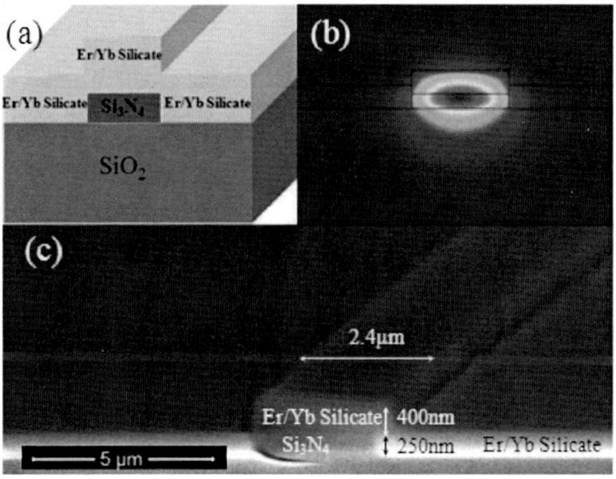

Figure 5.13 (a) Designed structure diagram of waveguide. (b) Calculated fundamental TE-mode profile of fabricated waveguide. (c) Scanning electron microscopy (SEM) image of the cleaved.

The gain characteristics of fabricated waveguides are depicted by applying different pumping powers to measure the transition spectrum for fixed signal intensity. Figure 5.14a shows the transition spectrum through a 5.9 mm-long waveguide at wavelength range from 1520 to 1580 nm for 1480 nm pump power of 0 mW (pump off), 39, 101, 161, 272 and 372, respectively. Under the pump-off condition, the maximum Er absorption is found to be 10.3 dB at 1532 nm, and α_{abs} the waveguide absorption loss is calculated to be 2.23 cm^{-1}. Given Γ of 0.33 and L of 5.9 mm, the maximum Er absorption cross section is 8.64×10^{-21} cm^2. This is same order of magnitude with $4\sim8 \times 10^{-21}$ cm^2 typical used for

Er-doped silica fibers,[52] and 6 × 10^{-21} cm^2 reported for Er-doped Al_2O_3.[53,54] With increasing pump power, the rip depth decreases quickly. The absorption reduction resulting from the decrease of ground state population is in accordance with anticipation of increasing signal enhancement. Figure 5.14b shows the 1532 nm signal light enhancement as a function of pump power. Despite large coupling loss, at the pump power of 372 mW, signal enhancement of 5.5 and 10.1 dB/cm was observed in the fabricated waveguides annealed at 750 and 1000°C, respectively. With higher annealing temperature, the signal enhancement of the waveguide became higher because more Er^{3+} ions were activated at high temperature. But still the emission do not suppress the absorption, the rip still remains in the transmission spectrum (data not shown) of the waveguide annealed at 1000°C, as shown in Fig. 5.14a. So no population inversion happens. Given higher pump power or pump at forward and backward of waveguide simultaneously, the signal enhancement can be further increased. To characterize the enhancement for different wavelength, an enhancement spectrum range of 1520–1580 nm for fixed pump power of 372 mW is shown in the inset of Fig. 5.14b.

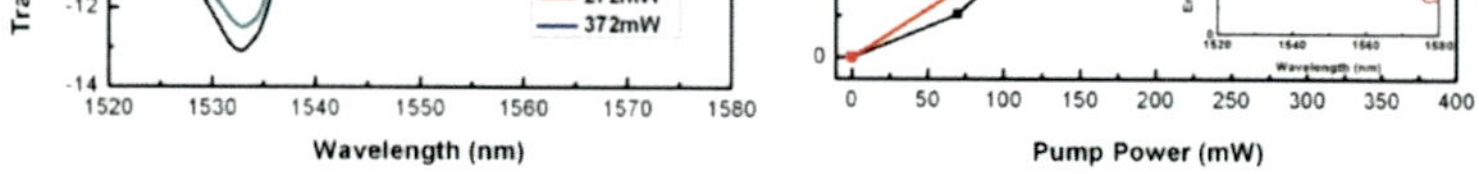

Figure 5.14 (a) Transmission spectrum of signal light through a 5.9 mm-long waveguide at different pump power. (b) Normalized enhancement at 1532 nm signal light as a function of pump power. Inset: enhancement spectrum for fixed pump power of 372 mW.

Upconversion of Er ions also has an important effect on optical gain of Er/Yb silicate materials system. To characterize this property, a light emission spectrum at visible light range was measured, the strong light emission at 660 nm was observed, corresponding naked-eye observed red light in the experiment, which limits the optical gain of Er/Yb silicate waveguide. Figure 5.15a shows the Er^{3+} energy levels involved in the cooperative upconversion (CU) process. One excited Er^{3+} ions at $^4I_{13/2}$ state interact with another one by energy transfer, and decay non-radiatively to the ground state with the other jump to the higher excited level. Based on the first-order upconversion process, and derived from the rate equation of Er^{3+} ions, a simplified 1530 nm optical gain model was established with CUC as follows:

$$g=\sigma_{\mathrm{em}}\times\left[\frac{-\left(\frac{1}{\tau_2}+\sigma_{\mathrm{ex}}\varphi_{\mathrm{e}}\right)+\sqrt{\left(\frac{1}{\tau_2}+\sigma_{\mathrm{ex}}\varphi_{\mathrm{e}}\right)^2+8C_{\mathrm{up}}\sigma_{\mathrm{ex}}\varphi_{\mathrm{e}}N_{\mathrm{total}}}}{2C_{\mathrm{up}}}-N_{\mathrm{total}}\right], \quad (5.2)$$

where σ_{em}, σ_{ex}, φ_{e}, τ_2, C_{up}, and N_{total} are the emission cross section, excitation cross section, pump photon flux, Er^{3+} decay lifetime, CUC, and concentration of Er, respectively. As shown in Fig. 5.15b, we obtain a good fit with a CUC value of $(3.0 + 1) \times 10^{-17}$ cm^3/sec. This value is nearly one order of magnitude larger than that reported from $Er_xY_{2-x}SiO_5$ nanocrystal aggregates fabricated using Si nanowires of $(2.2 + 1.1) \times 10^{-18}$ cm^3/sec with comparable Er concentration,[40] and that reported from Er-doped Al_2O_3 of $(4 + 1) \times 10^{-18}$ cm^3/sec at a Er concentration of $\sim 1 \times 10^{20}$ cm^{-3}.[54] It is suggested that such high value of CUC is the reason that population inversion cannot be reached. On the other hand, Polman et al. reports that CUC from thin film can be reduced by as much as two orders of magnitude, if special care is taken to atomically disperse the Er^{3+} ions.[53,55] So, it should be possible to reduce the CUC from Er silicate thin film to levels that would enable net optical gain.

In the past few years, slot waveguides have attracted increasing attention in photonics community.[56–60] In this structure, low-index material is sandwiched by high-index silicon to form high-index-contrast configuration. Due to the continuity of normal electric displacement ($D = \varepsilon E$) at dielectric interface, the electric field for a TM-like mode can be enhanced by a factor of $n_{\mathrm{Si}}^2/n_{\mathrm{slot}}^2$ in the low-index slot region. Furthermore, this structure can provide relatively high electric field confinement, which is expected to produce high modal gain.[61] For active material embedded in the slot region, the concentrated electric field can promote the light-matter interaction. According to Fermi's Golden Rule, the spontaneous emission rate is expected to increase due to enhanced local density of optical states (LDOS). Despite wide research on passive SiO_2 slot waveguide and spontaneous emission of Er material embedded in slot structure,[62,63] the active devices such as amplifiers and lasers based on this kind of high-index-contrast configuration remain unreported.

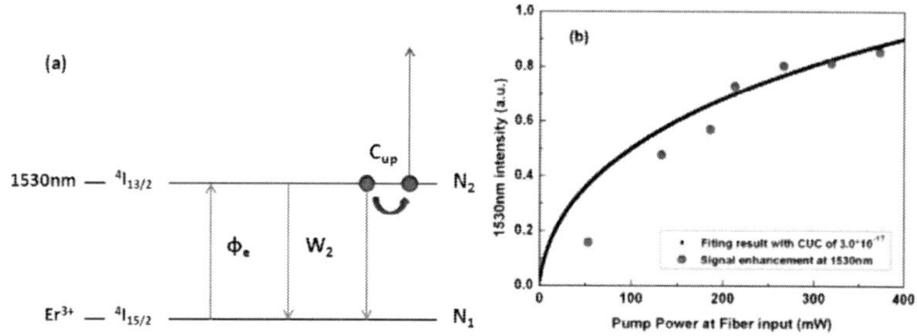

Figure 5.15 (a) Simplified Er^{3+} energy levels involved in the cooperative upconversion (CU) process. (b) 1530 nm intensity fitting with CUC.

We also fabricated slot structure waveguides embedding active erbium/ytterbium (Er/Yb) silicate, which can provide relatively high Er/Yb concentration.[48] Slot waveguides were fabricated from a commercial Si-on-insulator (SOI) wafer, which has a buried oxide of 3 μm and upper crystal silicon layer of 250 nm. Through thermal oxidation and hydrofluoric acid (HF) etching, the crystal silicon layer was thinned to a thickness of 191 nm. Then sputtering process was executed to form a 110 nm-thick Er/Yb silicate layer using a mixed target, which is a mixture of Er_2O_3, Yb_2O_3, and SiO_2 with a molar ratio of 1:19:20.[48] Then a 225 nm-thick Si was deposited as the upper high index layer. The sputtered film was annealed for 30 min at a temperature of 750°C to activate the Er ions and partly eliminate the defects of the upper Si layer. Through photolithography and Si etching, 2.7 μm-wide and 55 nm-thick silicon strips were formed. Finally, the fabricated slot waveguides were cleaved to a length of

6 mm for optical amplification testing. During the measurement, a tunable laser was used as signal light source, which has fixed output power of 5.4 mW at wavelength range of 1440 to 1640 nm. The TM-like mode propagating along the slot waveguide was assured by using a polarization controller. The active material was excited by a fiber-coupled continuous-wave 1476 nm laser with a maximum power of 372 mW in the fiber output port.

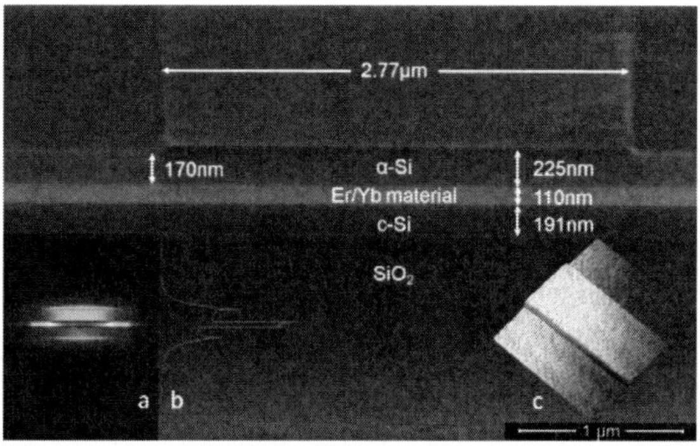

Figure 5.16 (a) SEM image of as-fabricated slot waveguide. Inset: (a) the calculated fundamental TM-like mode profile, (b) the plot of normalized electric field along vertical direction at waveguide center and (c) AFM image of the fabricated slot waveguide.

Figure 5.16a shows the cross-sectional image of the fabricated slot waveguide by scanning electron microscope. The white part is the active material, which has a refractive index of about 1.65 measured by ellipsometer. The bottom and upper layers are c-Si and α-Si, which have similar index of about 3.45. The thickness of the bottom c-Si layer, active material, upper α-Si slab and α-Si strip are 191, 110, 170 and 55 nm, respectively. The inset (a) shows the calculated fundamental TM-like modal profile for 1.53 μm signal light. It can be seen that the electric field is enhanced in the slot region. Inset (b) shows the normalized electric field along vertical direction at the horizontal center of the waveguide. For 1.48 μm pump and 1.53 μm signal light, the mode profiles are similar. The confinement factor Γ can be described as[61]

$$\Gamma = \frac{n_g}{n_{slot}} \frac{\iint_{slot} \varepsilon |E|^2 \, dxdy}{\iint_{\infty} \varepsilon |E|^2 \, dxdy}, \tag{5.3}$$

where n_g is the group index and n_{slot} is the refractive index in the slot region. For pump and signal light, the calculated Γ is 0.237 and 0.250, respectively. Inset (c) gives the atomic force microscope (AFM) image of the fabricated waveguide. Despite two depositing and one annealing process, the waveguide has a relatively smooth surface with a root mean square (RMS) roughness of around 1.8 nm.

We measure the propagation loss and the coupling loss at 1640 nm where the Er absorption is relatively weak and could be neglected. The detailed method could be found in our previously published paper.[48] By averaging data from five 6 mm-long waveguides and five 11 mm-long ones, the coupling loss and propagation loss were determined to be 6.4 dB/facet and 14.7 dB/cm, respectively. Compared with the published propagation loss of 7 and 15 dB/cm in similar single horizontal slot waveguides.[14,60] the propagation loss in our structure is moderate. Due to the annealing process, the slight crystallization of upper silicon partly contributes to the propagation loss. Furthermore, the formed small particles in Er/Yb material during the annealing process serve as scattering center, which leads to part of the propagation loss. Of course, scattering at dielectric interfaces, surface and sidewall of the silicon strip also bring propagation loss. The coupling loss partly comes from the modal mismatch between the lensed fiber and fabricated slot waveguide, and partly comes from facet roughness, which can be decreased by facet polishing.[64]

Figure 5.17a shows the transmission spectrum for signal light at wavelength range of 1515 to 1560 nm under pump power of 0 and 372 mW. Under pump-off condition, the maximum absorption dip (α_{abs}) at 1.53 μm is about 4.9 dB (1.13 cm^{-1}). It can be seen that the signal was enhanced at a large wavelength range. The wide range corresponds to the broad Er absorption, which covers wavelength range of 1440 to 1640 nm. Figure 5.17b gives the enhancement spectrum at pump power of 372 mW. From this figure, it can be clearly seen that the enhancement peak appears at 1.53 μm, which corresponds to the PL peak and absorption dip. The high-frequency oscillation noise comes from the Fabry–Perot (F–P) cavity effect, which is formed by the reflection of waveguide facets.

At 1.53 μm wavelength, the waveguide absorption loss can be written as $\alpha_{abs} = \Gamma \times \sigma_{abs} \times N_{tot} \times L$, where Γ is the confinement factor, σ_{abs} is the peak absorption cross section at 1.53 μm, N_{tot}

is the total Er concentration and L is the waveguide length. For the measured waveguide, α_{abs}, Γ, N_{tot}, and L are 1.13 cm^{-1}, 0.25, 1 × 1021/cm^3 and 0.6 cm, respectively. Thus, we can get the peak absorption cross section of $\sigma_{abs} = 7.53 \times 10^{-21}$ cm^2, which is equal to the peak emission cross section σ_{em}. This value is close to the reported Er concentration in the similar Er/Y silicate- and Er-doped silicon-rich silicon nitride.[64,65] Based on the estimated Er absorption cross section, the excited Er fraction can be calculated by assuming a two-level model $N_{tot} = N_0 + N_1$, where N_0 and N_1 are the population on the ground and excited states. The observed signal enhancement can be described as $\beta = 2 \times \Gamma \times \sigma_{em} \times N_1 \times L$. Given β, Γ, σ_{em}, and L of 0.39, 0.25, 7.53 × 10^{-21} cm^2 and 0.6 cm, the excited Er concentration is determined as 0.17 × 10^{21}/cm^3, which corresponds to a excited Er fraction of 0.17 ($f = N_1/N_{tot}$).

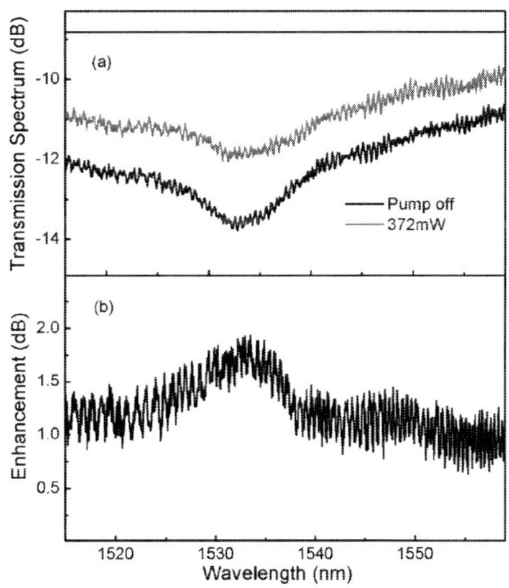

Figure 5.17 (a) Transmission spectrum for signal light range from 1515 nm to 1560 nm at pump power of 0 mW and 372 mW. (b) The signal enhancement spectrum for the pump power of 372 mW.

5.3.3 Light Emitter Diode

An efficient electrical-pumped silicon-based light source compatible with silicon technology is a key component for the rapid developing silicon photonics.[5] The Er silicate compounds provide much higher optically active Er concentrations than the conventional Er-doped silicon-based materials due to the stoichiometric nature.[8–11,37–43,48–50] Recently, we have demonstrated extraordinary PL efficiency and optical amplification at 1.53 μm in ErYb silicate.[42,43,48–50] In contrast to the intense investigation on the PL and optical amplification of this material system, EL is difficult to obtain due to the insulating nature of the silicate compound, which prevents efficient current injection. Up to now, the realization of EL from Er silicate materials has to be assisted by additional silicon-rich oxide layer or excess silicon contained in the Er silicate film.[11,66] Although the current injection was improved and sensitization effect of excess silicon on Er might function, the excess silicon simultaneously brings luminescence quenching of Er due to the Auger effect induced energy back-transfer from the excited Er ions to silicon.[68] Direct EL from Er silicate materials without the assistance of excess Si has not been realized yet.

One possible way to get EL directly from Er silicate materials is to use the metal–insulator–semiconductor (MIS) structure, which has been successfully used in the Er-doped silicon dioxide light emitting devices.[67–69] There the Er ions are directly impact excited by the energetic hot electrons produced by high electric field in the conduction band of silicon dioxide. This method is not suffered from the energy back-transfer or free carrier absorption that is usually appeared in the Er-doped silicon rich materials. High EL efficiency has been reported.[70] However, this method has never been applied to the Er silicate compounds. In this chapter, we fabricated MIS light emitting device containing ErYb silicate thin film as optical active material using the standard silicon technology. The direct 1.53 μm EL from the ErYb silicate was observed the first time at an ultralow current density of 0.3 mA/cm^2 due to the direct impact excitation of Er ions by hot electrons produced in the silicate. The current conduction mechanism of the device was revealed to be the Fowler–Nordheim (FN) tunneling.

The ErYb silicate thin film was deposited on a 4′ p-type c-Si (100) wafer (resistance: 0.05~0.2 Ωcm) by radio-frequency sputtering of a compound target which was fabricated from fully mixed oxide powders with a molar ratio of Er_2O_3:Yb_2O_3:SiO_2 = 1:9:10. The thickness of the deposited film is 60 nm, measured by spectroscopic ellipsometer. The as-deposited film was then annealed in furnace at 1000°C for half an hour, under flow N_2 protection. The schematic cross section view of the device structure is shown in Fig. 5.18. After the post-annealing, a layer of 60 nm-thick SiN_x ($n \sim 2.02$) was deposited on the ErYb silicate film by Plasma Enhanced Chemical Vapor Deposition. Aluminum (Al) was sputtered on the backside of the Si substrate and furnace annealed at 520°C in N_2 for 10 minutes to form Ohmic contact. Finally, 80 nm-thick indium tin oxide (ITO) film was sputtered on the top of the SiN_x film and patterned by the standard photolithography to form square shape electrodes. The size of the electrode is 1 × 1 mm. Two reference samples, A (without SiN_x layer) and B (without ErYb layer) were fabricated at the same time in order to compare with the ErYb silicate MIS light-emitting device.

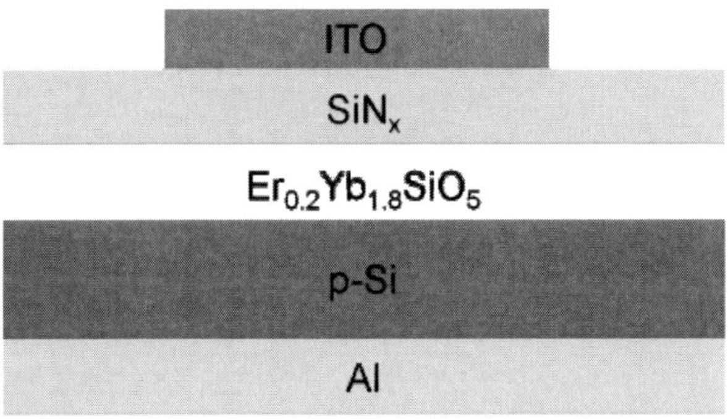

Figure 5.18 A schematic cross section view of the ErYb silicate MIS device.

The current–voltage (I–V) characterization of the device was carried out by a Keithley 2611 sourcemeter. The EL spectrum was collected by a 1 mm-diameter multimode fiber and analyzed by a thermoelectric-cooled InGaAs photodetector mounted on a Horiba JY iHR550 monochromator. During the EL operation of the device, positive bias was applied on the Si substrate side. All the measurements were performed at room temperature.

Figure 5.19 shows the ErYb silicate MIS device's I–V curve measured on a 1 mm^2 ITO electrode. To analyze the current transportation mechanism, the I–V curve is re-plotted as the FN tunneling manner shown in the inset. The equation describing the FN tunneling is $J/E^2 = A \exp(-B/E)$, where J is the current density, E is the electric field, and A and B are constants. A semilog plot of the equation gives a linear relation between the J/E^2 and $1/E$. The inset of Fig. 5.19 demonstrates that the I–V curve follows the FN tunneling behavior very well. This conduction mechanism is very similar to the Er-doped SiO_2 MIS devices,[68] but different from the reported Er silicate EL devices.[11,66] The FN tunneling happens when the electric field is higher than about 60 V, corresponding to the $1/E$ smaller than 2×10^{-7} cm/V. As a comparison, reference sample A without the SiN_x layer shows no current before the happening of destructive breakdown of the ErYb silicate layer when the voltage bias is higher than 30 V. On the other hand, reference sample B without the ErYb silicate layer shows very similar I–V curve with the EL device (data not shown), and the conduction mechanism is determined to be also FN tunneling where the tunneling current started from about 37 V.

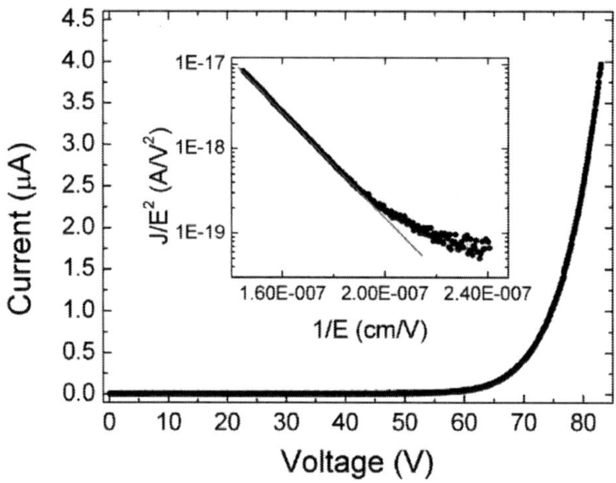

Figure 5.19 I–V characterization of the MIS device. The inset shows the FN plot of the I–V curve.

The FN tunneling transportation mechanism and the similar behavior between the EL–I–V and I–V curves strongly indicates that the Er ions are excited by the direct impact of energetic hot electrons. At high electric field, electrons from the ITO electrode enter the conduction band of the underneath insulators through the FN tunneling and are accelerated by the high electric field. Those electrons obtaining kinetic energy larger than 0.8 eV (~1.53 μm) are able to impact excite the Er ions to the first excited state. The excited Er ions then relax to the ground state and emit 1.53 μm photons. This EL work principle is different from the Er silicate EL devices with additional Si-rich oxide layer or excess Si contained in the Er silicate.[11,66] There the current transportation was improved by excess Si, and the sensitization effect of Si nanoclusters on Er luminescence may function. However, it was shown the excess Si brings Er luminescence quenching due to the Auger effect at the same time. In our device, no similar phenomenon was observed.

It should be noted that the additional SiN_x layer is critical for the stable EL operation. The SiN_x layer functions as a current limiting layer to prevent destructive breakdown of the underneath dielectric layer, similar to the function of the SiON layer used in the rare earth–doped SiO_2 MIS devices.[69–70] A single ErYb silicate layer is not able to conduct a stable current, while the SiN_x has a better conductivity via the FN tunneling. When stacking the ErYb silicate with a thin layer of SiN_x, the sudden increase of the breakdown current of the ErYb silicate layer is limited by the SiN_x layer. The destructive breakdown of the ErYb silicate layer is therefore prevented and a stable EL operation becomes possible. The 0.3 mA/cm^2 current density is not only much lower than the Si-rich Er silicate EL device,[66] but also about one order of magnitude smaller than the Er-doped SiO_2 MIS devices working in the same principle.[68] Considering the Er concentration of our device is about one order of magnitude higher than the Er-doped SiO_2 MIS device, it is reasonable to obtain Er EL at a smaller current density. This means the ErYb silicate materials have a good potential to be used for silicon-based light source.

Figure 5.20a shows the EL spectra of the device, measured using current pulse with different amplitude from 1.5 to 5 μA. The EL spectra profiles are identical to the 1.53 μm PL spectra (PL not shown). This indicates that the EL is due to the radiative transition of Er ion from $^4I_{13/2}$ to $^4I_{15/2}$ level. The EL spectra are peaked at 1531 nm and the spectra structure is due to the Stark splitting of the referred Er ions' energy levels.[71] The EL intensities are linearly increased versus the current, as shown in Fig. 5.20b. No saturation of EL intensity was observed, which means the EL intensity can be further increased. It is well known that under optical pumping Yb and Er ions have efficient coupling. Using electrical pumping and impact excitation, in Er and Yb co-doped

SiO_2 MISLED the excitation of Yb ions and Er upconversion emission can be also observed.[72] However, in the measured injection range of our device, neither visible Er upconversion nor Yb-related EL emission was observed. It seems the presence of Yb does not have influence on the excitation/de-excitation of Er. This may indicate that (1) the electrical pumping level was not high enough to induce Er-Er upconversions or Er-Yb interactions, or (2) the hot carriers' kinetic energies were not high enough to either directly excite Yb (excitation energy is about 1.27 eV) or to excite higher energy levels of Er. Further enhancement of the hot carrier's energy and/or injection level may help us to learn more about the interaction of Yb and Er in silicate under electrical pumping.

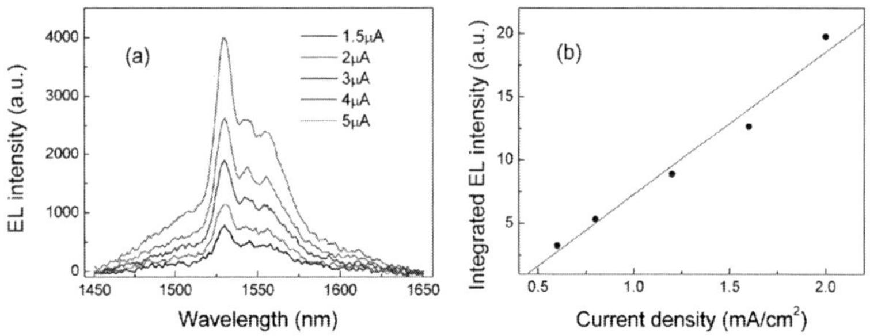

Figure 5.20 (a) EL spectra under different current pulse bias. (b) Integrated EL intensity versus current density.

Figure 5.21a shows time-resolved EL signals recorded at 1.53 μm using different current pulse bias. Applying single-exponential decay to fit the decay part of each curve, a $1/e$ time of about 1.5 ms can be figured out for all the EL decay curves. Differently, the rise part of each curve was characterized by different rise time as shown in the inset. When the device is pumped by pulsed 980 nm laser light, a 1.42 ms $1/e$ PL decay time can be obtained, close to the EL decay time. This means no extra decay path was introduced in the EL process. The EL excitation cross section σ can be obtained from the time-resolved EL signals using the following relation:

$$\frac{1}{\tau_{\text{rise}}} = \sigma\varphi + \frac{1}{\tau_{\text{decay}}}, \tag{5.4}$$

where τ_{rise} means the EL rise time, τ_{decay} means the EL decay time, and φ means the injected carrier flux. Figure 5.21b shows the reciprocal of EL rise time $1/\tau_{rise}$ versus the injected carrier flux φ. The slope of a linear fit gives the excitation cross section $\sigma = 3 \times 10^{-14}$ cm^2 of Er ions. This is three times of the highest reported Er^{3+} impact excitation cross section in Er-doped SiO_2 MISLED.[73] This large cross section may be related to the microstructure surrounding Er ions and the high Er concentration.[74] Further work is under way to investigate the detailed reason. The large impact excitation cross section of ErYb silicate indicates a high excitation efficiency of Er ions.

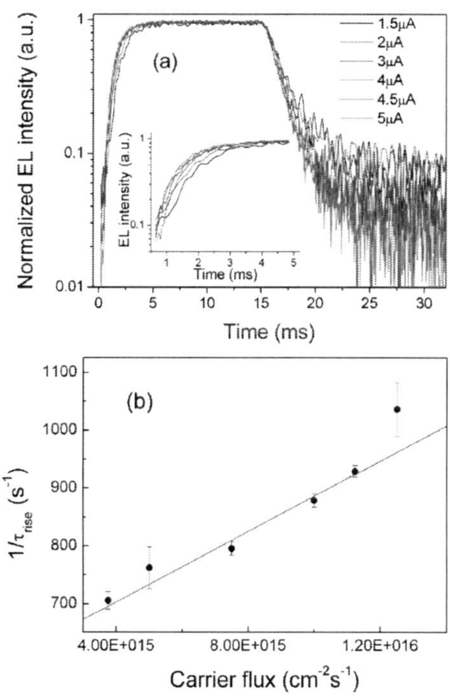

Figure 5.21 (a) Time-resolved EL signals versus current pulse bias, the inset shows the enlarged rise part of each curve; (b) The reciprocal of the EL rise time versus the carrier flux.

Since only the lowest energy's Er emission was observed, the excited Er fraction can be evaluated from rate equations based on a simplified two energy level model, where only the Er ion's ground state ($^4I_{15/2}$ level) and first excited state ($^4I_{13/2}$ level) are considered. The excitation process is depicted by $\sigma(N_{tot} - N_1)J/q$, where N_{tot} is the total Er concentration and N_1 is the excited Er concentration. For the decay process, except for the spontaneous decay τ_{decay}, we also consider upconversions, which happen between two neighboring excited Er ions and have been studied in previous work using optical pumping.[49,50] Although no upconversion was observed in present EL device, we speculate that taking it into account is reasonable because (1) the EL device was not pumped hard, which means the excited Er fraction should be small and the possible upconversion effect may be too weak to be observed; and (2) even if there is no upconversion effect under electrical-pumping, the inclusion of such an effect would only give a more conservative estimation on the population inversion. With these considerations, the rate equation describing the EL process is expressed as

$$\frac{dN_1}{dt} = \sigma\varphi(N_{tot} - N_1) - \frac{N_1}{\tau_{decay}} - 2C_{up}N_1^2, \tag{5.5}$$

where C_{up} is the upconversion coefficient. In steady state, using the evaluated excited Er concentration N_1 and identical emission and absorption cross section at 1.53 μm, the material's gain can be obtained through:

$$g = 4.343\sigma_{em}[N_1 - (N_{tot} - N_1)] \tag{5.6}$$

Table 5.1 lists the selected parameters used for the evaluation. The upconversion coefficient 3×10^{-17} cm^3s^{-1} and emission cross section 7×10^{-21} cm^2 of ErYb silicate are from previous waveguide measurements.[48–50] For the Er-doped SiO_2, the Er concentration, emission cross section and upconversion coefficient are selected from typical values.[75] The excitation cross section and decay time are obtained from electrical-pumped MISLED.[73]

Table 5.1 Selected parameters used for population inversion calculation

Material	N_{tot}(cm^{-3})	τ_2 (ms)	σ_{ex} (cm^2)	σ_{em} (cm^2)	C_{up} (cm^3s^{-1})
Er-SiO_2	2×10^{20}	1.5	1×10^{-14}	6×10^{-21}	3×10^{-18}
$Er_{0.2}Yb_{1.8}SiO_5$	1.9×10^{21}	1.5	3×10^{-14}	7×10^{-21}	3×10^{-17}

Figure 5.22a shows the invertible Er fractions *n* of ErYb silicate and Er-doped SiO_2 under electrical pumping. When *n* equals to 0.5, a half of Er ions are excited and the absorption and emission of Er ions are equal to each other. For Er-doped SiO_2, population inversion appears at current density of 21 mA/cm^2. For ErYb silicate, it requires 308 mA/cm^2 to invert a half of Er ions. At 1 A/cm^2 injection level, 97% of Er ions in Er-doped SiO_2 are inverted, while only about 70% of Er ions in ErYb silicate are inverted. The reason is that the ErYb silicate has as high as 10 times Er concentration and upconversion coefficient than the Er-doped SiO_2. As a result it is easier to reach population inversion in Er-doped SiO_2. Figure 5.22b shows the calculated material's gain. At 387 mA/cm^2, ErYb silicate and Er-doped SiO_2 show the same gain of about 4.5 dB/cm. Using lower injection current density, Er-doped SiO_2 has higher gain than ErYb silicate. However, further increase the current density to be higher than 0.39 A/cm^2, the gain achievable from Er-doped SiO_2 is severely saturated and the highest gain obtained is about 5 dB/cm. This value is comparable with previous estimation.[14] In contrast, the gain obtained from ErYb silicate continues to increase with the current density. At 1 A/cm^2, the gain is already higher than 23 dB/cm, corresponding to 70% of inverted Er ions. Further evaluation indicates that 44 dB/cm gain can be achieved at current density of 4 A/cm^2 (not displayed in Fig. 5.4), corresponding to 88% of inverted Er ions.

The evaluation indicates that the material's gain achievable from the ErYb silicate is exciting, in expense of high injection level. The highest current density reached in measurement was about 0.1 A/cm^2 (data not shown), still about one order of magnitude smaller than the required injection level. Further increase of the injection current density was prevented by the destructive breakdown of the devices mainly due to the silicate thin film's quality. Previously in Tb-doped SiO_2 MISLED a high injection level of 4 A/cm^2 was achieved.[70] Improvements are under way to improve the material quality, to increase the breakdown electric field strength and the injection level of ErYb silicate. In addition, it should be noted that in our evaluation the current density for population inversion may be overestimated due to the included upconversion effect. Therefore, the current density may need not be as high as we estimated. Recently we observed a 1.7 dB optical signal enhancement at 1.53 μm in ErYb silicate embedded silicon slot waveguide using 1480 nm optical pumping. The propagation loss was about 15 dB/cm and the waveguide structure is suitable for current injection. Considering the evaluated highest achievable material gain of 44 dB/cm under electrical-pumping and a confinement factor Γ of 0.3 in the slot region, about 13 dB/cm mode gain can be supplied. This is already close to the propagation loss of 15 dB/cm we measured. We estimate that the loss was due to the scattering in the polycrystalline upper silicon, in ErYb silicate, and the roughness of the surface and sidewall of the silicon stripe. Further optimization of the design, processing, and fabrication should be able to reduce the waveguide loss and net gain may be achieved in ErYb silicate embedded slot waveguide using electrical-pumping.

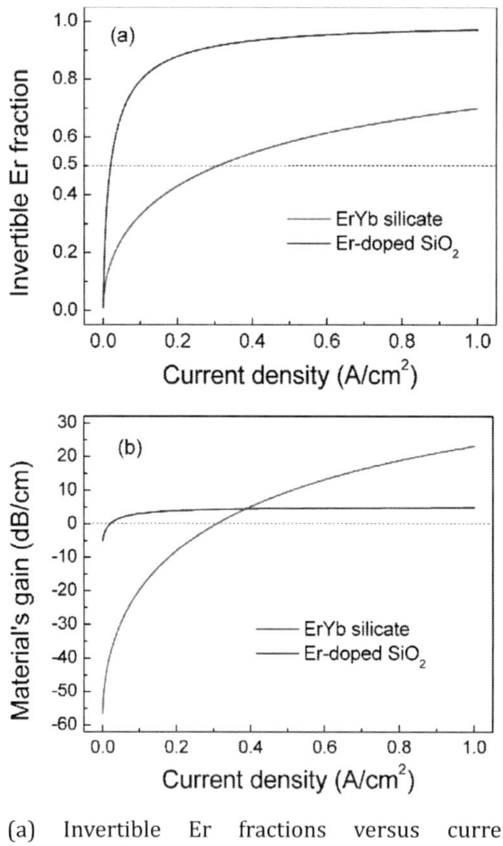

Figure 5.22 (a) Invertible Er fractions versus current density; (b) Material's gain versus current density.

5.4 Conclusion

In conclusion, we have introduced Er-doped silicon-rich silicon oxide (SRSO:Er) and silicon nitride (SRSN:Er), and Er silicates as novel materials that can provide high and stable Er concentrations and a high refractive index for compact on-chip light sources. The SRSO:Er and SRSN:Er allows sensitization of Er by nc-Si without optical de-activation of Er. For Er silicates, the two orders of magnitude photoluminescence intensity enhancement was observed for yttrium and ytterbium co-doped Er silicates compared with that of pure Er silicate. Three kinds of waveguide structure, the strip-loaded, slot and hybrid ErYb/Y silicates waveguides have been fabricated, and the optical amplification was observed in these waveguide structures. 1.53 μm electroluminescence in ErYb silicates was also realized using hot carriers' impact excitations of Er ions. The actual realization of light sources using these materials still needs more research, and many promising on-chip photonic devices can already be realized with them.

Acknowledgments

This work is supported by Peking University's 985 startup fund, the National Natural Science Foundation of China Grant Nos. 60907024, and 61036011, 61377056, and Program for New Century Excellent Talents in University.

References

1. R. Soref (2006). The past, present, and future of silicon photonics, *IEEE J. Selected Top. Quantum Electron.*, **12** (6), 1678–1687.
2. Z. Zhou, Z. Tu, B. Yin, W. Tan, L. Yu, H. Yi, and X. Wang (2013). Development trends in silicon photonics, *Chin. Opt. Lett.*, **11**(1), 12501.
3. E. Kasper, M. Kittler, M. Oehme, and T. Arguirov (2013). Germanium tin: Silicon photonics toward the mid-infrared [Invited], *Photon. Res.*, **1**(2), 69.
4. P. Boucaud, M. El Kurdi, A. Ghrib, M. Prost, M. de Kersauson, S. Sauvage, F. Aniel, X. Checoury, G. Beaudoin, L. Largeau, Sagnes, I., Ndong, G., M. Chaigneau, and R. Ossikovski (2013). Recent advances in germanium emission [Invited], *Photon. Res.*, **1**(3), 102.

5. L. Pavesi, L. Dal Negro, C. Mazzoleni, G. Franzò, and F. Priolo (2000). Optical gain in silicon nanocrystals, *Nature*, **408**(6811), 440–444.
6. J. Palm, F. Gan, B. Zheng, J. Michel, and L. Kimerling (1996). Electroluminescence of erbium-doped silicon, *Phys. Rev. B*, **54**(24), 17603–17615.
7. P. G. Kik, and A. Polman (2000). Exciton–erbium interactions in Si nanocrystal-doped SiO_2, *J. Appl. Phys.*, **88**(4), 1992.
8. H. Isshiki, M. J. A. de Dood, A. Polman, and T. Kimura (2004). Self-assembled infrared-luminescent Er–Si–O crystallites on silicon, *Appl. Phys. Lett.*, **85**(19), 4343.
9. M. Miritello, R. Lo Savio, F. Iacona, G. Franzò, A. Irrera, A. M. Piro, C. Bongiorno, and F. Priolo (2007). Efficient luminescence and energy transfer in erbium silicate thin films, *Adv. Mater.*, **19**(12), 1582–1588.
10. R. Lo Savio, M. Miritello, F. Iacona, A. M. Piro, M. G. Grimaldi, and F. Priolo (2008). Thermal evolution of Er silicate thin films grown by RF magnetron sputtering, *J. Phys. Condensed Matter*, **20**(45), 454218.
11. Y. Yin, K. Sun, W. J. Xu, G. Z. Ran, G. G. Qin, S. M. Wang, and C. Q. Wang (2009). 1.53 μm photo- and electroluminescence from Er^{3+} in erbium silicate, *J. Phys. Condensed Matter Inst. Phys. J.*, **21**(1), 12204.
12. T. Kippenberg, J. Kalkman, A. Polman, and K. Vahala (2006). Demonstration of an erbium-doped microdisk laser on a silicon chip, *Phys. Rev. A*, **74**(5), 51802.
13. R. Sun, P. Dong, N. Feng, C. Hong, J. Michel, M. Lipson, and L. Kimerling (2007a). Horizontal single and multiple slot waveguides: Optical transmission at λ = 1550 nm, *Opt. Express*, **15**(26), 17967.
14. K. Preston, and M. Lipson (2009). Slot waveguides with polycrystalline silicon for electrical injection, *Opt. Express*, **17**(3), 1527.
15. D. Olaosebikan, S. Yerci, A. Gondarenko, K. Preston, R. Li, L. Dal Negro, and M. Lipson (2011). Absorption bleaching by stimulated emission in erbium-doped silicon-rich silicon nitride waveguides, *Opt. Lett.*, **36**(1), 4–6.
16. G. Franzò, S. Coffa, F. Priolo, and C. Spinella (1997). Mechanism and performance of forward and reverse bias electroluminescence at 1.54 μm from Er-doped Si diodes, *J. Appl. Phys.*, **81**(6), 2784.
17. F. Priolo, G. Franzò, S. Coffa, and A. Carnera (1998). Excitation and nonradiative deexcitation processes of Er^{3+} in crystalline Si, *Phys. Rev. B*, **57**(8), 4443–4455.

18. A. J. Kenyon, P. F. Trwoga, M. Federighi, and C. W. Pitt (1994). Optical properties of PECVD erbium-doped silicon-rich silica: Evidence for energy transfer between silicon microclusters and erbium ions, *J. Phys. Condensed Matter*, **6**(21), L319–L324.

19. M. Fujii, M. Yoshida, Y. Kanzawa, S. Hayashi, and K. Yamamoto (1997). 1.54 μm photoluminescence of Er^{3+} doped into SiO_2 films containing Si nanocrystals: Evidence for energy transfer from Si nanocrystals to Er^{3+}, *Appl. Phys. Lett.*, **71**(9), 1198.

20. D. Pacifici, G. Franzò, F. Priolo, F. Iacona, and L. Dal Negro (2003). Modeling and perspectives of the Si nanocrystals–Er interaction for optical amplification, *Phys. Rev. B*, **67**(24), 245301.

21. G., Franzò, S. Boninelli, D. Pacifici, F. Priolo, F. Iacona, and C. Bongiorno (2003). Sensitizing properties of amorphous Si clusters on the 1.54 μm luminescence of Er in Si-rich SiO_2, *Appl. Phys. Lett.*, **82**(22), 3871.

22. F. Priolo, G. Franzò, D. Pacifici, V. Vinciguerra, F. Iacona, and A. Irrera (2001). Role of the energy transfer in the optical properties of undoped and Er-doped interacting Si nanocrystals, *J. Appl. Phys.*, **89**(1), 264.

23. N. Daldosso, D. Navarro-Urrios, M. Melchiorri, L. Pavesi, F. Gourbilleau, M. Carrada, R. Rizk, C. García, P. Pellegrino, B. Garrido, and L. Cognolato (2005). Absorption cross section and signal enhancement in Er-doped Si nanocluster rib-loaded waveguides, *Appl. Phys. Lett.*, **86**(26), 261103.

24. N. Daldosso, D. Navarro-Urrios, M. Melchiorri, L. Pavesi, C. Sada, F. Gourbilleau, and R. Rizk (2006). Refractive index dependence of the absorption and emission cross sections at 1.54 μm of Er^{3+} coupled to Si nanoclusters, *Appl. Phys. Lett.*, **88**(16), 161901.

25. P. G. Kik, and A. Polman (2002). Gain limiting processes in Er-doped Si nanocrystal waveguides in SiO_2, *J. Appl. Phys.*, **91**(1), 534.

26. J. Lee, J. Shin, and N. Park (2004). Optical gain at 1.5 μm in Si-nanocrystal sensitized, Er-doped silica waveguide using top-pumping 470 nm LED-Technical Digest (CD). Los Angeles, California: Optical Society of America, D19.

27. V. Toccafondo, F. Di Pasquale, S. Faralli, N. Daldosso, L. Pavesi, and H. E. Hernandez-Figueroa (2007). Study of an efficient longitudinal multimode pumping scheme for Si-nc sensitized EDWAs, *Opt. Express*, **15**(22), 14907.

28. O. Jambois, Y. Berencen, K. Hijazi, M. Wojdak, A. J. Kenyon, F. Gourbilleau, R. Rizk, and B. Garrido (2009). Current transport and

electroluminescence mechanisms in thin SiO_2 films containing Si nanocluster-sensitized erbium ions, *J. Appl. Phys.*, **106**(6), 63526.

29. O. Jambois, F. Gourbilleau, A. J. Kenyon, J. Montserrat, R. Rizk, and B. Garrido (2010). Towards population inversion of electrically pumped Er ions sensitized by Si nanoclusters, *Opt. Express*, **18**(3), 2230.
30. A. Kanjilal, L. Rebohle, W. Skorupa, and M. Helm (2009). Correlation between the microstructure and electroluminescence properties of Er-doped metal-oxide semiconductor structures, *Appl. Phys. Lett.*, **94**(10), 101916.
31. A. Polman, A. Lidgard, D. C. Jacobson, P. C. Becker, R. C. Kistler, G. E. Blonder, and J. M. Poate (1990). 1.54 μm room-temperature luminescence of MeV erbium-implanted silica glass, *Appl. Phys. Lett.*, **57**(26), 2859.
32. S. Saini, J. Michel, and L. C. Kimerling (2003). Index contrast scaling for optical amplifiers, *J. Lightwave Technol.*, **21**(10), 2368–2376.
33. T. Kippenberg, J. Kalkman, A. Polman, and K. Vahala (2006b). Demonstration of an erbium-doped microdisk laser on a silicon chip, *Phys. Rev. A*, **74**(5), 51802.
34. J. Michel, J. L. Benton, R. F. Ferrante, D. C. Jacobson, D. J. Eaglesham, E. A. Fitzgerald, Y. H. Xie, J. M. Poate, and L. C. Kimerling (1991). Impurity enhancement of the 1.54 μm Er^{3+} luminescence in silicon, *J. Appl. Phys.*, **70**(5), 2672.
35. K. S. Cho, N. Park, T. Kim, K. Kim, G. Y. Sung, and J. H. Shin (2005). High efficiency visible electroluminescence from silicon nanocrystals embedded in silicon nitride using a transparent doping layer, *Appl. Phys. Lett.*, **86**(7), 71909.
36. L. Dal Negro, R. Li, J. Warga, and S. N. Basu (2008). Sensitized erbium emission from silicon-rich nitride/silicon superlattice structures, *Appl. Phys. Lett.*, **92**(18), 181105.
37. K. Masaki, H. Isshiki, and T. Kimura (2005). Erbium-Silicon-Oxide crystalline films prepared by MOMBE, *Opt. Mater.*, **27**(5), 876–879.
38. K. Suh, J. H. Shin, S. Seo, and B. Bae (2006). Large-scale fabrication of single-phase Er_2SiO_5 nanocrystal aggregates using Si nanowires, *Appl. Phys. Lett.*, **89**(22), 223102.
39. T. Kimura, Y. Tanaka, H. Ueda, and H. Isshiki (2009). For mation of highly oriented layer-structured Er_2SiO_5 films by pulsed laser deposition, *Phys. E Low-Dimens. Syst. Nanostruct.*, **41**(6), 1063–1066.
40. K. Suh, J. H. Shin, S. Seo, and B. Bae (2008). Er^{3+} luminescence and cooperative upconversion in $Er_xYb_{2-x}SiO_5$ nanocrystal aggregates fabricated using Si nanowires, *Appl. Phys. Lett.*, **92**(12), 121910.

41. M. Vanhoutte, B. Wang, J. Michel, and L. C. Kimerling (2009). Processing and properties of ytterbium-erbium silicate thin film gain media, *Group IV Photonics, 2009. GFP '09. 6th IEEE International Conference on*, 63–65.

42. X. J. Wang, G. Yuan, H. Isshiki, T. Kimura, and Z. Zhou (2010). Photoluminescence enhancement and high gain amplification of $Er_xY_{2-x}SiO_5$ waveguide, *J. Appl. Phys.*, **108**(1), 13506.

43. X. J. Wang, B. Wang, L. Wang, R. M. Guo, H. Isshiki, T. Kimura, and Z. Zhou (2011). Extraordinary infrared photoluminescence efficiency of $Er_{0.1}Yb_{1.9}SiO_5$ films on SiO_2/Si substrates, *Appl. Phys. Lett.*, **98**(7), 71903.

44. X. Orignac, D. Barbier, X. Min Du, R. M. Almeida, O. McCarthy, and E. Yeatman (1999). Sol-gel silica/titania-on-silicon Er/Yb-doped waveguides for optical amplification at 1.5 μm, *Opt. Mater.*, **12**(1), 1–18.

45. B. Dong, D. P. Liu, X. J. Wang, T. Yang, S. M. Miao, and C. R. Li (2007). Optical thermometry through infrared excited green upconversion emissions in Er^{3+}-Yb^{3+} codoped Al_2O_3, *Appl. Phys. Lett.*, **90**(18), 181117.

46. L. Zhang, H. Hu, C. Qi, and F. Lin (2001). Spectroscopic properties and energy transfer in Yb^{3+}/Er^{3+}-doped phosphate glasses, *Opt. Mater.*, **17**(3), 371–377.

47. G. N. van den Hoven, E. Snoeks, A. Polman, J. W. M. van Uffelen, Y. S. Oei, and M. K. Smit (1993). Photoluminescence characterization of Er-implanted Al_2O_3 films, *Appl. Phys. Lett.*, **62**(24), 3065.

48. R. Guo, X. Wang, K. Zang, B. Wang, L. Wang, L. Gao, and Z. Zhou (2011). Optical amplification in Er/Yb silicate strip loaded waveguide, *Appl. Phys. Lett.*, **99**(16), 161115.

49. L. Wang, R. M. Guo, B. Wang, X. J. Wang, and Z. Zhou (2012). Hybrid Si_3N_4-Er/Yb silicate waveguides for amplifier application, *IEEE Photon. Technol. Lett.*, **24**(11), 900.

50. R. M. Guo, B. Wang, X. J. Wang, L. Wang, and Z. Zhou (2012). Optical amplification in Er/Yb silicate slot waveguide, *Opt. Lett.*, **37**(9), 1427.

51. D. McCumber (1964). Einstein relations connecting broadband emission and absorption spectra, *Phys. Rev.*, **136**(4A), A954–A957.

52. W. J. Miniscalco (1991). Erbium-doped glasses for fiber amplifiers at 1500 nm, *J. Lightwave Technol.*, **9**(2), 234–250.

53. P. G. Kik, and A. Polman (2003). Cooperative upconversion as the gain-limiting factor in Er doped miniature Al_2O_3 optical waveguide amplifiers, *J. Appl. Phys.*, **93**(9), 5008.

54. G. N. van den Hoven, E. Snoeks, A. Polman, C. van Dam, J. W. M. van Uffelen, and M. K. Smit (1996). Upconversion in Er-implanted Al_2O_3 waveguides, *J. Appl. Phys.*, **79**(3), 1258.

55. J. S. Chang, I. Y. Kim, G. Y. Sung, and J. H. Shin (2011). Population inversion and low cooperative upconversion in Er-doped silicon-rich silicon nitride waveguide, *Opt. Express*, **19**(9), 8406–8412.

56. V. R. Almeida, Q. Xu, C. A. Barrios, and M. Lipson (2004). Guiding and confining light in void nanostructure, *Opt. Lett.*, **29**(11), 1209.

57. Q. Xu, V. R. Almeida, R. R. Panepucci, and M. Lipson (2004). Experimental demonstration of guiding and confining light in nanometer-size low-refractive-index material, *Opt. Lett.*, **29**(14), 1626.

58. R. Sun, P. Dong, N. Feng, C. Hong, J. Michel, M. Lipson, and L. Kimerling (2007). Horizontal single and multiple slot waveguides: Optical transmission at λ = 1550 nm, *Opt. Express*, **15**(26), 17967.

59. R. M. Briggs, M. Shearn, A. Scherer, and H. A. Atwater (2009). Wafer-bonded single-crystal silicon slot waveguides and ring resonators, *Appl. Phys. Lett.*, **94**(2), 21106.

60. J. T. Robinson, K. Preston, O. Painter, and M. Lipson (2008). First-principle derivation of gain in high-index-contrast waveguides, *Opt. Express*, **16**(21), 16659.

61. Y. C. Jun, R. M. Briggs, H. A. Atwater, and M. L. Brongersma (2009). Broadband enhancement of light emission in silicon slot waveguides, *Opt. Express*, **17**(9), 7479.

62. C. Creatore and L. Andreani (2008). Quantum theory of spontaneous emission in multilayer dielectric structures, *Phys. Rev. A*, **78**(6), 63825.

63. K. Suh, M. Lee, J. S. Chang, H. Lee, N. Park, G. Y. Sung, and J. H. Shin (2010). Cooperative upconversion and optical gain in ion-beam sputter-deposited $Er_xY_{2-x}SiO_5$ waveguides, *Opt. Express*, **18**(8), 7724–7731.

64. D. Olaosebikan, S. Yerci, A. Gondarenko, K. Preston, R. Li, L. Dal Negro, and M. Lipson, *Opt. Lett.*, **36**, 4 (2011).

65. L. C. Kimerling, L. Dal Negro, S. Saini, Y. Yi, D. Ahn, S. Akiyama, D. Cannon, J. Liu, J. G. Sandland, D. Sparacin, J. Michel, K. Wada, and M. R. Watts, *Silicon Photonics*, Lockwood, D. J. and Pavesi, L. New York (2004).

66. Y. Yin, W. J. Xu, F. Wei, G. Z. Ran, G. G. Qin, Y. F. Shi, Q. G. Yao, and S. D. Yao (2010). Room temperature Er^{3+} 1.54 μm electroluminescence

from Si-rich erbium silicate deposited by magnetron sputtering, *J. Phys. D Appl. Phys.*, **43**(33), 335102.

67. S. Wang, A. Eckau, E. Neufeld, R. Carius, and C. Buchal (1997). Hot electron impact excitation cross section of Er^{3+} and electroluminescence from erbium-implanted silicon metal-oxide-semiconductor tunnel diodes, *Appl. Phys. Lett.*, **71**(19), 2824.

68. M. E. Castagna, S. Coffa, M. Monaco, A. Muscara, L. Caristia, S. Lorenti, and A. Messina (2003). High efficiency light emitting devices in silicon, *Mater. Sci. Eng. B*, **105**(1–3), 83–90.

69. J. M. Sun, W. Skorupa, T. Dekorsy, M. Helm, L. Rebohle, and T. Gebel (2005). Bright green electroluminescence from Tb^{3+} in silicon metal-oxide-semiconductor devices, *J. Appl. Phys.*, **97**(12), 123513.

70. J. M. Sun, L. Rebohle, S. Prucnal, M. Helm, and W. Skorupa (2008). Giant stability enhancement of rare-earth implanted SiO_2 light emitting devices by an additional SiON protection layer, *Appl. Phys. Lett.*, **92**(7), 71103.

71. B. Wang, R. M. Guo, X. J. Wang, L. Wang, L. Y. Hong, B. Yin, L. F. Gao, and Z. Zhou (2012). Near-infrared electroluminescence in ErYb silicate based light-emitting device, *Opt. Mater.*, **34**(8), 1371–1374.

72. S. Prucnal, L. Rebohle, and W. Skorupa (2011). Electroluminescence from Er and Yb co-doped silicon dioxide layers: The excitation mechanism, *J. Non-Crystalline Solids*, **357**(3), 915–918.

73. M. E. Castagna, S. Coffa, M. Monaco, A. Muscara, L. Caristia, S. Lorenti, and A. Messina (2003). High efficiency light emitting devices in silicon, *Mater. Sci. Eng. B*, **105**(1–3), 83–90.

74. J. M. Sun, S. Prucnal, W. Skorupa, T. Dekorsy, A. Müchlich, M. Helm, L. Rebohle, and T. Gebel (2006). Electroluminescence properties of the Gd^{3+} ultraviolet luminescent centers in SiO_2 gate oxide layers, *J. Appl. Phys.*, **99**(10), 103102.

75. E. Snoeks, van den G. N. Hoven, and A. Polman (1996). Optimization of an Er-doped silica glass optical waveguide amplifier, *IEEE J. Quantum Electron.*, **32**(9), 1680–1684.

Chapter 6

Nonclassical Light Sources and Frequency Converters with Integrated Opto-Mechanical Systems

Zhang-qi Yin,[a] Yong-Chun Liu,[b] and Yun-Feng Xiao[b]

[a]*Center for Quantum Information, Institute for Interdisciplinary Information Sciences, Tsinghua University, Beijing 100084, China*
[b]*State Key Laboratory for Mesoscopic Physics and School of Physics, Peking University, Beijing 100871, China*

yinzhangqi@tsinghua.edu.cn, ycliu@pku.edu.cn

6.1 Introduction

In this chapter, we discuss the theoretical and experimental progress in integrated opto-mechanical systems. Opto-mechanics is a quickly developing new field that has applications in precise measurements, classical-quantum transitions, etc. (Kippenberg and Vahala, 2008; Aspelmeyer et al., 2013). The study of opto-mechanics was back to Kepler, who found that the tails of comet were always way from the sun, and explained the phenomena as radiation pressure of light. In 1960s, Braginsky studied the role of radiation pressure in the context of interferometers (Braginsky and Manukin, 1967). He also proposed to cool the motion of the

Integrated Nanophotonic Resonators: Fundamentals, Devices, and Applications
Edited by Ya Sha Yi

ISBN 978-981-4613-78-1 (Hardcover), 978-981-4613-79-8 (eBook)
www.panstanford.com

mirror by radiation pressure. The quantum ground state cooling of mechanical oscillator has been proposed in 2007 (Wilson et al., 2007). Marquardt et al., (2007) and realized in 2011 (Chan et al., 2011). The experimental study of classical-quantum transitions in the macroscopic regime will be the next goal in this field.

Beside the fundamental interests, it is also highly demanded to study the applications of opto-mechanical systems. The output light phase and frequency are highly sensitively related with the motion of the mechanical oscillator in opto-mechanics, which is widely studied as ultra-sensitive sensors. Recently, more and more attention has been paid to the nonlinear optics and optical devices in opto-mechanical systems. Here we focus on the applications of opto-mechanics as nonclassical light sources, and the frequency converters.

This review is organized as follows. In Section 6.1, we discuss the basic model of opto-mechanics and reviews the applications and progress in the field. Section 6.2 introduces the nonlinear optics and nonclassical light sources in opto-mechanical systems. Section 6.3 discusses how to realize the two-mode squeezed light source and the light frequency converter with integrated opto-mechanical systems. Section 6.4 summarizes the chapter.

6.1.1 The Basic Model of Opto-Mechanics

The typical opto-mechanics studies the systems that contain a mechanical oscillator and an optical cavity, as shown in Fig. 6.1. The motion of the mechanical oscillator induces the frequency shift on optical cavity. Usually, the frequency shift is proportional to the oscillation amplitude of the mechanical mode. The frequency of the cavity mode can be expressed in the following linear equation:

$$\omega_c(t) = \omega_0 + g_0 x(t), \tag{6.1}$$

where ω_0 denotes the cavity frequency without opto-mechanical coupling, and $g_0 = \partial\omega_c/\partial x$ is the opto-mechanical coupling constant. For the case of Fabry–Pérot cavity with length L, we have $g_0 = \omega_0/L$, and for a whispering-gallery mode (WGM) resonator with radius R, we have $g_0 = \omega_0/R$.

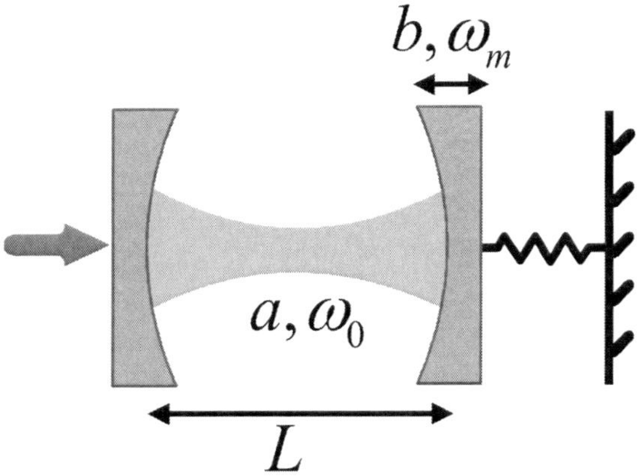

Figure 6.1 The opto-mechanical system that contains a cavity mode a and the mechanical oscillator mode b. The frequencies are ω_0 and ω_m, respectively. The cavity and mechanical modes couple with each other with strength $g_0 = \omega_0/L$.

The Hamiltonian of the system can be written as (Law, 1995)

$$H = \hbar\omega_0 a^\dagger a + \hbar\omega_m b^\dagger b + \hbar g_0 x_0 a^\dagger a(b + b^\dagger) + H_d, \tag{6.2}$$

where $a(b)$ and $a(b^\dagger)$ are the annihilation and the creation operators of the optical cavity (mechanical oscillator) mode, $x_0 = \sqrt{\hbar/2m\omega_m}$ is the zero field fluctuation for the mode b, and m is the effective mass of the mechanical oscillator. $H_d = \hbar\sqrt{\eta\kappa}(s_{in}a^\dagger e^{-i\omega_l t} + \text{h.d.})$ is the driving Hamiltonian, where $\kappa = \kappa_0 + \kappa_c$, s_{in} is the driving amplitude with $|s_{in}|^2 = P_{in}/\hbar\omega_0$. $\eta = \kappa_{ex}/(\kappa_0 + \kappa_{ex})$ describes the laser driving efficiency, κ_0 and κ_{ex} being the cavity decay rates due to intrinsic losses and the coupling to the taper waveguide. We denote $S = 2\sqrt{\eta\kappa}s_{in}$ as the laser driving strength.

6.1.2 Effective Hamiltonian and Corresponding Applications

We define the detuning between the driving laser and the cavity mode is $\Delta' = \omega_l - \omega_0$. We assume that the driving strength is so strong that the steady state of the cavity mode amplitude $|\alpha|$ is much larger than 1. The steady-state amplitude can be written as

$\alpha = \Omega'/(2\Delta' + ik)$. The Hamiltonian of the system can be expanded with $a - \alpha$, and the effective Hamiltonian is

$$H_{\mathrm{om}} = \Delta' a^\dagger a + \omega_{\mathrm{m}} b^\dagger b + G(a^\dagger + a)(b + b^\dagger), \tag{6.3}$$

where $G = \alpha g_0 x_0$. The main advantage of the opto-mechanical coupling Hamiltonian is that we can tune the detuning of the driving laser, and form the different type of effective Hamiltonian. Here we briefly discuss several situations: Kerr nonlinear Hamiltonian, beam splitter Hamiltonian, and two-mode squeezing Hamiltonian.

If we choose $\Delta' = 0$, under the condition $G \ll \omega_{\mathrm{m}}$, the mechanical mode can be adiabatically eliminated, the effective Hamiltonian takes the form

$$H_{\mathrm{Kerr}} = (G^2/\omega)\,(a^\dagger a^\dagger + aa), \tag{6.4}$$

also called Kerr nonlinear Hamiltonian, which can be used for generating squeeze lights (Fabre et al., 1994; Mancini and Tombesi, 1994). We will discuss its application in Section 6.2.

If we set $\Delta' = -\omega_{\mathrm{m}}$ and assume that $\omega_{\mathrm{m}} \gg G$, under the rotating wave approximation, we get the effective Hamiltonian of

$$H_{\mathrm{bs}} = Gab^\dagger + \mathrm{h.c.}, \tag{6.5}$$

which is also called beam-splitter Hamiltonian. This effective Hamiltonian is useful for laser sideband cooling of mechanical oscillators (Wilson et al., 2007; Marquardt et al., 2007), and light frequencies converter, which we will discuss in Section 6.3.

We take the detuning as $\Delta' = \omega_{\mathrm{m}}$, and assume $\omega_{\mathrm{m}} \gg G$, under the rotating wave approximation, the effective Hamiltonian becomes

$$H_{\mathrm{sq}} = Gab + \mathrm{h.c.}, \tag{6.6}$$

which can generate the two-mode squeezed state between cavity and mechanical modes. Combining both beam splitter and two-mode squeezing Hamiltonian, we may generate two-mode squeezed lights with opto-mechanical systems (Yin and Han, 2009), which will be discussed in Section 6.3.

6.2 Nonlinear Quantum Opto-Mechanics and Nonclassical Light Source

The opto-mechanical interaction Hamiltonian has the form $ga^\dagger a(b + b^\dagger)$ (Law, 1995), which is cubic in photonic and phononic field operators. In this case, the equations of motion (quantum Langevin equations) for the field operators are nonlinear. Therefore, such cubic-type interaction corresponds to the nonlinear nature of the opto-mechanical coupling. The nonlinearity is crucial for generating nonclassical states of the optical and mechanical modes. Generally speaking, the nonclassical states correspond to negative Wigner functions. Typically examples of nonclassical states include squeezed state, Fock state, and Schrödinger cat state. These nonclassical states are important for applications ranging from high-precision measurement and quantum information processing (Kippenberg and Vahala, 2008; Aspelmeyer et al., 2013). For example, squeezed state of light enables the measurement to surpass the standard quantum limit.

6.2.1 Basic Physics of Nonlinear Quantum Opto-Mechanics

To explore the physics of nonlinear quantum opto-mechanics, let us start from the system Hamiltonian

$$H = \omega_0 a^\dagger a + \omega_{\mathrm{m}} b^\dagger b + g a^\dagger a(b + b^\dagger) + \Omega(a^\dagger e^{i\omega t} + ae^{-i\omega t}). \qquad (6.7)$$

Here the first (second) term describes the energy of the optical (mechanical) mode, and the third term represents the opto-mechanical interaction, with the single-photon opto-mechanical coupling rate $g = g_0 x_0$. The last term describes the driving of the input laser.

The nonlinear properties depend on both the opto-mechanical coupling term $ga^\dagger a(b + b^\dagger)$ and the driving term $\Omega(a^\dagger e^{i\omega t} + ae^{-i\omega t})$. In the following we consider the two regimes of weak driving (Rabl, 2011; Nunnenkamp et al., 2011; Safavi-Naeini et al., 2012; Stannigel et al., 2012) and strong driving (Liu et al., 2013; Lemonde et al., 2013; Børkje et al., 2013; Kronwald and Marquardt, 2013), respectively.

6.2.1.1 Weak driving

In the weak driving regime, the driving term can be regarded as a perturbation. Starting from the system Hamiltonian, we use the polaron transformation $U = e^{-S}$ with $S = (g/\omega_m)a^\dagger a(b^\dagger - b)$ to diagonalize the opto-mechanical coupling term (Mahan, 2000). This can be understood by completing the square for the mechanical part

$$H = \omega_0 a^\dagger a + \omega_m \left(b^\dagger + \frac{g}{\omega_m} a^\dagger a\right)\left(b + \frac{g}{\omega_m} a^\dagger a\right) - \frac{g^2}{\omega_m} a^\dagger a a^\dagger a. \tag{6.8}$$

To simplify the expression, we denote a new operator

$$c = b + \frac{g}{\omega_m} a^\dagger a. \tag{6.9}$$

Then the Hamiltonian is written in the form

$$H = \omega_0 a^\dagger a + \omega_m c^\dagger c - \frac{g^2}{\omega_m} a^\dagger a a^\dagger a. \tag{6.10}$$

The above Hamiltonian reveals two important features. First, it can be verified that c satisfies the bosonic commutation relation $[c, c^\dagger] = 1$, and thus it is a bosonic operator. We can find that it represents a displaced phonon mode, with the equilibrium position of the harmonic oscillator shifted by an amount proportional to $(g/\omega_m)a^\dagger a$. Note that the displacement depends linearly on the photon number $a^\dagger a$. Second, in the Hamiltonian (6.10), the third term $-(g^2/\omega_m)a^\dagger a a^\dagger a$ represents an effective photon–photon interaction, with effective Kerr nonlinearity

$$X = -\frac{g^2}{\omega_m}. \tag{6.11}$$

After diagonalizing the Hamiltonian, the energy eigenvalues can be obtained as

$$E_{n,m} = \omega_0 n - \frac{g^2}{\omega_m} n^2 + \omega_m m, \tag{6.12}$$

where n and m denote the photon number and shifted phonon number, respectively. The energy level diagram for the nonlinear quantum opto-mechanics is plotted in Fig. 6.2. It shows (1) equally spaced phonon ladders whose offset shifts depend on the photon number and (2) anharmonic photon ladders with decreasing photon level spacing for higher energy levels.

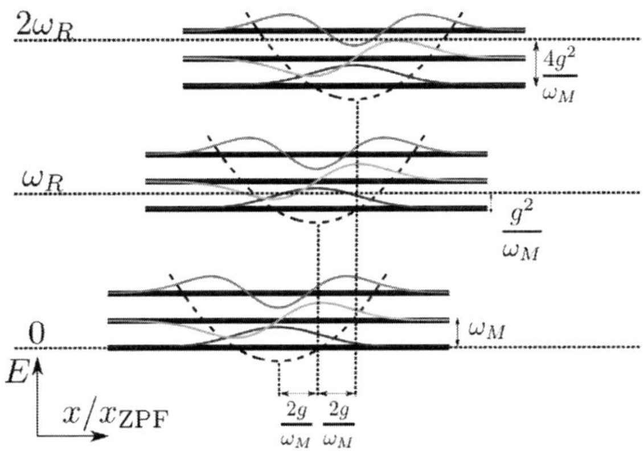

Figure 6.2 Eigenenergy spectrum (thick black lines) and mechanical eigenfunctions (blue, green, and red curves) under weak driving. Photon number n = 0, 1, 2 and phonon number m = 0, 1, 2. The energy axis is not to scale. The parabolas indicate the displaced harmonic oscillator potentials. Here ω_R and ω_M represent the optical and mechanical resonance frequencies, respectively. x_{ZPF} denotes the zero-point fluctuation of the mechanical resonator. Figure adapted with permission from Nunnenkamp et al. (2011), copyright (2011) by the American Physical Society.

With a weak driving as a perturbation, after the polaron transformation the driving term represents rich transition processes. Except for the common single mechanical sideband structure in the cavity response spectrum, it is possible to achieve multiple mechanical sidebands, which corresponds to the multiple-phonon transition processes (Fig. 6.3). This enables the application of multiple-phonon cooling, i.e., one photon interacts with two phonons simultaneously.

The effective Kerr nonlinearity leads to the nonclassical photon correlations. Under appropriate conditions, we can observe the photon blockade effect, where the first photon blocks the second photon from entering the opto-mechanical system. This phenomenon is quantitatively described by the second-order correlation function (Scully and Zubairy, 1997).

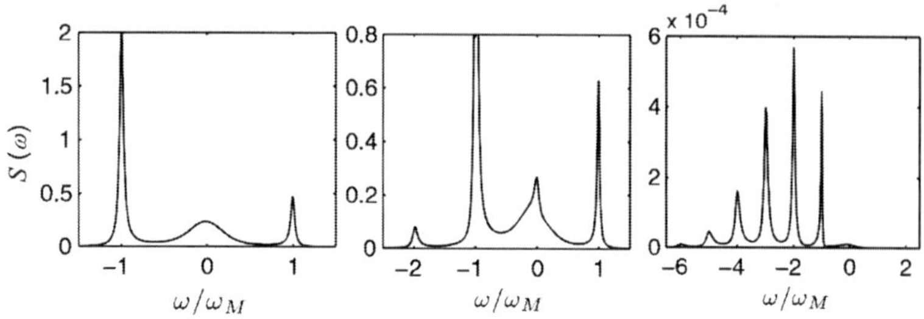

Figure 6.3 Crossover from single to multiple mechanical sidebands as the driving strength increases. $S(\kappa)$ represents the cavity output spectrum. Figures adapted with permission from Nunnenkamp et al. (2011), copyright (2011) by the American Physical Society.

$$g^{(2)}(t) = \frac{\langle a^\dagger(t)a^\dagger(t+\tau)a(t+\tau)a(t)\rangle}{\langle a^\dagger(t)a(t)\rangle\langle a^\dagger(t+\tau)a(t+\tau)\rangle} = \frac{\langle a^\dagger(t)a^\dagger(t+\tau)a(t+\tau)a(t)\rangle}{\langle a^\dagger a\rangle^2}. \tag{6.13}$$

The definition of $g^{(2)}(\tau)$ shows that it is a measure of the photon correlations between some time t and a later time $t + \tau$. If $g^{(2)}(\tau) < g^{(2)}(0)$, then the photons exhibit more correlations when they come together. This is called photon bunching, since the photons tend to distribute themselves preferentially in bunches. On the other hand, $g^{(2)}(\tau) > g^{(2)}(0)$ corresponds to photon antibunching, where the photons tend to distribute further apart rather than close together. Photon antibunching is important for realizing single-photon sources, where photons are emitted one by one. Another classification of photon statistics depends on the value of $g^{(2)}(0)$. If $g^{(2)}(0) = 1$, the filed satisfies the Poissonian distribution, where the photons arrive randomly. Fields for $g^{(2)}(0) < 1$ and $g^{(2)}(0) > 1$ are referred to as sub-Poissonian distribution and super-Poissonian distribution, respectively.

Figure 6.4 plots the dependence of $g^{(2)}(0)$ as functions of the laser detuning, the opto-mechanical coupling strength and the cavity decay rate. It reveals that $g^{(2)}(0) < 1$ can be realized for suitable parameter ranges, which enable the application of single photon generation.

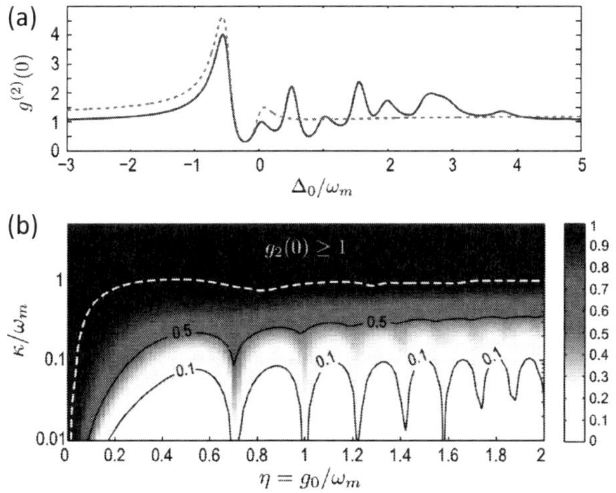

Figure 6.4 (a) Second-order photon correlation function $g^{(2)}(0)$ as a function of laser detuning Δ_0. (b) Minimum of $g^{(2)}(0)$ with respect to the laser detuning as functions of cavity decay rate κ and single-photon opto-mechanical coupling strength g_0. Figures adapted with permission from Rabl (2011), copyright (2011) by the American Physical Society.

6.2.1.2 Strong driving

In the strong driving regime, usually the opto-mechanical interaction can be linearized, while the nonlinear part is negligible. However, under appropriate condition, the nonlinear term can be resonantly enhanced. Starting from the system Hamiltonian (6.7), in the frame rotating at the input laser frequency ω, we obtain

$$H = -(\omega - \omega_0)a^\dagger a + \omega_{\mathrm{m}} b^\dagger b + g a^\dagger a(b + b^\dagger) + \Omega(a^\dagger + a). \tag{6.14}$$

Under the displacement transformation $a \to a_1 + \alpha$, $b \to b_1 + \beta$, we obtain

$$H = H_{\mathrm{L}} + H_{\mathrm{NL}},$$

$$H = H_{\mathrm{L}} + H_{\mathrm{NL}},$$

$$\begin{aligned} H_{\mathrm{L}} &= -\Delta a_1^\dagger a_1 + \omega_{\mathrm{m}} b_1^\dagger b_1 + G(a_1^\dagger + a_1)(b_1 + b_1^\dagger), \\ H_{\mathrm{L}} &= g a_1^\dagger a_1 (b_1 + b_1^\dagger). \end{aligned} \tag{6.15}$$

Here α and β represent the c-number steady-state values of the optical and mechanical modes, while a_1 and b_1 are the corresponding fluctuation operators. $G \equiv g\alpha$ describes the light-enhanced opto-mechanical coupling strength, which is assumed to be real and positive without loss of generality; $\Delta \equiv \omega_{\mathrm{L}} - \omega_{\mathrm{c}} + 2|G|^2/\omega_{\mathrm{m}}$ is the detuning where opto-mechanical coupling has been taken into account; H_{L} represents the quadratic Hamiltonian that leads to the linear dynamics, and H_{NL} denotes the nonlinear term.

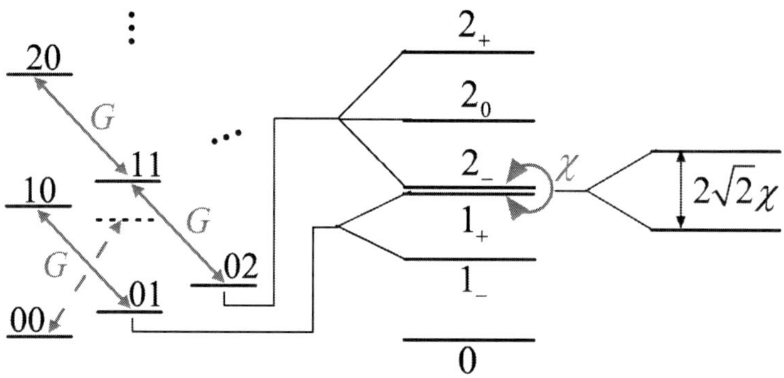

Figure 6.5 Energy level diagram of the system. Left: $|nm\rangle$ represents a state of n photons and m phonons. Right: $|0\rangle$ denotes the ground state, $|1\pm\rangle$ represents 1-excitation subspace, $|2\pm\rangle$ and $|20\rangle$ represents 2-excitation subspace. Figure adapted with permission from Liu et al. (2013), copyright (2013) by the American Physical Society.

$$H_{\mathrm{L}} = \omega_+ c_+^\dagger c_+ + \omega_- c_-^\dagger c_-, \tag{6.16}$$

where $\omega_\pm$ are the eigenfrequencies given by

$$\omega_\pm = \sqrt{\frac{\Delta^2 + \omega_{\mathrm{m}}^2 \pm \sqrt{(\omega^2 - \omega_{\mathrm{m}}^2)^2 - 16G^2\Delta\omega_{\mathrm{m}}}}{2}}, \tag{6.17}$$

and $c_\pm$ are the eigenmodes (polaritons) with the transformation

$$\begin{pmatrix} a_1 \\ b_1 \end{pmatrix} = \begin{pmatrix} \mu_1\,\mu_2\,\mu_3\,\mu_4 \\ \nu_1\,\nu_2\,\nu_3\,\nu_4 \end{pmatrix} \begin{pmatrix} c_+ \\ c_- \\ c_+ \\ c_- \end{pmatrix}. \tag{6.18}$$

As a typical simple example, consider the red-detuned resonant case $\Delta = -\omega_m$, which yields $\mu_{1,3} = \nu_{1,3} = \left(\sqrt{\omega_m/\omega_1} + \sqrt{\omega_{1+}/\omega_m}\right)/(2\sqrt{2})$ with $\omega_\pm = \sqrt{\omega_m^2 + 2G\omega_m}$, corresponding to the equal-weight mode mixing. In Fig. 6.5, the energy levels for this resonant case are plotted. It shows that the linear coupling G results in two independent sets of sub-levels n_+ and n_-, corresponding to the two normal modes c_+ and c_-. In the figure, 1_+ represents the first excited state of mode c_+, and 2_- denotes the second excited state of mode c_-. With frequency matching condition $\omega_+ = 2\omega_-$, these two states are degenerate, while H_{NL} leads to nonlinear interaction between them.

Under the condition $g \ll \omega_m$, H_{NL} can be re-expressed in the normal mode basis described by $H_{NL} = f(c_+, c_-, c^\dagger, c^\dagger_-)$. By using the rotating-wave approximation, the high-frequency off-resonant terms are neglect. In this case, H_{NL} can be approximated as

$$H_{NL} \simeq -\chi(c_+^\dagger c_-^2 + c_+ c_-^{\dagger 2}), \tag{6.19}$$

which corresponds to the parametric down conversion Hamiltonian, i.e., the annihilation of a c_+ mode polariton leads to the generation of a pair of c_- mode polariton. The parameter χ corresponds to the nonlinear coefficient, given by

$$\chi = [(\mu_1\mu_2 + \mu_3\mu_4)(\nu_2 + \nu_4) + \mu_2\mu_4\,(\nu_1 + \nu_3)]\,g. \tag{6.20}$$

In Fig. 6.6, the nonlinear parameter χ as a function of Δ is plotted, showing that χ is on the order of g for detuning $-2\omega_m < \Delta < -\omega_m/2$. This is different from the weak driving case where the nonlinear parameter scales as g^2/ω_m. This is because in the strong driving case, the resonant interaction conserves the energy as shown by Eq. (6.19). Noted that the major off-resonant nonlinear interaction terms include $\tilde{\chi}_{+,\omega-}c_+^\dagger c_+(c_- + c_+^\dagger)$, $\tilde{\chi}_{-,\omega-}c_-^\dagger c_-(c_- + c_-^\dagger)$, $\tilde{\chi}_{-,\omega+}c_-^\dagger c_-(c_+ + c_+^\dagger)$, and $\tilde{\chi}_{+,\omega+}c_+^\dagger c_+(c_+ + c_+^\dagger)$. The corresponding effective nonlinear parameters are described by $\chi_{\pm,\omega_-} = \tilde{\chi}^2_{\pm,\omega-}/\omega_-$ and $\chi_{\pm,\omega_+} = \tilde{\chi}^2_{\pm,\omega_+}/\omega_+$, which are plotted in the inset of Fig. 6.6 for comparison with χ. It reveals that these off-resonant nonlinear parameters are strongly reduced due to the far detuned interaction, typically on the order of $(g/\omega_m)\chi$.

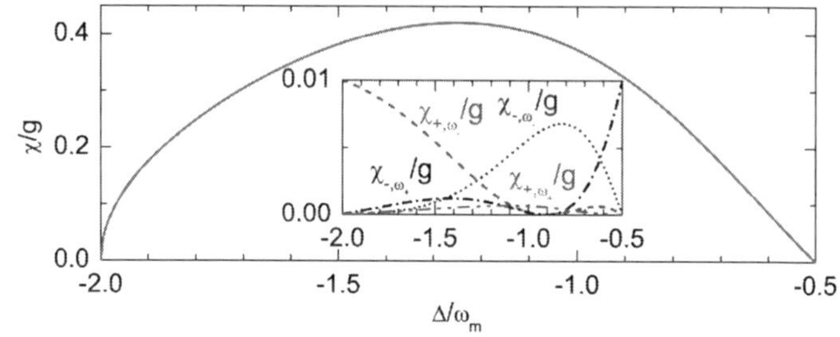

Figure 6.6 χ/g as a function of Δ, given by Eq. (6.20). Inset: $\chi_{+,\omega_-}/g$(green dashed curve), $\chi_{-,\omega_-}/g$ (blue dotted curve), $\chi_{-,\omega_+}/g$ (black dash-dotted curve) and $\chi_{+,\omega_+}/g$ (magenta dash-dot-dotted curve) as functions of Δ. Figure adapted with permission from Liu et al. (2013), copyright (2013) by the American Physical Society.

The parametric downconversion Hamiltonian (6.19) enables polariton pair generation. For example, photon pairs or phonon pairs can be generated through optimizing the detuning Δ. For $\Delta \simeq -2\omega_m$, c_+ mode is photon-like while c_- mode is phonon-like, and thus it is able to create phonon pairs. Similarly, for $\Delta \simeq -\omega_m/2$, photon pairs can be obtained. This holds potential for nonclassical light sources with entangled photon pairs.

To demonstrate the photon and phonon pair generation, numerical simulations are performed using the master equation

$$\dot{\rho} = i[\rho, H] + \kappa D[a_1]\rho + \gamma(n_{\text{th}} + 1)D[b_1]\rho + \gamma n_{\text{th}} D[b_1^\dagger]\rho, \tag{6.21}$$

where ρ is the density operator of the system; $D[\hat{o}]\rho = \hat{o}\rho\hat{o}^\dagger - (\hat{o}^\dagger\hat{o}\rho + \rho\hat{o}^\dagger\hat{o})/2$ denotes the standard dissipator in Lindblad form; κ and γ represent the decay rates of the optical and mechanical modes; $n_{\text{th}} = 1/[e^{\hbar\omega_m/(k_B T)} - 1]$ is the thermal phonon number at the environmental temperature T. Figure 6.7a plots the dynamics of photon pair generation. The laser detuning is set as $\Delta = -1.9\omega_m$. In this case, the photon (phonon) almost behaves the same as the $c_+(c_-)$ polariton, since the $c_+(c_-)$ mode consists of 97% optical (mechanical) component. With a single photon input, the photon-like polariton has effectively converted to a pair of phonon-like polaritons at $t \sim 200/\omega_m$. The maximum phonon number being slightly less than 2 is a result of dissipations. In Fig. 6.7b, photon pair generation is demonstrated for $\Delta = -0.6\omega_m$ with single-phonon initial state.

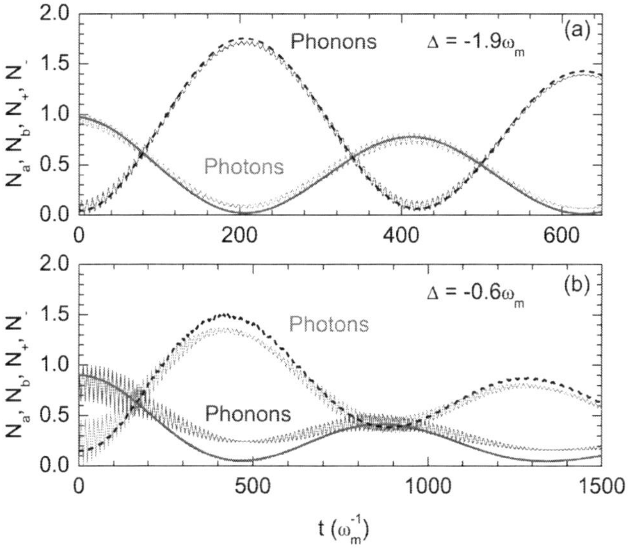

Figure 6.7 Time evolution of + polariton number N_+ (green thick curves), − polariton number N_- (black dashed curves), mean photon number N_a (red thin curves) and phonon number N_b (blue thin curves), for $\Delta = -1.9\omega_m$ (a) and $-0.6\omega_m$ (b). Figure adapted with permission from Liu et al. (2013), copyright (2013) by the American Physical Society.

6.2.2 Opto-Mechanical Squeezed Light Source

Squeezed light is a typical example of nonclassical light that allows quantum limits to be overcome in precision metrology. In the quantum theory, the Heisenberg uncertainty principle imposes a fundamental limit on the precision of measurement. For example, the amplitude and phase of an optical field cannot be measured precisely at the same time. Commonly used laser light is in the coherent state, whose quantum noise is described by a circle in the phasor diagram, representing the uncertainty of measurement. The minimum quantum noise achievable using a coherent laser source determines the shot noise limit, or standard quantum limit. The squeezed state of light reduces the quantum noise in one direction with increased quantum noise in the other direction, while the Heisenberg uncertainty principle still maintains. In order to generate the squeezed light, nonlinear process is a must. One typical example is the optical Kerr effect, where the phase shift of light depends on the light intensity. As shown in Fig. 6.8, the quantum noise circle of an initial coherent state will skew under such intensity-dependent phase shift, leading to the squeezing of quantum noise below the shot noise limit. Experimental demonstration of squeezed light was realized using nonlinear optical media (atomic sodium) (Slusher et al., 1985), optical fibers (Shelby et al., 1986) and nonlinear crystals (Wu et al., 1986). Large degree of quantum noise suppression up to 10 dB has been achieved in recent years (Vahlbruch et al., 2008; Eberle et al., 2010).

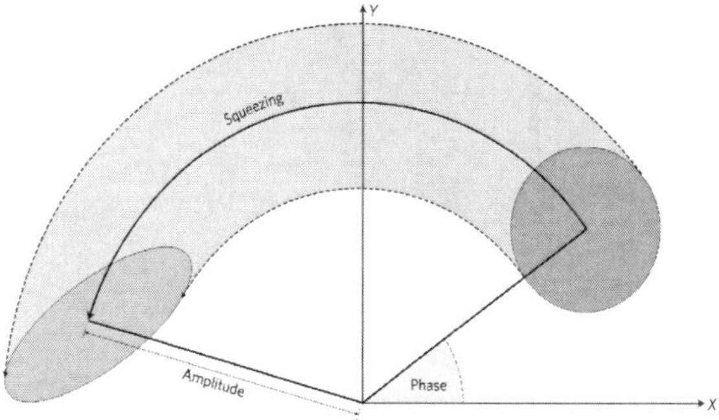

Figure 6.8 Squeezed state generation with optical Kerr effect. Figure adapted with permission from Brawley et al. (2013), copyright (2013) by Nature Publishing Group.

Squeezed light can be generated in opto-mechanical systems by making use of the nonlinearity induced by opto-mechanical coupling (Braginsky and Manukin, 1967; Fabre et al., 1994; Mancini and Tombesi, 1994). This is because opto-mechanical systems have the effective Kerr-type interaction $-(g_2/\omega_m)a^\dagger a a^\dagger a$ which correlates the amplitude and phase quadratures. From the quantum Langevin equations

$$\dot{a} = \left(i\Delta - \frac{\kappa}{2}\right)a - iga(b + b^\dagger) - i\Omega - \sqrt{\kappa}a_{\text{in}}, \tag{6.22}$$

$$\dot{b} = \left(-i\omega_{\text{m}} - \frac{\gamma}{2}\right)b - iga^\dagger a - \sqrt{\gamma}b_{\text{in}}, \tag{6.23}$$

in the frequency domain, it is found that the optical field $a(\omega)$ depends on the noise operators $a_{\text{in}}(\omega)$, $a_{\text{in}}^\dagger(\omega)$, $b_{\text{in}}(\omega)$ and $b_{\text{in}}^\dagger(\omega)$. As a result, the optical field can be squeezed or anti-squeezed in some particular quadrature.

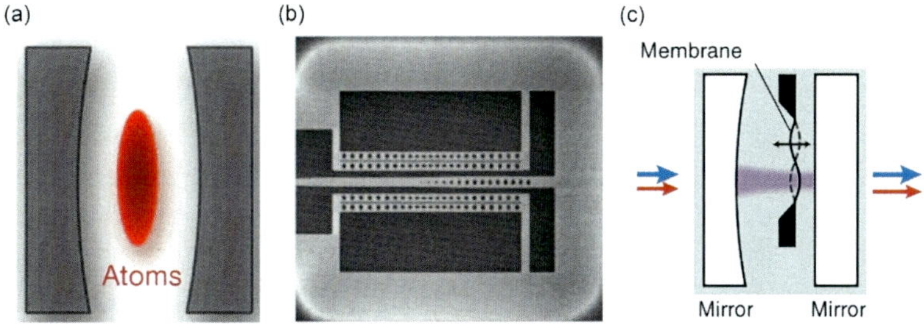

Figure 6.9 Various opto-mechanical systems that have realized squeezed light generation. (a) Cold atom opto-mechanical system. (b) Zipper-type photonic crystal opto-mechanical system. (c) Membrane-in-the-middle opto-mechanical system. Figures adapted with permission from (a) Brooks et al. (2013), copyright (2012) by Nature Publishing Group; (b) Safavi-Naeini et al. (2013), copyright (2013) by Nature Publishing Group; (c) Purdy et al. (2013), copyright (2013) by American Physical Society.

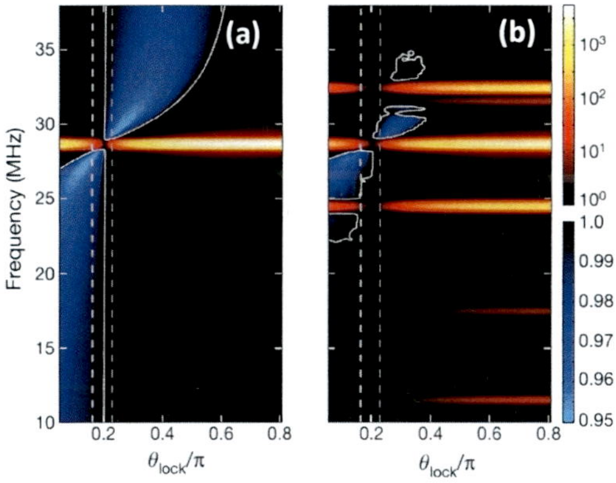

Figure 6.10 Noise power spectral density (PSD) showing the squeezing of light. Areas with noise below shot-noise are shown in blue shades on a linear scale. Areas with noise above shot-noise are shown in orange shades on a log scale. (a) Theoretical model. (b) Experimental data. Figures adapted with permission from Safavi-Naeini et al. (2013), copyright (2013) by Nature Publishing Group.

Recent experimental efforts have demonstrated squeezed light generation in opto-mechanical systems (Brawley et al., 2013; Brooks et al., 2013; Safavi-Naeini et al., 2013; Purdy et al., 2013; Mavalvala and Corbitt, 2013). In Brooks et al. (2013), quantum noise of light was reduced to 0.8% below the shot noise limit in a cold atom opto-mechanical system. In Safavi-Naeini et al. (2013), 4.5% squeezing of light is achieved in a photonic crystal opto-mechanical system. In Purdy et al. (2013), the experiments demonstrate squeezing of 32% (1.7 dB) by using a membrane-in-the-middle opto-mechanical system. These experiments show that opto-mechanical systems provides an architecture for integrating quantum light source, which will hold potential for applications in high-precision force and displacement sensing, and quantum information processing.

For example, it is important for improving the shot-noise-limited sensitivity of kilometer-scale laser interferometers, which are the basis of gravitational wave detectors, such as Laser Interferometer Gravitational-Wave Observatory (LIGO) (Caves, 1981; Kimble et al., 2001; LIGO, 2011, 2013). Squeezed states are also demonstrated to be useful in biological measurements for beating the quantum limit (Taylor et al., 2013).

6.2.3 Outlook

Nonlinear quantum opto-mechanics provides the unique characteristic to generate nonclassical states. By making use of the nonlinearity, opto-mechanical systems enable the construction of single-photon sources and squeezed light sources. These nonclassical light sources have applications in quantum communication, quantum computation, and high-precision measurement. The opto-mechanical nonclassical light sources possess several unique advantages. For example, they can be fabricated on a chip, and they can be easily integrated with other optical systems. With further development of device engineering, the single-photon opto-mechanical coupling strength is expected to be enhanced, so that the nonclassical characteristics can be increased. Advances are also expected on better environmental isolation so as to suppress the environmental thermal noise.

6.3 Two-Mode Squeezed State and Frequency Converter

In this section, we discuss the opto-mechanical systems that contain two cavity modes and one mechanical mode. We will extend the squeezed light source from single mode to two modes with different frequencies. Besides, by red sideband driving on both of two cavity modes, the frequency converter between two cavity modes can be realized. The frequency converter is the crucial component for the optical communication and quantum networks. The opto-mechanical frequency converter is in principle coherent, lossless, and noiseless. Therefore, it attracts a lot of attention.

6.3.1 Two-Mode Squeezed Light Source

In order to generate two-mode squeezed state between lights, we need to consider two different cavity modes, which couple with the same mechanical mode. The Hamiltonian of the system is

$$H'_{\mathrm{om}} = \sum_{i=1,2} \Delta_i a_i^\dagger a_i + \omega_{\mathrm{m}} b^\dagger b + \sum_{i=1,2} G_i (a_i^\dagger + a_i)(b + b^\dagger), \tag{6.24}$$

where a_i is the i-th cavity mode, Δ_i is the laser detuning. We assume that $\Delta_1 - \omega_{\mathrm{m}} = \delta > 0$, and $\Delta_2 + \omega_{\mathrm{m}} = \delta$.

In the rotating wave frame that $H_0 = \omega_{\mathrm{m}}(a_1^\dagger a_1 - a_2^\dagger a_2 + b^\dagger b)$, under the rotating wave approximation $G_i \ll \omega_{\mathrm{m}}$, we get the effective Hamiltonian is

$$H_{\mathrm{sq1}} = \sum_{i=1,2} \delta a_i^\dagger a_i + (G_1 a_1^\dagger b + G_2 a_2 b + \mathrm{h.c.}), \tag{6.25}$$

Equation (6.25) has very clear physical meaning, a_1 mode couples with b mode with beam splitter Hamiltonian Eq. (6.5), and a_2 mode couples with b mode through two-mode squeezing Hamiltonian Eq. (6.6). Mode b is treated as a quantum bus in the processing of entangling a_1 and a_2 modes. We can choose a specific time when the b is disentangled with both mode a_1 and a_2, but a_1 and a_2 modes are entangled (Tian 2013, Wang and Clerk 2013). The main noise of the scheme is from the thermal noise of mechanical mode. Therefore, we need to adiabatically eliminate the mechanical mode b (Yin and Han 2009). Under the condition that $\delta \gg G_i$, we get the effective coupling Hamiltonian between a_1 and a_2 modes

$$H_{\mathrm{sq2}} = \chi a_1^\dagger a_2^\dagger + \mathrm{h.c.}, \tag{6.26}$$

where $\chi = G_1 G_2/\delta$.

In opto-mechanical systems, the two-mode squeezed light source can have two different frequencies of output mode. The output lights frequencies can be chosen with large range of regime, e.g., from GHz to 100 THz. Therefore, it can be used to entangle distant quantum nodes with different frequencies. We will discuss the entanglement spectrum and the applications in quantum information processes, such as teleporation between two light modes.

6.3.1.1 Entanglement spectrum of the entangled light source

From Eq. (6.25), we get the Langevin equations are linearized as

$$\dot{a}_j = -iG_j(b+b^\dagger) + \left(i\Delta_j' - \frac{\gamma}{2}\right)a_j + \sqrt{\gamma}a_j^{\text{in}}, \tag{6.27}$$

$$\dot{b} = -i\sum_{p=1}^{2}(G_p^* a_p + G_p a_p^\dagger) + \left(i\omega_{\text{m}} - \frac{\gamma_{\text{m}}}{2}\right)b + \sqrt{\gamma_{\text{m}}}b^{\text{in}}, \tag{6.28}$$

where $j = 1, 2$ and $\Delta_j' = \Delta_j + 2(G_1^2 + G_2^2)/\omega_{\text{m}}$. We suppose Δ_1' and $\Delta_2' < 0$. We define $\delta = (\Delta_2' - \Delta_1')/2 - \omega_{\text{m}}$ and $d = -(\Delta_1' + \Delta_2')/2$. The thermal noise inputs are defined as correlation functions $\langle a_{\text{m}}^{\text{in}\dagger}(t), a_{\text{m}}^{\text{in}}(t)\rangle = n_{\text{m}}\delta(t - t')$, $\langle a_{\text{m}}^{\text{in}\dagger}(t), a_{\text{m}}^{\text{in}\dagger}(t')\rangle = \langle a_{\text{m}}^{\text{in}}(t), a_{\text{m}}^{\text{in}}(t')\rangle = 0$ and $\langle a_j^{\text{in}\dagger}(t), a_j^{\text{in}}(t')\rangle = \langle a_j^{\text{in}\dagger}(t), a_j^{\text{in}\dagger}(t')\rangle = \langle a_j^{\text{in}}(t), a_j^{\text{in}}(t')\rangle = 0$, with n_{m} the thermal occupancy number of thermal bath for mechanical oscillator mode. We suppose cavity modes couple with the vacuum bath.

In the limit $\omega_{\text{m}} \gg \delta, d, \gamma, \gamma_{\text{m}}$, Eqs. (6.27) and (6.28) can be simplified under rotating

$$\dot{a}_1 = -ida_1 - iG_1 a_{\text{m}} - \frac{\gamma}{2}a_1 + \sqrt{\gamma}a_1^{\text{in}},$$

$$\dot{a}_2 = -ida_2 - iG_2 a_{\text{m}}^\dagger - \frac{\gamma}{2}a_2 + \sqrt{\gamma}a_2^{\text{in}},$$

$$\dot{a}_{\text{m}} = i\delta a_{\text{m}} - i(G_1^* a_1 + G_2 a_2^\dagger) - \frac{\gamma_{\text{m}}}{2}b + \sqrt{\gamma_{\text{m}}}b^{\text{in}}. \tag{6.29}$$

Without losing generality, we can always tune the coupling strength $G_1 = G_2 = G$. Define the Fourier components of the intracavity field by $a(t)=\frac{1}{\sqrt{2\pi}}\int_{-\infty}^{\infty} e^{-\omega(t-t_0)}a(\omega)d\omega$. In the limit $\delta \gg \omega, \gamma_{\rm m}$, we can adiabatically eliminate the $a_{\rm m}$ mode. Then we have effective quantum Langevin equations for $a_1(\omega)$ and $a_2^\dagger(-\omega)$:

$$
\begin{aligned}
-i\omega a_1(\omega) &= -ig'a_1(\omega) - iga_2^\dagger(-\omega) - \frac{\gamma}{2}a_1(\omega) \\
&\quad + \sqrt{\gamma}a_1^{\rm in}(\omega) + \sqrt{\tilde{a}_{\rm m}}a_{\rm m}^{\rm in}(\omega), \\
-i\omega a_2^\dagger(-\omega) &= ig'a_2^\dagger(-\omega) + iga_1(\omega) - \frac{\gamma}{2}a_2^\dagger(-\omega) \\
&\quad + \sqrt{\gamma}a_2^{\dagger\rm in}(-\omega) - \sqrt{\tilde{\gamma}_{\rm m}}a_{\rm m}^{\rm in}(\omega),
\end{aligned}
\tag{6.30}
$$

where $g = G^2/\delta$, $\tilde{\gamma}_{\rm m} = (G/\delta)^2\gamma_{\rm m}$, and $g = g + d$. In Eq. (6.30), we neglect the phase of G because it is not important.

Denote $\vec{a}(\omega)=\begin{pmatrix} a_1(\omega) \\ a_2^\dagger(-\omega)\end{pmatrix}$, $\vec{a}^{\rm in}(\omega)=\begin{pmatrix} a_1^{\rm in}(\omega) \\ a_2^{\rm in\dagger}(-\omega)\end{pmatrix}$ and $\vec{a}_{\rm m}^{\rm in}(\omega)=\begin{pmatrix} a_{\rm m}^{\rm in}(\omega) \\ -a_{\rm m}^{\rm in}(\omega)\end{pmatrix}$.
We get the following matrix equation:

$$
\mathbf{A}\vec{a}(\omega) = \sqrt{\gamma}\vec{a}^{\rm in}(\omega) + \sqrt{\tilde{\gamma}_{\rm m}}\vec{a}_{\rm m}^{\rm in},
\tag{6.31}
$$

where $\mathbf{A} = \begin{pmatrix} -i\omega + \frac{\gamma}{2} + ig' & ig \\ -ig & -i\omega + \frac{\gamma}{2} - ig' \end{pmatrix}$.

Using boundary conditions $a_j^{\rm out}(\omega) = -a_j^{\rm in}(\omega) + \sqrt{\gamma}a_j(\omega)$ for j = 1, 2, we can calculate the output field as

$$
\begin{aligned}
a_1^{\rm out}(\omega) &= G(\omega)a_1^{\rm in}(\omega) - H(\omega)a_2^{\rm in\dagger}(-\omega) + I(\omega)a_{\rm m}^{\rm in}(\omega), \\
a_2^{\rm out\dagger}(\omega) &= G(\omega)^* a_2^{\rm in\dagger}(\omega) - H(\omega)^* a_1^{\rm in}(-\omega) - I(\omega)^* a_{\rm m}^{\rm in}(-\omega),
\end{aligned}
\tag{6.32}
$$

where $G(\omega) = \left(\omega^2 + \frac{\gamma^2}{4} + g^2 - g'^2 - ig'\gamma\right)/\Delta(\omega), H(\omega) = ig\gamma/\Delta(\omega)$, $I(\omega)$ = $\left(-i\omega + \frac{\gamma}{2} - ig' + ig\right)\sqrt{\gamma\tilde{\gamma}_{\rm m}}/\Delta(\omega)$, and $\Delta(\omega) = \left(-i\omega + \frac{\gamma}{2}\right)^2 + g'^2 - g^2$.

Let us define the dimensionless position and momentum operators of fields $X_j^{\rm out}(\omega)$ = $[a_1^{\rm out}(\omega)$ + $a_1^{\rm out\dagger}(-\omega)]$ and $P_j^{\rm out}(\omega)$

$= [a_j^{out}(\omega) + a_j^{out\dagger}(-\omega)]/i$ for $j = 1, 2$. We define the correlation matrix of the output field as $V_{ij} = \langle(\xi_i\xi_j + \xi_j\xi_i)/2\rangle$, where $x = (X_1^{out}, P_1^{out}, X_2^{out}, P_2^{out})$. We calculate the correlation matrix with Eq. (6.32). Up to local unitary transformation, the standard form of it is

$$V_S = \begin{pmatrix} n & 0 & k_x & 0 \\ 0 & n & 0 & -k_x \\ k_x & 0 & n & 0 \\ 0 & -k_x & 0 & n \end{pmatrix}, \tag{6.33}$$

where $n = \left\{\left(\omega^2 + \frac{\gamma^2}{4} + g^2 - g'^2\right)^2 + (g'^2 + g^2)\gamma^2 + \left[(\omega+\delta)^2 + \frac{\gamma^2}{4}\right]\gamma\tilde{\gamma}_m(n_m+1)\right.$
$\left. + \left[(\omega-\delta)^2 + \frac{\gamma^2}{4}\right]\gamma\tilde{\gamma}_m n_m\right\} / |\Delta(\omega)|^2, k_x = \sqrt{V_{14}^2 + V_{24}^2},$

where $V_{14} = \left[-2g\gamma\left(\omega^2 + \frac{\gamma}{4} + g^2 - g'^2\right) + \gamma^2\delta\tilde{\gamma}_m(2n_m+1)\right] / |\Delta(\omega)|^2$, $V_{24} = \left\{2g'^2 + \left(\omega^2 - \delta^2 + \frac{\gamma^2}{4}\right)\gamma\tilde{\gamma}_m(2n_m+1)\right\} / |\Delta(\omega)|^2$. This is the symmetric Gaussian state. The EOF for the symmetric Gaussian states is defined as (Giedke et al., 2003)

$$\begin{aligned} E_F = {} & C_+(n - k_x)\log_2[C_+(n - k_x)] \\ & - C_-(n - k_x)\log_2[C_-(n - k_x)], \end{aligned} \tag{6.34}$$

where $C_\pm(x) = (x^{-1/2} \pm x^{1/2})^2/4$. V describes an entanglement state if and only if $n - k_x < 1$. Based on the standard form of matrix (6.33), we also find that $\langle\delta^2(X_1 + X_2)\rangle = \langle\delta^2(P_1 - P_2)\rangle = n - k_x$. We define the two-mode squeezing as $S = -10\log_{10}(n - k_x)$. In Fig. 6.11, we estimate the bath noise influence in experimentally accessible conditions. It is found that even at room temperature, the output entanglement is larger than zero. The output entanglement between lights is in proportional to the driving strength of the laser.

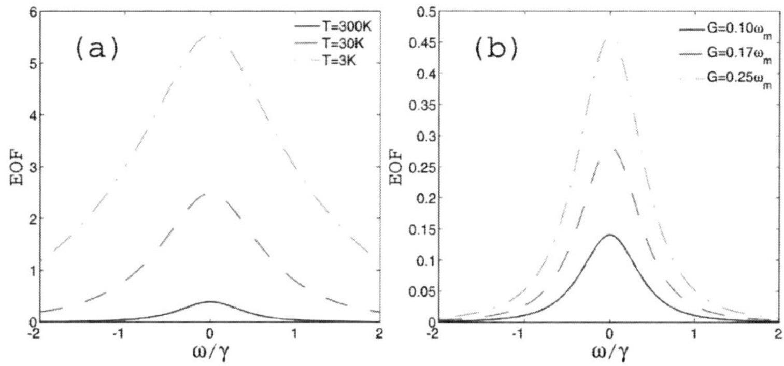

Figure 6.11 (a) EOF for different temperature. $\omega_{\rm m}/2\pi$ = 73.5 MHz, $G = 0.1\omega_{\rm m}$, $\delta/2\pi$ = 10 MHz, $\gamma/2\gamma$ = 3.2 MHz, d =0.07γ. (b) EOF for different G. The other parameters are the same as (a).

6.3.2 Frequency Transducer with Opto-Mechanical System

In modern optical communication networks, frequency converters play a center role in linking the notes with different frequencies. For example, converting the electrical signals to the optical signals enables the hybrid networks that contain both electric circuit and optical fibers. Unlike the conventional method, opto-mechanical systems can transduce the signals between two modes with arbitrary frequencies difference. Is also possible to convert quantum signals between different frequencies, maintain the coherence of the signals, and induce negligible noises. The microwave to optical frequencies transducer was proposed in 2011 (Regal and Lehnert, 2011), and realized recently (Andrews et al., 2014).

In order to convert signals between two modes with different frequencies, we also need to consider two different cavity modes $a_{1,2}$, that couple with the same mechanical mode b, as shown in Fig. 6.12. The Hamiltonian of the system was written in Eq. (6.24). Here we assume that $\Delta_1 = -\Delta_2 = \omega_{\rm m}$. Under the rotating wave approximation, the effective Hamiltonian reads

$$H_{\rm t} = \sum_{i=1,2} \delta_i a_i^\dagger a_i + G_1 a_1 b^\dagger + G_2 a_2 b^\dagger + \text{h.c.}, \tag{6.35}$$

where $G_{1,2}$ is time dependent, $\delta_i = \Delta_i - \omega_{\rm m}$.

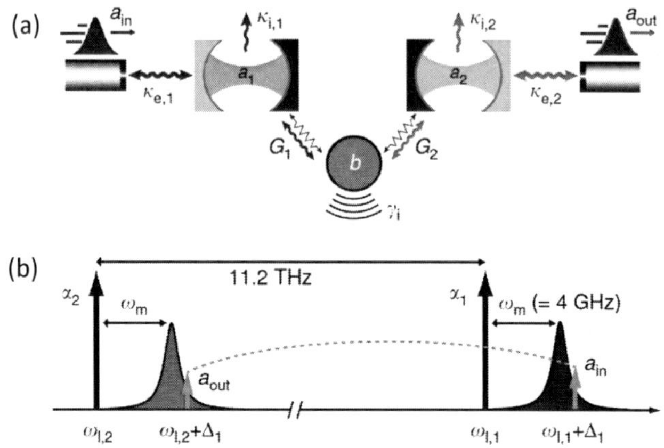

Figure 6.12 (a) Diagram of the wavelength transducing process as realized via two separate Fabry–Pérot cavities. The two optical cavity modes, a_1 and a_2, are coupled to the same mechanical mode b, with coupling strengths G_1 and G_2, respectively. Figure adapted with permission from Jeff et al. (2012), copyright (2012) by Nature Publishing Group.

If we want to transduce single photon pulse between mode a_1 and a_2, the simplest way contains 3 steps (Tian and Wang, 2010). Here we set $\delta_i = 0$. Step 1, cooling the mechanical mode down to the ground state; step 2, preparing a single photon state in mode a_1 and transfer it to the mechanical state by turning on the beam-splitter interaction between a_1 and b modes for time $t_1 = \pi/(2G_1)$; step 3, transfer the state from mode b to a_2 by turning off the interaction between a_1 and b and turning on the interaction between a_2 and b for time $t_2 = \pi/(2G_2)$. The main disadvantage of the scheme is that the thermal noises of mechanical mode greatly decrease the fidelity of the transducer (Jeff et al., 2012).

6.3.2.1 Frequency transducer by adiabatically eliminating mechanical mode

In order to decrease the thermal noise effects, we can adiabatically eliminate mode b. Here we briefly summarize the two different way of adiabatically elimination.

The first method is similar as the one we used in two-mode squeezing. In the large detuning limit $\delta_1 = \delta_2 = \delta \gg G_i$, the mechanical mode b can be adiabatically eliminated, and the effective Hamiltonian between a_1 and a_2 is

$$H_{t1} = \sum_{i=1,2} (G_1 G_2/\delta) a_1 a_2^\dagger + \text{h.c.}, \tag{6.36}$$

The quantum state is initially stored in the a_1 mode, a_2 mode is prepared in the vacuum state. Then we turn on the interaction Eq. (6.36). After some time $t = \pi\delta/(2G_1G_2)$, the state will be mapped into the mode a_2. During the process, the mechanical mode b is only virtually excited, because of adiabatical elimination. The thermal noise effects decrease $\delta^2/(G_1G_2)$ times.

The second method requires adiabatically tuning the strength ratio between G_1 and G_2 and maintaining the phase difference between them (Wang and Clerk, 2012; Tian, 2012). For simplicity, we assume $\delta = 0$, the Hamiltonian becomes

$$H_t = \sum_{i=1,2} G_1 a_1 b^\dagger + G_2 a_2 b^\dagger + \text{h.c.}, \tag{6.37}$$

Let us denote $G_0 = \sqrt{G_1^2 + G_2^2}$. The eigenvalues of the Hamiltonian are $E_1 = 0$ and $E_{2,3} = \pm G_0$. The eigenstate of the eigenvalue E_0 is $\psi_1 = (-G_1 a_1 + G_2 a_2)/G_0$, which is called dark mode, because it contains no mechanical mode. The energy gap between eigenvalue is G_0. The quantum state initially stores in the mode a_1. The mode a_2 are in arbitrary single photon state, and mechanical mode b is in thermal state. At time $t = 0$, G_2 is large but $G_1 = 0$. The dark mode is mode a_1 at this time. Then, we adiabatically increase G_1 from zero to large positive value, and adiabatically decrease G_2 to zero at final time T. The adiabatical condition requires that $T \gg 1/G_0$. At time T, the dark mode becomes mode a_2. During the process, the mechanical mode b is also only virtually excited. Therefore, the mechanical thermal noise effect can be greatly decreased. However, the drawback is that the transducing time T is much longer than the resonant evolution scheme in last subsection. The optical dark mode has been realized in experiment in 2012 (Dong et al., 2012). Therefore, the frequencies convertor that is based on dark mode is in principle possible.

6.3.2.2 Applications as quantum interfaces

Both two-mode squeezing light source and the frequency transducer can be used as a quantum interface between lights with different frequencies, let us say the microwave and optical frequencies. Here we briefly discuss the theoretical and experimental progress in this direction.

As we discussed before, the frequency difference between two output modes is in principle arbitrary. As shown in Fig. 6.13 (Barzanjeh et al., 2012), we can use the two-mode squeezing Hamiltonian to generate the entanglement between output light pulses a_1^{out} and the superconducting resonator output mode a_2^{out}. The quantum state that we want to map is stored in the input light pulse a_3^{in}. Alice with mix both a_1^{out} and a_3^{in} go through a beam splitter, then homodyne detection is performed. Based on the measurement results, Bob performs the local operation on the a_2^{out} mode and recovers the state of a_3^{in}. The scheme is reversible by exchanging the roles of Alice and Bob. In this way, Bob performs Bell measurements on the microwave field a_2^{out} and the input microwave field a_3^{in}, in which the quantum information in encoded, and will be teleported onto the optical output field a_1^{out} in Alice's hand, by conditional local operations based on the Bell measurements' results from Bob.

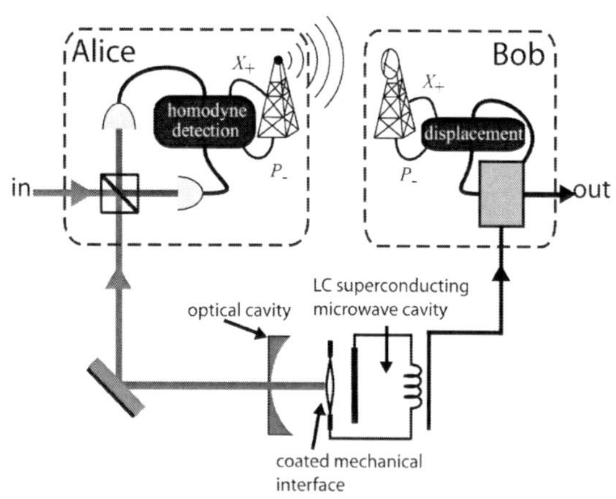

Figure 6.13 Schematic description of the proposed optical-microwave interface based on two-mode squeezing source. Figure adapted with permission from Barzanjeh et al. (2012), copyright (2012) by the American Physical Society.

If we can realize beam-splitter Hamiltonian between optical mode and microwave mode, the noiseless state transfer between them is capable directly. Such kind of proposal was firstly proposed by (Regal and Lehnert, 2011), then realized by the same group recently (Andrews et al., 2014). The experimental setup is shown in Fig. 6.14, A stoichiometric silicon nitride (Si_3N_4) membrane (light blue) that has been partially covered with niobium (dark blue) interacts with an inductor–capacitor (LC) circuit that forms the microwave resonator and a Fabry–Pérot cavity that forms the optical resonator (mode shown in red). Both the optical mode and microwave mode couple with mechanical oscillator of Si_3N_4 membrane. If we apply the fields on red sidebands of both optical and microwave modes, the beam-splitter like Hamiltonian (6.37) can be realized in this system.

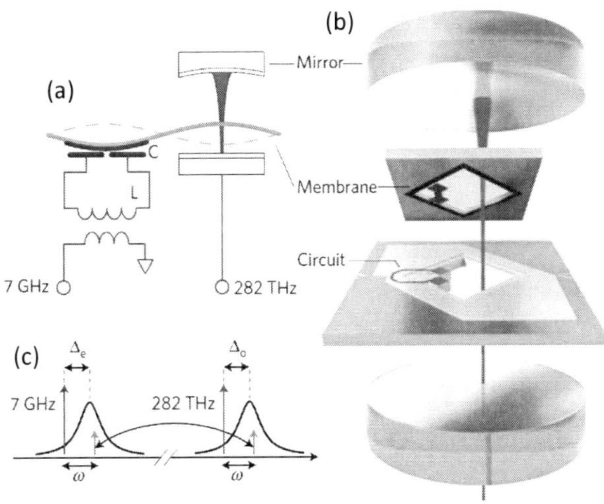

Figure 6.14 Layout and operation of microwave-to-optical converter. Figure adapted with permission from Andrews et al. (2014), copyright (2014) by Nature Publishing Group.

The experiment was performed in the temperature of 4.5 K. The conversion efficiency in the experiment is only around 0.086. The vibrational noise at such temperature is nonzero, as the mechanical oscillator is too hot. The noiseless conversion is only realizable at the temperature around 40 mK, which is challenging but possible by using the dilution refrigerator.

6.4 Summary and Outlook

The multi-mode opto-mechanical systems provide the unique opportunity that linking lights modes with different frequencies. In order to generate squeezed lights with high efficiency, we should consider the three cavity modes that couple with the same mechanical mode (Deng et al., 2014). The neighbor modes distance is similar to the mechanical mode frequency. By resonantly driving on the center cavity mode, the two sideband modes will resonantly emit the entangled lights. As all three cavity modes are nearly resonant, the efficiency of the two-mode squeezed light source is maximized. Recently, it is found that the thermal noises in the frequency conversion between optical microwave qubits can be decreased by using sideband cooling method to initialize the superconducting qubits (Yin et al., 2014). The noiseless and high efficiency quantum interface can be realized at the temperature 1 K, other than 40 mK, between the optical photons and the superconducting qubits.

References

R. W. Andrews, R. W. Peterson, T. P. Purdy, K. Cicak, R. W. Simmonds, C. A. Regal, and K. W. Lehnert, Bidirectional and efficient conversion between microwave and optical light, *Nat. Phys.*, **10**, 321 (2014).

M. Aspelmeyer, T. J. Kippenberg, and F. Marquardt, Cavity optomechanics, arXiv:1303.0733 (2013).

Sh. Barzanjeh, M. Abdi, G. J. Milburn, P. Tombesi, and D. Vitali, Reversible optical-to-microwave quantum interface, *Phys. Rev. Lett.*, **109**, 130503 (2012).

K. Børkje, A. Nunnenkamp, J. D. Teufel, and S. M. Girvin, Signatures of nonlinear cavity optomechanics in the weak coupling regime, *Phys. Rev. Lett.*, **111**, 053603 (2013).

V. B. Braginsky and A. B. Manukin, Ponderomotive effects of electromagnetic radiation, *Sov. Phys. JETP*, **25**, 653 (1967).

G. A. Brawley and W P. Bowen, Quantum nanomechanics: Feeling the squeeze, *Nature Photon.*, **7**, 854 (2013).

D. W. C. Brooks, T. Botter, S. Schreppler, T. P. Purdy, N. Brahms, and D. M. Stamper-Kurn, Non-classical light generated by quantum-noise-driven cavity optomechanics, *Nature*, **488**, 476 (2012).

C. M. Caves, Quantum-mechanical noise in an interferometer, *Phys. Rev. D.*, **23**, 1693 (1981).

J. Chan, T. P. Mayer Alegre, A. H. Safavi-Naeini, J. T. Hill, A. Krause, S. Gröblacher, M. Aspelmeyer, and O. Painter, *Nature* (London), **478**, 89 (2011).

Z. J. Deng, S. J. M. Habraken, and F. Marquardt, An entanglement rate for continuous variables and its application to a resonant optomechanical multimode setup, arXiv:1406.7815.

C. Dong, V. Fiore, M. C. Kuzyk, H. Wang, Optomechanical dark mode, *Science*, **338**, 6114 (2012).

T. Eberle, S. Steinlechner, J. Bauchrowitz, V. Händchen, H. Vahlbruch, M. Mehmet, H. Müller-Ebhardt, and R. Schnabel, Quantum enhancement of the zero-area Sagnac interferometer topology for gravitational wave detection, *Phys. Rev. Lett.*, **104**, 251102 (2010).

C. Fabre, M. Pinard, S. Bourzeix, A. Heidmann, E. Giacobino, and S. Reynaud, Quantum-noise reduction using a cavity with a movable mirror, *Phys. Rev. A*, **49**, 1337 (1994).

G. Giedke, M. M. Wolf, O. Krüger, R. F. Werner, and J. I. Cirac, Entanglement of formation for symmetric Gaussian states, *Phys. Rev. Lett.*, **91**, 107901 (2003).

J. T. Hill, A. H. Safavi-Naeini, J. Chan, O. Painter, Coherent optical wavelength conversion via cavity optomechanics, *Nat. Commun.*, **3**, 1196 (2012).

H. J. Kimble, Y. Levin, A. B. Matsko, K. S. Thorne, and S. P. Vyatchanin, Conversion of conventional gravitational-wave interferometers into quantum nondemolition interferometers by modifying their input and/or output optics, *Phys. Rev. D*, **65**, 022002 (2001).

T. J. Kippenberg and K. J. Vahala, Cavity optomechanics: Back-action at the mesoscale, *Science*, **321**, 1172 (2008).

A. Kronwald and F. Marquardt, Optomechanically induced transparency in the nonlinear quantum regime, *Phys. Rev. Lett.*, **111**, 133601 (2013).

C. K. Law, Interaction between a moving mirror and radiation pressure: A Hamiltonian formulation, *Phys. Rev. A*, **51**, 2537 (1995).

M.-A. Lemonde, N. Didier, and A. A. Clerk, Nonlinear interaction effects in a strongly driven optomechanical cavity, *Phys. Rev. Lett.*, **111**, 053602 (2013).

LIGO Scientific Collaboration, The, A gravitational wave observatory operating beyond the quantum shot-noise limit, *Nat. Phys.*, **7**, 962 (2011).

LIGO Scientific Collaboration, The, Enhanced Sensitivity of the LIGO Gravitational wave detector by using squeezed states of light, *Nat. Photon.*, **7**, 613 (2013).

Y.-C. Liu, Y.-F. Xiao, Y.-L. Chen, X.-C. Yu, and Q. Gong, Parametric down-conversion and polariton pair generation in optomechanical systems, *Phys. Rev. Lett.*, **111**, 083601 (2013).

M. Ludwig, A. H. Safavi-Naeini, O. Painter, and F. Marquardt, Enhanced quantum nonlinearities in a two-mode optomechanical system, *Phys. Rev. Lett.*, **109**, 063601 (2012).

G. M. Mahan, *Many-Particle Physics*, 3rd ed. (Kluwer Academic, New York, 2000).

S. Mancini and P. Tombesi, Quantum noise reduction by radiation pressure, *Phys. Rev. A*, **49**, 4055 (1994).

F. Marquardt, J. P. Chen, A. A. Clerk, and S. M. Girvin, *Phys. Rev. Lett.*, **99**, 093902 (2007).

N. Mavalvala and T. Corbitt, Vibrating membrane puts a squeeze on light, *Physics*, **6**, 95 (2013).

A. Nunnenkamp, K. Børkje, and S. M. Girvin, Single-photon optomechanics, *Phys. Rev. Lett.*, **107**, 063602 (2011).

T. P. Purdy, P.-L. Yu, R. W. Peterson, N. S. Kampel, and C. A. Regal, Strong optomechanical squeezing of light, *Phys. Rev. X*, **3**, 031012 (2013).

P. Rabl, Photon blockade effect in optomechanical systems, *Phys. Rev. Lett.*, **107**, 063601 (2011).

C. A. Regal and K. W. Lehnert, From cavity electromechanics to cavity optomechanics. *J. Phys. Conf. Ser.*, **264**, 012025 (2011).

A. H. Safavi-Naeini, S. Grøblacher, J. T. Hill, J. Chan, M. Aspelmeyer, and O. Painter, Squeezed light from a silicon micromechanical resonator, *Nature*, **500**, 185 (2013).

M. O. Scully and M. S. Zubairy, *Quantum Optics*, 1st ed. (Cambridge University Press, Cambridge, 1997).

R. M. Shelby, M. D. Levenson, S. H. Perlmutter, R. G. DeVoe, and D. F. Walls, Broad-band parametric deamplification of quantum noise in an optical fiber, *Phys. Rev. Lett.*, **57**, 691 (1986).

R. E. Slusher, L. W. Hollberg, B. Yurke, J. C. Mertz, and J. F. Valley, Observation of squeezed states generated by fourwave mixing in an optical cavity, *Phys. Rev. Lett.*, **55**, 2409 (1985).

K. Stannigel, P. Komar, S. J. M. Habraken, S. D. Bennett, M. D. Lukin, P. Zoller, and P. Rabl, Optomechanical quantum information processing with photons and phonons, *Phys. Rev. Lett.*, **109**, 013603 (2012).

M. A. Taylor, J. Janousek, V. Daria, J. Knittel, B. Hage, H.-A. Bachor, and W. P. Bowen, Biological measurement beyond the quantum limit, *Nat. Photon.*, **7**, 229 (2013).

L. Tian, Adiabatic state conversion and pulse transmission in optomechanical systems, *Phys. Rev. Lett.*, **108**, 153604 (2012).

L. Tian and H. Wang, Optical wavelength conversion of quantum states with optomechanics, *Phys. Rev. A*, **82**, 053806 (2010).

H. Vahlbruch, M. Mehmet, S. Chelkowski, B. Hage, A. Franzen, N. Lastzka, S. Goiler, K. Danzmann, and R. Schnabel, Observation of squeezed light with 10-dB quantum-noise reduction, *Phys. Rev. Lett.*, **100**, 033602 (2008).

Y. D. Wang and A. A. Clerk, Using interference for high fidelity quantum state transfer in optomechanics, *Phys. Rev. Lett.*, **108**, 153603 (2012).

I. Wilson-Rae, N. Nooshi, W. Zwerger, and T. J. Kippenberg, *Phys. Rev. Lett.*, **99**, 093901 (2007).

L.-A. Wu, H. J. Kimble, J. L. Hall, and H. Wu, Generation of squeezed states by parametric down conversion, *Phys. Rev. Lett.*, **57**, 2520 (1986).

Z.-Q. Yin and Y.-J. Han, Generating EPR beams in a cavity optomechanical system, *Phys. Rev. A*, **79**, 024301 (2009).

Z.-Q. Yin, W. L. Yang, L. Sun, and L. M. Duan, Quantum network of superconducting qubits through opto-mechanical interface, arXiv: 1407.4938.

Chapter 7

Scintillators Boosted by Nanophotonics

Bo Liu

School of Physics Science and Engineering, Tongji University, Shanghai 200092, PR China

lbo@tongji.edu.cn

This chapter provides a review on the scintillators improved by the methods based on the principles of nanophotonics. The main contents contain the enhanced light extraction efficiency of scintillator by photonic crystals, the controllable directivity of emission from scintillators by photonic crystals, and the fast decay component of scintillators by surface plasmon polariton with metal nanoparticles. The application of nanophotonics on scintillators will significantly promote the development in the field of radiation detection in a totally new way. In this chapter, the basic principles are described in detail and the experimental demonstrations are summarized.

7.1 Introduction to Scintillators

In 1896, the earliest scintillator $CaWO_4$ was discovered followed by the discovery of X-ray by Röentgen.[1] In the 1940s, the modern radiation detection system emerged due to the development of

Integrated Nanophotonic Resonators: Fundamentals, Devices, and Applications
Edited by Ya Sha Yi

ISBN 978-981-4613-78-1 (Hardcover), 978-981-4613-79-8 (eBook)
www.panstanford.com

the photomultiplier tube and the discovery of NaI:Tl scintillator, which is the most typical scintillator. Scintillators can be defined as materials that absorb incident particles (such as electrons, protons, neutrons, α-particle) or high-energy photons (such as X-ray, γ-ray) and subsequently convert the deposition energy efficiently into a large number of photons in the visible or ultra-violet range, which can be directly detected by photomultipliers, photodiodes or CCD.[2] In this way, the energy, position, and time of the incident radiation can be efficiently determined. Therefore, scintillators play an important role in radiation detection systems with various applications in nuclear medical imaging, high-energy physics experiments, nuclear physics experiments and national security areas.

The nature of scintillation process is the spontaneous emission under the excitation of ionizing radiation. The physical processes involved in scintillation can be divided into three stages. The first stage is the ionization event, which creates inner shell holes and primary electrons by the interactions of photoabsorption, Compton scattering, and electron–positron pair formation. The second stage is the thermalization of hot electrons and holes until the energy becomes less than the ionization threshold. The carries diffuse in the lattice and, subsequently are trapped on defects and impurities. This process can lead to the excitation of the luminescence centers. The last stage is the luminescent process. The excited luminescent center returns to the ground state by emitting a photon or nonradiative transition.

In general, an ideal scintillator should have the following properties: high light yield, short decay time, high density, no afterglow, good spectral match to photo detectors, good chemical stability and mechanical strength, and low cost.[3]

The light yield can be defined as the number of photoelectrons generated by radiation with unit energy. The measured light yield must also take into account the light extraction efficiency and the quantum efficiency of the photodetector. High light yield of a scintillator is vital to improve the sensitivity, spatial resolution and energy resolution in detection systems. For instance, with the improved sensitivity, we can apply a lower radiation dose during the medical imaging procedure (such as X-ray CT). As a result, the cancer risks due to the radiation exposure for patients would be correspondingly decreased.[4]

The decay time of a scintillator is defined as the duration of the emission of scintillation light. The decay time strongly depends on the property of luminescent center. If only one type of luminescence center appears in the scintillator, the decay is exponential with one time constant. Additionally, the decay time is also influenced by the type and energy of the incident particle. Most practical scintillators exhibit rather complicated time profile. Short decay time is critical for improving the time-resolution for particle detection. This is also particular important in the application of positron emission tomography (PET) scan with the coincidence time technique in which a shorter time window can be applied, which leads to a reduced background of random events and thus a better quality of imaging.[5]

Generally, a high-density scintillator composed of elements with large atomic numbers has a strong stopping power for the incident radiation, which makes the detector more compact. A compact detector can not only largely save the cost of detector but also improve the spatial resolution in imaging systems.

However, an ideal scintillator combining all advantageous properties does not exist in practice. Therefore, trade-off is necessary, selecting the most suitable combination of properties from available scintillator depending on the requirements as a function of the priority for a given detection system.[6] The properties of the commonly used scintillators are listed in Table 7.1.

Table 7.1 The properties of the most commonly used scintillators

Scintillator	Light yield Phe/MeV	Density/ g/cm^3	Decay constant/ns	Wavelength/nm	Z_{eff}
NaI(Tl)	38,000	3.7	230	415	51
CsI(Tl)	60,000	4.5	1000	560	54
$PbWO_4$	190	8.2	30	420	
BGO	8,000	7.13	300	480	74
LSO:Ce	25,000	7.35	11/36	420	66
GSO:Ce	8,000	6.7	56/600	440	59
YSO:Ce	10,000	4.54	37/82	420	
YAP:Ce	16,000	5.37	28	360	34
LuAP:Ce	9,600	8.34	11/28	355	
Gd_2O_2S:Tb	70,000	7.3	—	545	64
Y_2O_2S:Tb	60,000	4.9	—	545	

7.2 The Challenge for the Development of Scintillator

In the past tens of years, the development of scintillators has been strongly promoted by the requirement for the high-energy physics experiments and medical imaging systems. New scintillators with high light yield and fast decay time have been proposed and obtained many practical applications. For example, the LYSO ($(Lu,Y)_2SiO_5$:Ce) scintillator is a promising candidate in the PET system.[7] $PbWO_4$ (PWO) scintillator has been used in the Compact Muon Solenoid (CMS) of the Large Hadron Collider (LHC) at CERN.[8]

7.2.1 Light Yield

Light yield is the most important requirement in most applications. Exploring scintillator with a much higher light yield than usual has attracted much attention in the field of scintillator. For example, in recent years, LuI_3:Ce with a light yield of 76000 phe/MeV was proposed, which is the highest value among the inorganic scintillators.[9] For a given crystal structure and the element, the light yield can be defined as[10]

$$\mathrm{LY(phe/MeV)} = (10^6/\beta E_\mathrm{g})SQ, \tag{7.1}$$

where E_g is the bandgap of scintillator, β is a constant with a typical value of 2~3, which is related with the phonon energy the dielectric constant, S is the efficiency of transfer of electron–hole pair energy to luminescent center, and Q is the quantum efficiency of the luminescent center.

From formula (7.1), assuming the S and Q to be the unity, the maximum light yield mainly depends on the bandgap of scintillator. Theoretically, the maximum light yield only reaches 150000 phe/MeV calculated with wide-bandgap semiconductor (bandgap of 3 eV). In recent years, although many efforts have been devoted into the exploring new scintillator with high light yield based on the crystal structure and elemental constituent, it becomes more and more difficult to find new materials with unique properties.

7.2.2 Light Extraction

Most inorganic scintillators have a high refractive index (n = 1.8 to 2.2), which is much higher than the ambient medium. The light created in the internal of scintillator could be totally reflected at the crystal–air interface if the incident angle is larger than the critical angle. These photons cannot reach the detectors but emit at the side face or cannot be absorbed during multi-reflection.

In practical applications, only those photons that can come into detectors are useful. Therefore, it is necessary to define an effective light yield as

$$LY_{eff} = \eta_{ex} \cdot LY_{abs}, \tag{7.2}$$

where LY_{abs} is absolute light yield, which is defined as photons created by radiation with unit energy; η_{ex} is the efficiency of light extraction indicating the ratio of photons can emit out of scintillator; and LY_{eff} is the effective light yield.

The total internal reflection (TIR) is the main reason for the low extraction efficiency. According to Snell's law, light can emit out of the scintillator when the incident angle is smaller than the critical angle. If we neglect the Fresnel reflection, considering an infinite plane scintillator–air interface, the extraction efficiency from one side face can be determined as the following formula[11]:

$$\eta_{ex} = \frac{1}{2}\int_0^{\theta_c} \sin\theta d\theta \quad \theta_c = \sin^{-1}(n_{air}/n_{sc}) \tag{7.3}$$

Assuming that the refractive index of scintillator is 2, the extraction efficiency is about 6.7% calculated as formula (7.3). Although in some cases, the extraction efficiency is enhanced by the optical coupling with grease between the surface of the scintillator and the window of photodetector, there is a possibility to further improve the light extraction because the coupling cannot totally eliminate the difference in refractive indices. In some specific cases, the optical coupling cannot be applied, and thus the efficiently light extraction becomes extremely important.

7.2.3 Directional Emission

For high-index scintillators with a planar surface, assuming a point-like source located a short distance below the scintillator–

air interface, the far-field emission exhibits a Lambertian pattern, which follows a cosine dependence on the angle. The intensity of emission reaches maximum at the normal direction, and decays with the increasing angle. This emission profile is disadvantageous to the detection system because in general only a fraction of the total emitted light can enter into the detector, which is illustrated in Fig. 7.1. It would be better to design the directionality of emission according to the layout of systems, aimed at the maximum of photon collection.

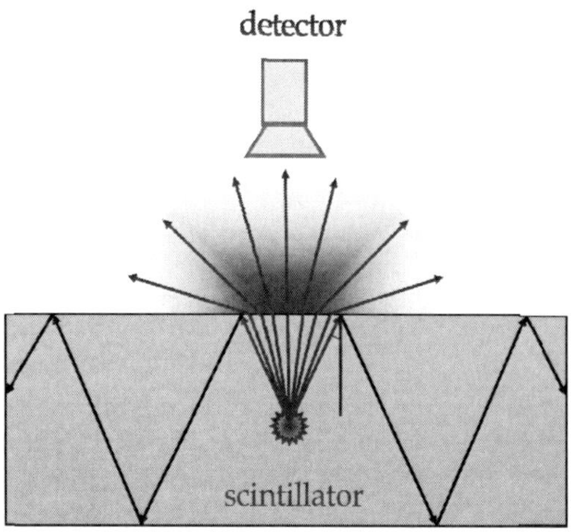

Figure 7.1 A general schematic sketch of the layout of scintillation detection system. It indicates the propagating of scintillation emission in the scintillator and out of the scintillator.

7.2.4 Decay Time

A strong demand on the timing performance appears in the application of particle identification in high energy physics, nuclear physics experiments, and time of flight techniques for PET systems. Fast timing characteristic is an important parameter in scintillation detection systems. Traditionally, thermal quenching can lead to a shorter decay time but at the expense of a reduction in light yield. PWO is a typical example.[12] Some direct-bandgap semiconductors such as ZnO, CuI, and CdS can act as fast scintillators, which exhibit subnanosecond radiative recombination due to the exciton emission.[13] Unfortunately, these scintillators suffer from the low luminescent efficiency and strong self-absorption. In the long term of exploring scintillators, high light yield and the fast decay time are always contradictory.

7.3 Application of Nanophotonics on Scintillator

7.3.1 The Concept of Nanophotonics

Nanophotonics deals with the interaction between light and material on nanoscale.[14] This interaction lead to the strong controlling of the propagation of electromagnetic waves and controlling of the distribution of the density of optical modes, which essentially influence the rate of spontaneous radiation. The type of nanostructured materials may include photonic crystal,[15] surface plasmon polariton,[16] optical microcavity.[17] The scale of structure is in the order of wavelength or subwavelength. Photonic crystal is a kind of nanostructured materials in which alternating dielectric components with higher and lower refractive indices create a periodic structure on the order of wavelength of light. Metallic nanoparticles exhibit unique optical property due to the excitation of surface plasmon polariton, which usually belong to the subwavelength scale. In the following subsection, some examples are provided to illustrate the principles and applications on the scintillators.

7.3.2 Enhancement of Light Extraction by Photonic Crystals

As discussed above, the high refractive index of the scintillator leads to light trapping in the internal of scintillator, which severely degrades the light extraction, thus giving rise to a low external efficiency. Using photonic crystals is a possible method to improve the extraction efficiency of scintillator, which is analogous to the application on light-emitting diodes.[18] It is natural to consider the so-called bandgap approach of photonic crystal to achieve the enhanced light extraction. The photonic crystal must penetrate through the scintillator layer and the structure requires to be tuned into the photonic bandgap, inhibiting light emission into guided modes, and thus increasing extraction efficiency.[19] However, the obvious drawback of this approach is that the internal quantum efficiency can be severely decreased due to the damage of the emission layer.[20]

Another approach is based on the diffraction gratings using photonic crystals.[21] Spontaneous emission into guided modes is allowed, which can be partially outcoupled by the diffractive structure of photonic crystal. In this scheme, the photonic crystal is covered onto the surface of the scintillator layer, and thus does not destroy the emission layer, avoiding the decrease of internal quantum efficiency. This is a more feasible method in practice. The drawback of this approach is that, in generally, not all of the guided modes can be totally extracted. A careful design is necessary for maximizing the extraction efficiency. At the same time, the extracted light by a photonic crystal is essentially direction dependent, which provides a possibility for the controlling the directivity of emission.

The principle of extraction process for scintillator film can be illustrated by Fig. 7.2. The in-plane wavevector of guided modes can be described as

$$k_{\|m} = (2\pi/\lambda)n_{sc}\sin\theta_{m}, \tag{7.4}$$

where λ is the wavelength of emission light in vacuum, n_{sc} is the refractive index, and θ_m is the angle corresponding to the mode number of m.

The diffraction condition of extraction needs satisfy the following formula[18]:

$$|k_{\|m} + pG_0| < \frac{2\pi}{\lambda} = k_0, \tag{7.5}$$

Where k_0 is the wavevector in air, G_0 is the primitive vector, which depends on the lattice constant of photonic crystal, and p is an integer representing the harmonic order.

The guided modes with different orders are possible to be extracted when they meet the photonic crystal at the surface of scintillator where the in-plane wavevectors are changed by the reciprocal lattice vectors pG_0 of photonic crystal, shown in Fig. 7.2a. The nature of extraction process is that the guided modes are changed by photonic crystal into leaky modes, which can propagate in the air.

The emission angle of extracted light by the photonic crystal is determined as

$$\theta_m = \sin^{-1}((\lambda/2\pi)\,|k_{||m} + pG_0|) \tag{7.6}$$

The angle-dependent of extracted light provides the possibility of the controlling of the directivity of emission.

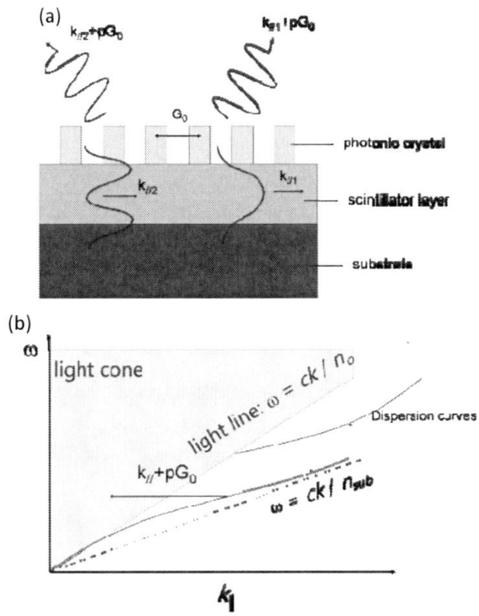

Figure 7.2 (a) Extraction process by photonic crystal for film scintillator. (b) Explanation with dispersion relation by photonic crystal.

In order to explain the process using the dispersion relation, Fig. 7.2b gives the understanding of extraction process. The guided modes that are originally trapped in the emission layer are altered by the reciprocal lattice vectors. If the in-plane wavevectors can come into the light cone shown in the blue color region, the corresponding modes become leaky modes, therefore propagating into the air.

The formulas concerning with the extraction process are explicit and easy to be understood. However, for the case with a large number of modes, the situation can become complicated. Not all of modes can be extraction. For a given structure of photonic crystal, only part of modes meet the diffraction condition as formula (7.5). Additionally, lattice parameters such as height, filling ratio have different effect on the extraction efficiency for different modes. Generally, the emission spectra for most of scintillators are broadband and therefore the wavelength-dependence should be taken into account. The emission angle of extracted light is determined as formula (7.6), which is different from the Lambertian profile, suggesting that the directionality of emission can be controlled by the photonic crystal.[22]

7.3.2.1 Light extraction enhancement for scintillator film by biologically inspired photonic structure

Biologically inspired moth-eye structure comprising an array of circular protuberances on the facet in a moth's eye provides a unique method to achieve broadband antireflection.[23] This bio-inspired moth-eye can be utilized to enhance the light extraction of the scintillator due to the periodicity. Yasha Yi's group reported that an improved moth-eye structure that was added some roughness on the sidewall of the original pyramid shape can achieve the enhancement of the extraction as large as 2.7 when the periodicity is 400 nm.[24] The structure was fabricated as the process illustrated in Fig. 7.3a. The SiO_2 nanoparticles as a mask layer were coated on the high index Si_3N_4 layer, which was previously deposited on the Lu_2SiO_5:Ce (LSO) scintillator thin film with a thickness of 500 nm. After the RIE etching and removing the remaining SiO_2 nanoparticles, the scanning electron microscope (SEM) image shown in Fig. 7.3b exhibits the improved bio-inspired moth-eye structure. The FDTD simulations in Fig. 7.3c show that the optimal structure parameter and the enhancement. The experimental characteristic of emission spectra with the reference and the structured sample under the X-ray excitation exhibits the significant enhancement with the broadband, shown in Fig. 7.3d. The method of using mask is superior to the conventional e-beam lithography because of its ability to fabricate samples with large-scale area, which is sometimes important in a practical application.

This moth-eye like structures are widely utilized as an antireflection coating in many applications, where the pyramid shape plays a role of gradual refractive index layer and therefore the Fresnel reflection can be significantly eliminated. However, the periodic structure is the most important factor for the light extraction because it provides the reciprocal lattice vector adding to the in-plane vector of guided modes. The individual pyramid shape may optimization the extraction efficiency.

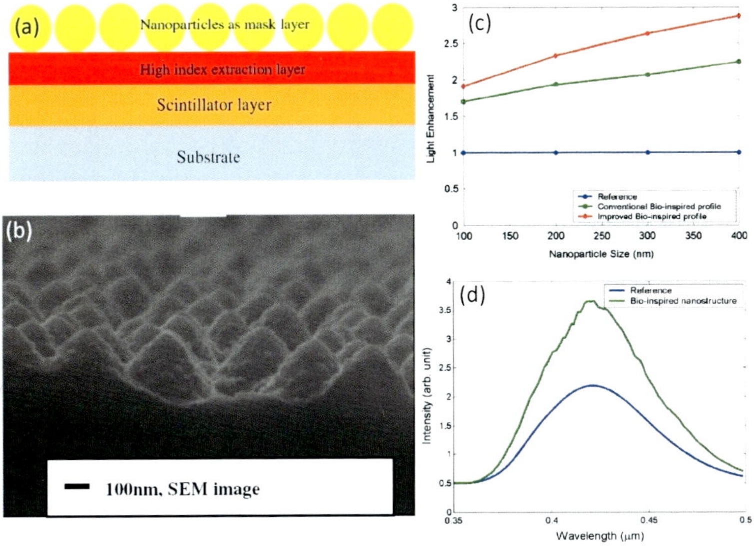

Figure 7.3 (a) The self assembly of SiO_2 nanoparticles on the top of high index light extraction layer Si_3N_4, which is deposited on Lu_2SiO_5:Ce thin film. (b) The SEM image of the improved bio-inspired moth-eye nanostructures with certain degree roughness on the sidewall, which shows interesting nano-on-nano features. (c) The FDTD simulation results, which shows light output enhancement for both bio-inspired light extraction structure, while the improved bio-inspired moth-eye structures shows more light output enhancement. (d) The comparison of light output between the two Lu_2SiO_5:Ce thin films with the excitation of X-ray mammographic unit with molybdenum target; the blue curve is the light output from the referenced sample, and the green curve is the light output from sample with improved bio-inspired moth-eye structure.

7.3.2.2 Light extraction enhancement for bulk scintillator by photonic crystal

When we design the nanostructured scintillator for practical applications, several important issues have to be taken into account. The bandwidth of the emission spectra for scintillators is generally rather broad. For instance, the full widths of half-maximum (FWHMs) of emission spectra of Lu_2SiO_5:Ce (LSO) and $Bi_3Ge_4O_{12}$ (BGO) are 80 and 150 nm, respectively. A large area of several square centimeters is often required. In most of applications for the high-energy particles detection, in order to fully absorb the energy of the particles, the scintillators require macroscopic thickness, which ranges from millimeters to several centimeters. In these cases, the allowed incident angle can vary continuously, which is different from the limited amount of guided modes in a thin film scintillator. The concept of guided mode is invalid. It is feasible to deal with the interaction of the incident light with a certain angle larger than the critical angel with the photonic crystals. Formulas (7.4)–(7.6) are still valid for the bulk scintillators.

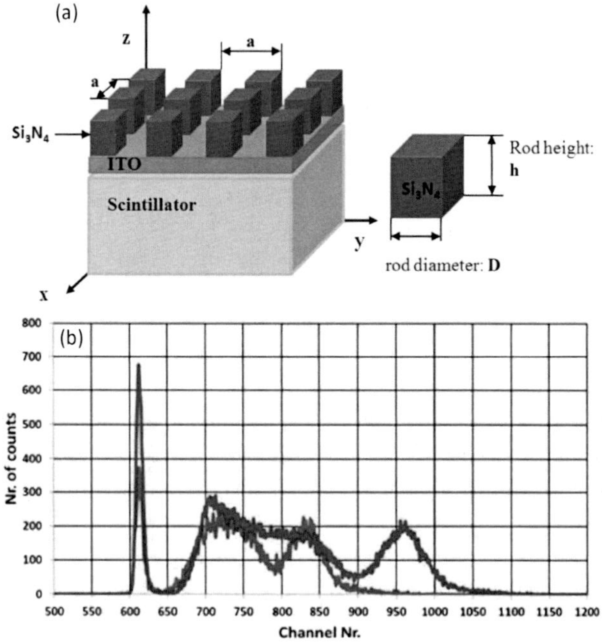

Figure 7.4 (a) Schematic diagram of the photonic crystal structure of silicon nitride pillars with a square cross section. (b) The measurement of light yield of the samples using gamma rays of 662 keV with a ^{137}Cs source. The blue and red curves represent, respectively, the reference sample and the sample with the photonic crystal structure shown in (a). The sampled were placed on top of a Q2020 photomultiplier tube without optical glue and wrapping. The PMT window has been masked to measure only the light coming from the photonic crystal surface of the scintillator.

Paul Lecoq's group from CERN has first proposed the concept of improved light extraction from heavy inorganic scintillators by photonic crystals as early as the year of 2008.[25] They fabricated two-dimensional photonic crystals slab with the area of 1.2 mm × 2.6 mm by electron beam lithography (EBL) on the surface of 1.2 mm × 2.6 mm × 5 mm LSO scintillators, shown in Fig. 7.4a.[26] The measurement of energy spectrum with ^{137}Cs showed that a maximum enhancement of light yield by 60% was obtained compared with the unstructured reference sample, shown in Fig. 7.4b.

7.3.2.3 Broadband light extraction enhancement of bulk scintillators by photonic crystals with monolayer periodic nanospheres

In recent years, photonic crystals forming with arrays of monolayer periodic nanospheres have attracted much attention due to their unique optical properties in the field of nanophotonics.[27] Compared with the conventional photonic crystal slab, the significant feature of an array of periodic nanospheres is based on the whispering gallery modes (WGMs) due to Mie resonance of individual spheres and Bragg diffraction arising from the periodic arrangement.[28] A wavelength-scale dielectric sphere with low-order resonances shows broadband characteristics for optical response, which is advantageous to the coupling with broadband emission of scintillators. An array of periodic nanospheres with WGMs has also shown an outstanding ability to enhance the absorption in a broad spectrum for solar cells.[29–31] The array of nanospheres with a large area can be easily prepared by self assembled methods,[32] which is particularly important for the applications in scintillation detection systems.

Bo Liu's group from Tongji university demonstrated the significant enhancement of light extraction for BGO[33] and $(Lu,Y)_2SiO_5$:Ce (LYSO)[34] scintillators in a broad emission spectrum using a monolayer of a periodic array of polystyrene (PS) nanospheres. The schematic diagrams and the top-view SEM images show in Figs. 7.5a–c. The SEM image of the self-assembled monolayer layer of PS nanospheres with diameter of 500 nm on the BGO scintillator surface exhibits a hexagonal-close-packed (HCP) structure with two directions of Γ–M and Γ–K.

In order to understand the effect of the monolayer of nanospheres on light extraction, the numerical simulations for transmission at the maximum emission wavelength of 510 nm are shown in Figs. 7.5d,e for s- and p-polarization, respectively. For the BGO with plain surface, the transmission drops to zero beyond the critical angle of 27.7°. After being covered with the monolayer of nanospheres, it is fortunate to find that two extra transmission peaks appear at about 30° and 40° for both polarizations. This implies that the incident light beyond the critical angle can be partly extracted, leading to an enhanced extraction efficiency.

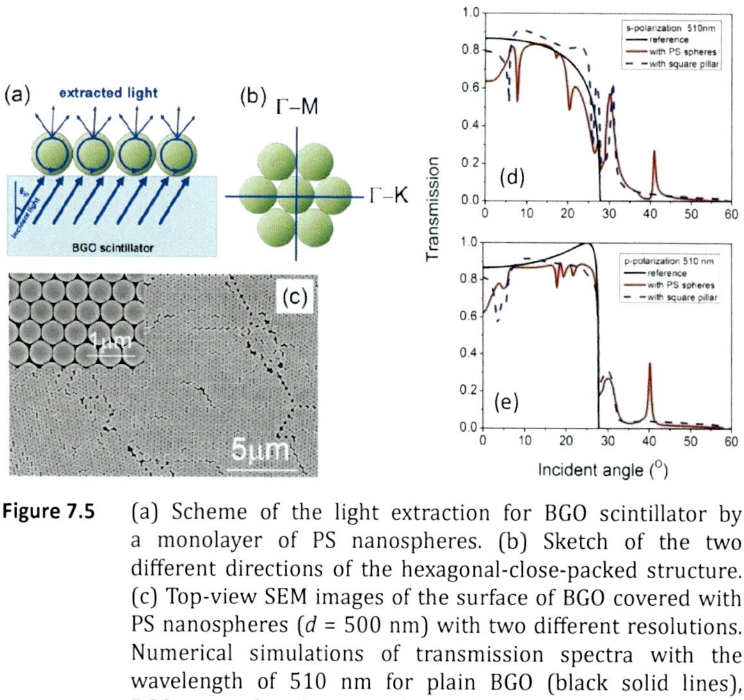

Figure 7.5 (a) Scheme of the light extraction for BGO scintillator by a monolayer of PS nanospheres. (b) Sketch of the two different directions of the hexagonal-close-packed structure. (c) Top-view SEM images of the surface of BGO covered with PS nanospheres (d = 500 nm) with two different resolutions. Numerical simulations of transmission spectra with the wavelength of 510 nm for plain BGO (black solid lines), BGO covered with a monolayer of PS nanospheres with the diameter of 500 nm (red solid lines), and BGO covered with an array of square pillars with the same structures as the monolayer of PS spheres (blue dashed lines). The simulations with s- and p-polarizations are shown in (d) and (e), respectively.

In Fig. 7.6, the simulated spatial distributions of electric field intensity for a cross section in the middle of a sphere in

Γ-K direction are presented in order to understand the extraction process. Figures 7.6a–c exhibit the strong confinement of electric field intensity, which is the evident characteristics of WGMs. In contrast, as shown in Fig. 7.6d, the distribution of electric field intensity with wavelength and incident angle of 510 nm and 30°, respectively, does not show any confinement inside a sphere. As a consequence, it is proved that the transmission peak at about 40° is associated with the WGMs. Combining with the simulation of the spatial distribution of electric field intensity, it is revealed that the WGMs for the individual sphere and the periodic arrangement are responsible for the enhancement of light extraction, and WGMs play an important role in providing extra enhancement compared with a traditional two-dimensional photonic crystal. Low-order modes of WGMs with the character of high energy leakage and low wavelength selectivity are beneficial to the wideband emission of scintillators. Additionally, it is worthwhile to note that below the critical angle several dips appear in the transmission spectra shown in Fig. 7.5. These dips result in some extra diffracted light back into the crystal. At a first glance, the back-diffracted light below the critical angle harms the light extraction. However, the situation is not severe because

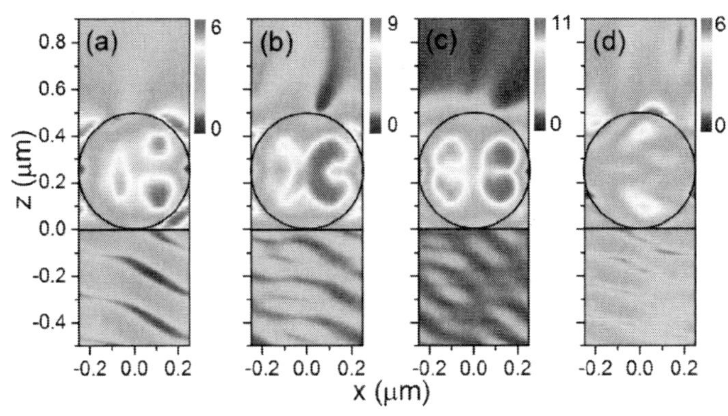

Figure 7.6 Simulations of spatial distributions of electric field intensity for BGO covered with a monolayer of PS nanospheres for p-polarization. The wavelengths and incident angles are (a) 694 nm, 37.4°, (b) 558 nm, 39.2°, (c) 500 nm, 40°, and (d) 510 nm, 30°. The black circle represents the site of nanosphere and the negative value in *z* coordination represents the scintillator.

the back-diffracted light can be reflected by the back side of the crystal and then be re-extracted by the structures, which could strengthen the ultimate extraction efficiency.

The photo-luminescence (PL) spectra of BGO in Figs. 7.7a,b show that the enhancements after covering with the monolayer PS spheres can be observed evidently, which is also strongly dependent on the wavelength and emergence angle. The wavelength- and angle-integrated enhancement reaches 72%. In order to obtain an optimal extraction efficiency, it is required that the diameter should be small enough for ensuring that only low order WG modes can be excited, and at the same time it should be large enough for ensuring that the diffraction can take place effectively. As a result, experimental and simulated results suggest that the optimal diameter is close to the wavelength involved. The angular profile of emission intensities at 510 and 620 nm shown in Figs. 7.7c,d, respectively, is significantly increased in the whole emergence angle with the strongest enhancement at normal direction, making the angular profile deviate from the Lambertian pattern. The directivity of emission is dependent on the diffraction of the structure of periodic array. This angle-dependent enhancement would be advantageous to the practical applications because only these photons emitted along with the direction of the window of a detector can contribute to the detection. Therefore, from the viewpoint of application, it is necessary to carefully design the structure for scintillators according to the layout of a system for a certain purpose of radiation detection. The measurements of energy spectra for γ-ray from a ^{137}Cs source were shown in Fig. 7.7e. The energy spectrum reflects the wavelength- and angle-integration of scintillation photons. For the photopeak, the channel number (the light output) of the structured BGO is larger by 34% than that of the reference sample. It should be noted that the structured BGO is covered with the PS nanospheres in half of area. Therefore, a corrected enhancement of light output is expected to be about 68%. It is interesting to observe that the energy resolution for the photopeak is improved from 33.6% (reference sample) to 30.3% (sample with PS nanospheres) deduced from the experiment data using Gaussian fitting.

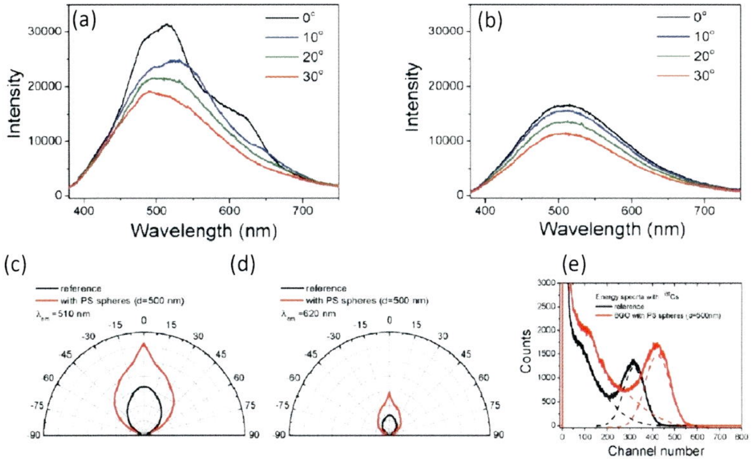

Figure 7.7 Emission spectra of BGO covered with a monolayer of PS nanospheres (a) and plain BGO (b) with different emergence angles. Angular profiles of emission at 510 nm (c) and 620 nm (d) for plain BGO (black lines) and BGO with nanospheres (red lines). The light output (e) under the excitation of γ-ray for plain BGO (black lines) and BGO with nanospheres (red lines). Gaussian fittings shown with dashed lines are used to analyze the energy resolution.

Similar with the case of BGO, the array of monolayer of PS nanospheres also increases the light extraction of LYSO single crystal, shown in Fig. 7.8. The enhancement of extraction efficiency up to 38% relative to the reference case without polystyrene spheres is achieved. Wavelength-dependence and the sphere-diameter-dependence are obvious. The optimal diameter of nanospheres is approximately the center wavelength of emission spectrum for both BGO and LYSO scintillators. The simulated results taking into account the multi-reflection shown in Fig. 7.8b exhibit the good agreement with the experiment results.

The obvious advantage of self-assembled method is the ability to fabricate the samples with large-scale area, which is important to the application of scintillation detection. Furthermore, using the self-assembled patterns as templates, more complicated photonic structures can be fabricated, which expands the compatibility and the ability to control the propagation of light. This work is under progress.

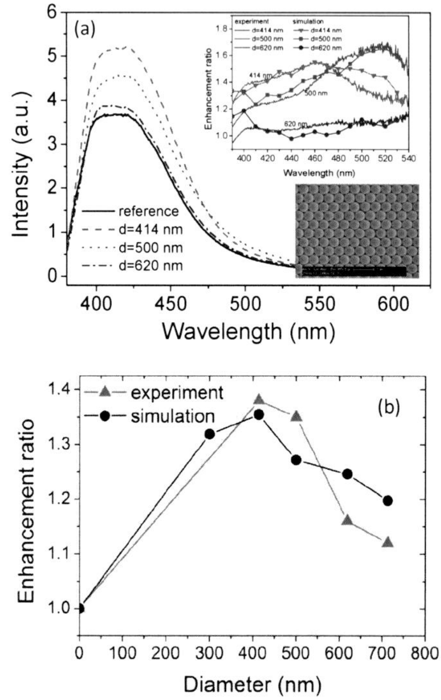

Figure 7.8 (a) Emission spectra of LYSO excited by an ultraviolet LED for the structures with different diameters in the normal direction. The top inset represents the enhancement ratio with respect to the reference sample, which shows a good consistence between experiment and calculation. The bottom inset is the SEM image of the array of monolayer PS nanospheres. (b) Experimental and simulated enhancement ratio of light extraction at 415 nm emission with various diameters of spheres.

7.3.3 Faster Luminescence Decay of Thin-Film Scintillator by Surface Plasmons

Surface plasmons (SPs), which propagate at the interface between a metal and a dielectric and evanescently confined in the perpendicular direction, can be used to control the spontaneous radiation by the coupling of SPs to the luminescent centers by changing the recombination rate.[35] Due to the significantly large photonic density of states (DOS) of SPs at near the resonance energy according to Fermi's golden rule, the spontaneous recombination transition rate into SPs modes can be significantly enhanced, which leads to a very large Purcell factor. Two necessary conditions need to be fulfilled. The first one is the spectral overlap of the resonance wavelength of SPs with the emission wavelength of luminescent center. Another one is the high efficiency of remission of free photons from SPs modes.

Many efforts are devoted to the enhancement of light-emitting diodes (LEDs) with low internal quantum efficiency by the competition from the non-radiative process.[36–40] However, for a luminescent material with high internal quantum efficiency, this method can provide a new approach to obtain faster component of scintillation light. The fundamental principle is based on the increased spontaneous recombination rate through the coupling of SPs, which leads to a faster luminescence decay.

Figure 7.9 exhibits the luminescence decay of LSO thin films coating with Ag nanoparticles with different dimensions for sample A and B. After fitting the curves, a significant fasten component about several nanoseconds appears, in contrast to the reference sample of 26 nanoseconds.[41]

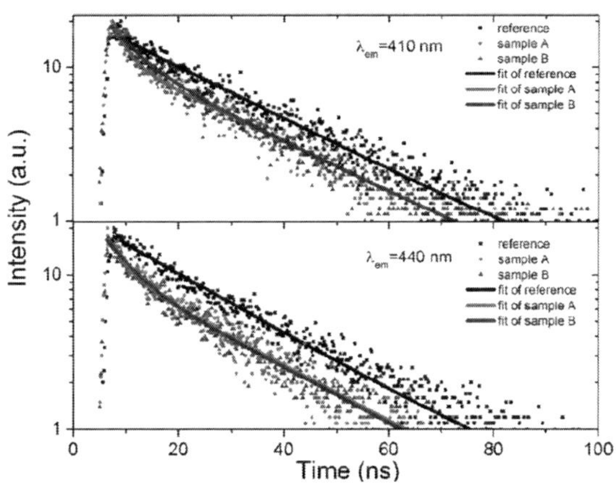

Figure 7.9 Time-resolved photoluminescence spectra with emission wavelength at 410 and 440 nm excited by 355 nm for sample A, sample B, and the reference sample. Fitting results are also presented with the solid lines.

For the reference sample without Ag-coating, the radiative and nonradiative recombinations can be described by the recombination rate of k_r and k_{nr}, which depend on the intrinsic electromagnetic characteristic of LSO. The nonradiative recombination with a rate of k_{nr} will give rise to heat by emitting phonons. The corresponding decay time constant can be described as $\tau_0 = \frac{1}{k_r + k_{nr}}$. With Ag-coating with the luminescent layer, the excited states can return to the ground states by exciting SPs modes in Ag nanoparticles due to the SP-scintillator coupling. This is a new recombination channel with a rate of k_{SP}. Thus, the decay time constant can be described as $\tau_1 = \frac{1}{k_r + k_{nr} + k_{SP}}$. The process is illustrated by Fig. 7.10. Since the k_{SP} is significantly large at resonance wavelength of SPs, the decay time of τ_1 is strongly dependent on k_{SP}.

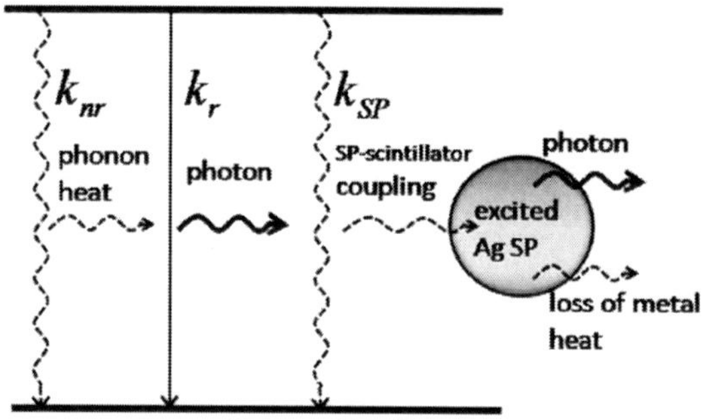

Figure 7.10 Recombination processes involving with SP-scintillator coupling.

In order to quantify the increase of spontaneous recombination rate, the Purcell factor as a function of wavelength can be defined as

$$F_p(\lambda) = \frac{k_r(\lambda) + k_{nr}(\lambda) + k_{SP}(\lambda)}{k_r(\lambda) + k_{nr}(\lambda)} = \frac{\tau_0(\lambda)}{\tau_1(\lambda)} \tag{7.7}$$

The new fast component of scintillation light will benefit to the radiation detection system with the requirement of high time-resolution. However, one severe problem of this method is the distance of SPs with luminescent center is very small. The effective thickness within the interaction is about several tens of nanometers, which gives rise to a weak scintillation efficiency.

7.4 Summary

Scintillators are the kind of traditional functional materials in the field of radiation detection systems. The combination of the principles of nanophotonics with the traditional materials will greatly improved the development of radiation detection. We summary the enhanced light extraction and the controllable emission directionality of scintillators by photonic crystals, and the fastening the decay time of scintillators by surface plasmon polariton. The effort in this field is just a beginning. However, there is strong evidence that nanophotonics innovations will have evolutionary and revolutionary impacts in the field of radiation detection.

References

1. W. C. Röntgen (1896). On a new kind of rays, *Science*, **3**, 227–231.
2. B. Liu and C. Shi (2002). Development of medical scintillator, *Chinese Sci. Bull.*, **47**, 1057–1063.
3. W. W. Moses (2002). Current trends in scintillator detectors and materials, *Nucl. Instrum. Methods Phys. Res. A*, **487**, 123–128.
4. D. J. Brenner, and E. J. Hall (2007). Computed tomography: An increasing source of radiation exposure, *N. Eng. J. Med.*, **357**, 2277–2284.
5. M. Conti (2011). Focus on time-of-flight PET: The benefits of improved time resolution, *Eur. J. Nucl. Med. Mol. Imaging*, **38**, 1147–1157.
6. M. J. Weber (2002). Inorganic scintillators: Today and tomorrow, *J. Luminescence*, **100**, 35–45.
7. I. Sarasola, R. Cuerdo, J. Navarrete, P. de Acilu, P. R. Mendes, J. M. Cela, J. C. Oller, L. Romero, and C. Willmott (2011). Performance evaluation of a PET demonstrator for PET-MR imaging based on monolithic LYSO: Ce scintillators, *J. Instrum.*, **6**, C12041.
8. P. Lecoq, I. Dafinei, E. Auffray, M. Schneegans, M. V. Korzhik, O. V. Missevitch, V. B. Pavlenko, A. A. Fedorov, A. N. Annenkov, and V. L. Kostylev (1995). Lead tungstate $PbWO_4$ scintillators for LHC EM calorimetry, *Nuclear Instruments and Methods in Physics Research Section A: Accelerators, Spectrometers, Detectors Assoc. Equipment*, **365**, 291–298.
9. J. Glodo, K. S. Shah, M. Klugerman, P. Wong, B. Higgins, and P. Dorenbos (2005). Scintillation properties of LuI3: Ce, *Nuclear Instruments and*

Methods in Physics Research Section A: Accelerators, Spectrometers, Detectors Assoc. Equipment, **537**, 279–281.

10. G. Blasse, and B. C. Grabmaier (1994). *Luminescent Materials*, Springer. p. 85.

11. A. Knapitsch, E. Auffray, C. W. Fabjan, J.-L. Leclercq, P. Lecoq, X. Letartre, and C. Seassal (2011). Photonic crystals: A novel approach to enhance the light output of scintillation based detectors, *Nuclear Instruments and Methods in Physics Research Section A: Accelerators, Spectrometers, Detectors Assoc. Equipment*, **628**, 385–388.

12. M. Nikl, P. Bohacek, E. Mihokova, M. Kobayashi, M. Ishii, Y. Usuki, V. Babin, A. Stolovich, S. Zazubovich, and M. Bacci (2000). Excitonic emission of scheelite tungstates AWO_4 (A = Pb, Ca, Ba, Sr), *J. Luminescence*, **87**, 1136–1139.

13. M. Nikl, A. Yoshikawa, A. Vedda, and T. Fukuda (2006). Development of novel scintillator crystals, *J. Crystal Growth*, **292**, 416–421.

14. R. Kirchain and L. Kimerling (2007). A roadmap for nanophotonics, *Nat. Photon.*, **1**, 303–305.

15. P. Lodahl, A. F. Van Driel, I. S. Nikolaev, A. Irman, K. Overgaag, D. Vanmaekelbergh, and W. L. Vos (2004). Controlling the dynamics of spontaneous emission from quantum dots by photonic crystals, *Nature*, **430**, 654–657.

16. W. L. Barnes, A. Dereux, and T. W. Ebbesen (2003). Surface plasmon subwavelength optics, *Nature*, **424**, 824–830.

17. H. Yokoyama (1992). Physics and device applications of optical microcavities, *Science*, **256**, 66–70.

18. J. J. Wierer, A. David, and M. M. Megens (2009). III-nitride photonic-crystal light-emitting diodes with high extraction efficiency, *Nat. Photon.*, **3**, 163–169.

19. M. Fujita, S. Takahashi, Y. Tanaka, T. Asano, and S. Noda (2005). Simultaneous inhibition and redistribution of spontaneous light emission in photonic crystals, *Science*, **308**, 1296–1298.

20. S. Noda, and M. Fujita (2009). Light-emitting diodes: Photonic crystal efficiency boost, *Nat. Photon.*, **3**, 129–130.

21. N. Ganesh, W. Zhang, P. C. Mathias, E. Chow, J. Soares, V. Malyarchuk, A. D. Smith, and B. T. Cunningham (2007). Enhanced fluorescence emission from quantum dots on a photonic crystal surface, *Nat. Nanotechnol.*, **2**, 515–520.

22. A.-L. Fehrembach, S. Enoch, and A. Sentenac (2001). Highly directive light sources using two-dimensional photonic crystal slabs, *Appl. Phys. Lett.*, **79**, 4280–4282.

23. S. A. Boden and D. M. Bagnall (2010). Optimization of moth-eye antireflection schemes for silicon solar cells, *Prog. Photovoltaics Res. Appl.*, **18**, 195–203.

24. P. Pignalosa, B. Liu, H. Chen, H. Smith, and Y. Yi (2012). Giant light extraction enhancement of medical imaging scintillation materials using biologically inspired integrated nanostructures, *Opt. Lett.*, **37**, 2808–2810.

25. M. Kronberger, E. Auffray, and P. Lecoq (2008). Probing the concepts of photonic crystals on scintillating materials, *IEEE Trans. Nucl. Sci.*, **55**, 1102–1106.

26. A. Knapitsch, E. Auffray, C. W. Fabjan, J. L. Leclercq, X. Letartre, R. Mazurczyk, and P. Lecoq (2012). Results of photonic crystal enhanced light extraction on heavy inorganic scintillators, *IEEE Trans. Nucl. Sci.*, **59**, 2334–2339.

27. H. T. Miyazaki, H. Miyazaki, K. Ohtaka, and T. Sato (2000). Photonic band in two-dimensional lattices of micrometer-sized spheres mechanically arranged under a scanning electron microscope, *J. Appl. Phys.*, **87**, 7152–7158.

28. Y. Yao, J. Yao, V. K. Narasimhan, Z. Ruan, C. Xie, S. Fan, and Y. Cui (2012). Broadband light management using low-Q whispering gallery modes in spherical nanoshells, *Nat. Commun.*, **3**, 664.

29. G. Kang, H. Park, D. Shin, S. Baek, M. Choi, D. Yu, K. Kim, and W. J. Padilla (2013). Broadband light-trapping enhancement in an ultrathin film a-Si absorber using whispering gallery modes and guided wave MODES with dielectric surface textured structures, *Adv. Mater.*, **25**, 2617–2623.

30. J. Grandidier, R. A. Weitekamp, M. G. Deceglie, D. M. Callahan, C. Battaglia, C. R. Bukowsky, C. Ballif, R. H. Grubbs, and H. A. Atwater (2013). Solar cell efficiency enhancement via light trapping in printable resonant dielectric nanosphere arrays, *Phys. Status Solidi A,* **210**, 255–260.

31. J. Grandidier, D. M. Callahan, J. N. Munday, and H. A. Atwater (2011). Light absorption enhancement in thin-film solar cells using whispering gallery modes in dielectric nanospheres, *Adv. Mater.*, **23**, 1272–1276.

32. J. F. Galisteo-López, M. Ibisate, R. Sapienza, L. S. Froufe-Pérez, Á. Blanco, and C. López (2011). Self-assembled photonic structures, *Adv. Mater.*, **23**, 30–69.

33. Z. Zhu, B. Liu, C. Cheng, H. Chen, M. Gu, Y. Yi, and R. Mao (2014). Broadband light output enhancement for scintillator using whispering-gallery modes in nanospheres, *Phys. Stat. Solidi A*, **211**, 1583–1588.

34. Z. Zhu, B. Liu, C. Cheng, Y. Yi, H. Chen, and M. Gu (2013). Improved light extraction efficiency of cerium-doped lutetium-yttrium oxyorthosilicate

scintillator by monolayers of periodic arrays of polystyrene spheres, *Appl. Phys. Lett.*, **102**, 071909.

35. W. L. Barnes (2004). Light-emitting devices: Turning the tables on surface plasmons, *Nat. Mater.*, **3**, 588–589.
36. A. Fujiki, T. Uemura, N. Zettsu, M. Akai-Kasaya, A. Saito, and Y. Kuwahara (2010). Enhanced fluorescence by surface plasmon coupling of Au nanoparticles in an organic electroluminescence diode, *Appl. Phys. Lett.*, **96**, 043307.
37. P. A. Hobson, S. Wedge, J. A. E. Wasey, I. Sage, and W. L. Barnes (2002). Surface plasmon mediated emission from organic light emitting diodes, *Adv. Mater.*, **14**, 1393–1396.
38. K. Okamoto, I. Niki, A. Shvartser, Y. Narukawa, T. Mukai, and A. Scherer (2004). Surface-plasmon-enhanced light emitters based on InGaN quantum wells, *Nat. Mater.*, **3**, 601–605.
39. A. A. Toropov, T. V. Shubina, V. N. Jmerik, S. V. Ivanov, Y. Ogawa, and F. Minami (2009). Optically enhanced emission of localized excitons in In_xGa_{1-x} N films by coupling to plasmons in a gold nanoparticle, *Phys. Rev. Lett.*, **103**, 037403.
40. Y. Xiao, J. P. Yang, P. P. Cheng, J. J. Zhu, Z. Q. Xu, Y. H. Deng, S. T. Lee, Y. Q. Li, and J. X. Tang (2012). Surface plasmon-enhanced electroluminescence in organic light-emitting diodes incorporating Au nanoparticles, *Appl. Phys. Lett.*, **100**, 013308.
41. B. Liu, Z. Zhu, J. Zhu, S. Wu, H. Chen, M. Gu, Q. Cheng, C. Cheng, Z. Wang, and Y. Zhao (2014). An approach to achieve significantly faster luminescence decay of thin-film scintillator by surface plasmons, *Appl. Phys. Lett.*, **104**, 061902.

Chapter 8

Optical Trapping of Nanoparticles

Shiyun Lin[a] **and Juejun Hu**[b]

[a]*Oracle Inc.*
[b]*Massachusetts Institute of Technology*

jjhu@mit.edu

8.1 Introduction and Scope

Ever since its first demonstration in the 1970s by Ashkin and co-workers,[1,2] optical trapping has become a mainstream technology extensively used for confinement and manipulation of micro- and nanoscale objects. In its most popular form, an optical trap, or "optical tweezers," is created by a single laser beam tightly focused through a high numerical aperture (NA) microscope objective. This relatively simple configuration is capable of exerting pico-Newton level forces in a non-contact manner and simultaneously measuring the trapped particle displacement with nanometer level precision, thereby enabling a plethora of applications in cell biology, immunology, rheology, biophysics, and colloid science.

Given the apparent practical importance of optical trapping technologies, there has been a large body of existing literature covering their historic development, underlying physics, instrument engineering, and various applications.[3–13] The readers may refer to the cited articles herein for more details.

Integrated Nanophotonic Resonators: Fundamentals, Devices, and Applications
Edited by Ya Sha Yi

ISBN 978-981-4613-78-1 (Hardcover), 978-981-4613-79-8 (eBook)
www.panstanford.com

While the majority of literature focus on manipulation of particles of a few microns in size (e.g., micro-beads or biological cells), in this chapter we will specifically address optical trapping of nanoscale objects with radii $r < 100$ nm, ranging from nanoparticles to single protein molecules. Trapping of nanoparticles is challenging mainly due to two reasons: (1) the trapping potential diminishes drastically for small nanoparticles, as typically the optical force scales with the target particle volume (or more precisely the polarizability $\alpha \propto r^3$) for a given trap; and (2) viscous dragging force also decreases for small particles, which contributes to increased probability of particle escape from the trap due to thermal motion. As a consequence, strong optical field is mandated to maintain stable trapping of nanoparticles. The size scaling leads to several important implications in both optical physics and experimental implementations: for example, nanophotonic or plasmonic resonant structures with tight optical confinement become a preferred trapping platform over free-space optical tweezers as they can readily generate strong optical gradient forces. Introduction of these new optical trapping schemes also entails new theoretical formalism and modeling tools to enhance our understanding into the optomechanical interactions in the nanoscale.

The chapter therefore emphasizes new optical physics and device designs specifically adapted for single nanoparticle and molecule manipulation. We start with a review of classical theoretical treatment and modeling tools of optical forces and highlight their limits when applied to new optical trapping schemes. Section 8.2.3 then examines from a theoretical perspective the benefits of emerging photonic and plasmonic platforms for nanoparticle trapping. Section 8.3 is dedicated to reviewing the latest advancements in experimental techniques that have recently extended optical manipulation to the unprecedented single molecule regime.

8.2 Theory of Optical Force

8.2.1 Classical Formulations

8.2.1.1 Overview

The general description of optical force is given by the Maxwell stress tensor, which is formulated in terms of traction over an enclosed surface (in SI units):

$$\ddot{\boldsymbol{T}} = \left[\varepsilon_0 \boldsymbol{EE} + \mu_0 \boldsymbol{HH} - \frac{1}{2}(\varepsilon_0 E^2 + \mu_0 H^2)\ddot{\boldsymbol{I}} \right] \quad (8.1)$$

where E and H are and electric and magnetic fields, ε_0 and μ_0 represent vacuum permittivity and permeability, and the notation $\boldsymbol{EE}$ denotes the dyadic product. The time-averaged optical force exerted on an object is given by integrating the Maxwell stress tensor over a surface S enclosing the object:

$$< \boldsymbol{F} > = \int_S < \ddot{\boldsymbol{T}}(r,t) > \cdot \, \boldsymbol{n} dS, \quad (8.2)$$

where $\boldsymbol{n}$ is the unit vector along the surface normal and $< \dots >$ denotes the time average. The Maxwell stress tensor formalism can be directly derived from the Maxwell equations and the Lorentz Force Law. The readers are encouraged to follow the derivation that can be found in many textbooks on electromagnetism.[14]

In general, the electromagnetic fields E and H in Eq. 8.1 are solved using finite difference time domain (FDTD) or the finite element method (FEM). It is worth noting that the field components used in the Maxwell tensor correspond to the fields when the trapped object is present rather than the unperturbed fields. To implement the Maxwell tensor for optical force calculation, it is mandatory to compute the field distribution for each particle size and location, which present a computationally intensive task. Therefore, while the Maxwell tensor presents a complete and rigorous solution to calculate optical force, alternative modeling techniques are often implemented to improve the computation efficiency.

In the sections that follow, two classes of modeling techniques are elaborated: the dipole approximation and its variants; and the T-matrix method, which originates from the generalized Lorenz–Mie theory. There is an apparent connection between optical scattering theories and optical force modeling, as optical force fundamentally stems from the conservation of momentum—the transfer of electromagnetic momentum to trapped particles via optical scattering.

8.2.1.2 Approximation models

The dipole approximation applies to scattering in the Rayleigh regime, i.e., when the particles are much smaller than the light wavelength λ. When such a Rayleigh particle is excited by a monochromatic electromagnetic wave with an angular frequency ω, the electric field of the radiation polarizes the particle and induces an electric dipole oscillating at the same frequency ω. The electric field E and the induced dipole moment P of the particle are generally complex numbers with a time dependence represented by an $\exp(i\omega t)$ factor. Assuming linear polarizability, they are connected through

$$\boldsymbol{P} = \boldsymbol{P}_0 \exp(\boldsymbol{i}\omega \boldsymbol{t}) = \varepsilon_2 \alpha \boldsymbol{E} = \varepsilon_2 \alpha \boldsymbol{E}_0 \exp(\boldsymbol{i}\omega \boldsymbol{t}), \tag{8.3}$$

where the subscript 0 denotes the time-independent complex amplitude, and ε_2 is the permittivity of the medium the particle resides in.[15] The electric field E in Eq. 8.3 is the unperturbed field from the light source; in other words, contribution from induced polarization of the particle (perturbation) is excluded. The electromagnetic field exerts an optical force F on the dipole. The time-averaged force is given by[16]

$$<\boldsymbol{F}> = \frac{1}{4}\varepsilon_2 \operatorname{Re}(\alpha)\cdot \nabla \boldsymbol{E}_0^2 + \frac{1}{2}\varepsilon_2 \operatorname{Im}(\alpha)\cdot \boldsymbol{E}_0^2 \nabla \varphi \tag{8.4}$$

where φ gives the phase angle of the electric field. The first term in Eq. 8.4 is customarily termed "gradient force." The gradient force results from polarization energy change of the particle: when the particle is placed at different locations with different electric field intensity, the electric polarization energy of the particle also varies, giving rising to a conservative force that scales with the gradient of polarization energy density. The latter term, called "scattering force," is non-conservative as it cannot be expressed as the gradient of a scalar potential. The scattering force describes the momentum transfer from electromagnetic radiation to the particle due to optical scattering and absorption. The connection becomes apparent when the magnitude of the scattering force is written in the following alternative form[17]:

$$|<\boldsymbol{F}_{\text{sca}}>| = \frac{1}{2}\varepsilon_2 \operatorname{Im}(\alpha)\cdot \boldsymbol{E}_0^2 \nabla \varphi = \frac{1}{2}\varepsilon_2 \boldsymbol{E}_0^2 \cdot \sigma_{\text{ext}} = \frac{1}{2}\varepsilon_2 \boldsymbol{E}_0^2 \cdot (\sigma_{\text{abs}} + \sigma_{\text{sca}}) \tag{8.5}$$

here σ_{ext}, σ_{abs}, and σ_{sca} represent the extinction, absorption, and scattering cross sections of the particle, respectively. The cross sections are given as follows (k gives the wave vector):

$$\sigma_{\text{sca}} = \frac{1}{6\pi} \cdot |\alpha|^2 k^4 \tag{8.6}$$

$$\sigma_{\text{abs}} = \text{Im}(\alpha) \cdot k - \frac{1}{6\pi} \cdot |\alpha|^2 k^4 \tag{8.7}$$

$$\sigma_{\text{ext}} = \sigma_{\text{sca}} + \sigma_{\text{abs}} = \text{Im}(\alpha) \cdot k \tag{8.8}$$

The advantage of the dipole approximation over direct force computation using the Maxwell stress tensor is evident from Eq. 8.4, as perturbation effects from the particle to the electromagnetic field are not necessary information for force calculation and hence repeated FDTD or FEM field simulations for each particle location are not required. For a given optical trap configuration, the optical force and trapping potential can be readily computed using Eq. 8.4 provided that the particle polarizability is determined.

In the quasi-static limit ($\lambda >> r$), the polarizability of a spherical Rayleigh particle with radius r is given by

$$\alpha = 4\pi \boldsymbol{r}^3 \cdot \frac{\varepsilon_1 - \varepsilon_2}{\varepsilon_1 + 2\varepsilon_2}, \tag{8.9}$$

where ε_1 is particle's permittivity. Validity of the quasi-static approximation depends on the r/λ ratio, the particle refractive index, as well as the specific trap configuration, and thus has to be evaluated on a case-by-case basis. For instance, Gaugiran et al. numerically calculated optical force applied on glass microspheres (index n = 1.55) trapped on a silicon nitride waveguide operating at λ = 1.064 μm wavelength, and they found good agreement between the quasi-static approximation with the Maxwell stress tensor method for particles with a radius up to 120 nm, corresponding to a r/λ ratio of ~11%.[18]

According to Eq. 8.9, the polarizability of a dielectric particle free of optical absorption is a real number. Consequently, Eq. 8.8 suggests that the extinction cross section of the particle will vanish in the classical quasi-static limit. This is apparently not physical as scattering always exists even for particles much smaller than wavelength. The paradox can be resolved by introducing the modified long wavelength approximation (MLWA), which specifies the particle polarizability as

$$\alpha = 4\pi\left(\frac{1}{\boldsymbol{r}^3}\cdot\frac{\varepsilon_1+2\varepsilon_2}{\varepsilon_1-\varepsilon_2} - \boldsymbol{i}\frac{2}{3}\boldsymbol{k}^3 - \frac{\boldsymbol{k}^2}{\boldsymbol{r}}\right)^{-1} \tag{8.10}$$

Compared to Eq. 8.9, the MLWA includes two additional correction terms: The imaginary part (the radiative reaction term) is dictated by the condition $\sigma_{abs} = 0$ in the case of a lossless dielectric particle (i.e., when both ε_1 and ε_2 are real positive numbers), whereas the second term results from dynamic depolarization.[19] Equation 8.10 in conjunction with Eq. 8.5 is routinely used in optical force calculations.

For non-spherical particles or large particles that cannot be treated within the dipole approximation framework, alternative formalisms have to be employed to solve the optical force. The Mie theory yields exact analytical solutions however only for spherical particles.[20] For particles of arbitrary shape and size, the discrete dipole approximation (DDA)[21–23] is a common method used to calculate their scattering properties and hence optical force. The DDA represents an arbitrary-shaped particle as a cubic lattice of N polarizable small subunits ($<<\lambda$) with polarizability α_{su} (Fig. 8.1), where α_{su} is given by the Clausius–Mosotti relation with the radiative reaction term.[24]

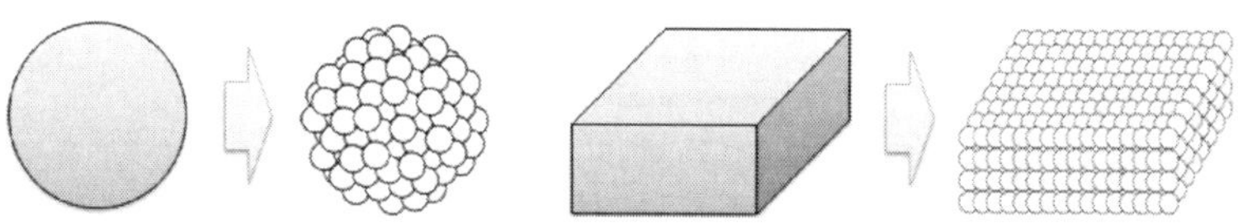

Figure 8.1 The DDA approximates a particle of arbitrary size and shape as an array of polarizable small subunits. Two examples are shown for a spherical particle (left) and a cuboid shaped particle (right).

When excited by an electromagnetic wave, each subunit induces an electric dipole moment p_i:

$$\boldsymbol{p}_i = \alpha_{su}\boldsymbol{E}_i \quad (i = 1, \ldots, 3N), \tag{8.11}$$

where E_i is the local electric field. There are N subunits and the electric field at each subunit has three components (x, y, z), and hence Eq. 8.11 represents a total of $3N$ equations. The field at each subunit located at r_i can be expressed as the sum of the incident field E_{inc} and field scattered by other subunits:

$$\boldsymbol{E}_i(\boldsymbol{r}_i) = \boldsymbol{E}_{\text{inc}} + \sum_{i \neq j} \boldsymbol{E}_{ij} \quad (i \neq j) \tag{8.12}$$

$$\boldsymbol{E}_{ij} = \frac{1}{4\pi\varepsilon_2}\left\{-\frac{\boldsymbol{p}_j}{\boldsymbol{r}_{ij}^3} + \frac{3\boldsymbol{r}_{ij}[\boldsymbol{r}_{ij}\cdot\boldsymbol{p}_j]}{\boldsymbol{r}_{ij}^5} + \frac{ik\boldsymbol{p}_j}{\boldsymbol{r}_{ij}^2} - \frac{3ik\boldsymbol{r}_{ij}[\boldsymbol{r}_{ij}\cdot\boldsymbol{p}_j]}{\boldsymbol{r}_{ij}^4} - \frac{k^2}{\boldsymbol{r}_{ij}^3}\boldsymbol{r}_{ij}\times[\boldsymbol{r}_{ij}\times\boldsymbol{p}_j]\right\}\cdot e^{ikr_{ij}}, \tag{8.13}$$

where $\boldsymbol{r}_{ij} = \boldsymbol{r}_i - \boldsymbol{r}_j$ and $\boldsymbol{p}_j = \alpha_{\text{su}}\boldsymbol{E}_j$. Solving Eq. 8.12, which contains $3N$ coupled linear equations, yields the complete electric field distribution. The total optical force is obtained by summing the contributions from each subunit given by Eq. 8.4.[25,26] Efficacy of the DDA applied to calculating optical force on inhomogeneous and anisotropic particles has been numerically validated[27] and a number of computer codes for the DDA are available in the public domain.[28]

8.2.1.3 The T-matrix method

The T-matrix method is another example highlighting the close connection between optical force and scattering. The classical Mie scattering theory rigorously solves optical scattering by a spherical particle. The theory expands the fields outside and inside the particle into a sum of spherical vector wave functions, and the expansion coefficients are determined by matching the boundary conditions at the sphere surface. The T-matrix method extends the approach to non-spherical particles. The electromagnetic waves incident on (E_{inc}) and scattered by (E_{sca}) the particle are expanded as series of vector spherical wave functions, respectively[29,30]:

$$\boldsymbol{E}_{\text{inc}}(\boldsymbol{r}) = \sum_{n=1}^{\infty}\sum_{m=-n}^{n}\left[a_{mn}\boldsymbol{RgM}_{mn}(k\boldsymbol{r}) + b_{mn}\boldsymbol{RgN}_{mn}(k\boldsymbol{r})\right] \tag{8.14}$$

$$\boldsymbol{E}_{\text{sca}}(\boldsymbol{r}) = \sum_{n=1}^{\infty}\sum_{m=-n}^{n}\left[p_{mn}\boldsymbol{RgM}_{mn}(k\boldsymbol{r}) + q_{mn}\boldsymbol{RgN}_{mn}(k\boldsymbol{r})\right] \tag{8.15}$$

where $\boldsymbol{RgM}_{mn}$ and $\boldsymbol{RgN}_{mn}$ are vector spherical wave functions:

$$\boldsymbol{RgM}_{mn}(k\boldsymbol{r}) = (-1)^m d_n \exp(im\phi) j_n(kr)\cdot\boldsymbol{C}_{mn}(\theta) \tag{8.16}$$

$$\boldsymbol{RgN}_{mn}(k\boldsymbol{r}) = (-1)^m d_n \exp(im\phi) \times \left\{\frac{n(n+1)}{kr}\cdot j_n(kr)\boldsymbol{P}_{mn}(\theta) + \left[j_{n-1}(kr) - \frac{n}{kr}j_n(kr)\right]\cdot\boldsymbol{B}_{mn}(\theta)\right\} \tag{8.17}$$

$$\boldsymbol{C}_{mn}(\theta) = \hat{\theta}\frac{im}{\sin\theta}d^{n}_{0m}(\theta) - \hat{\phi}\frac{d}{d\theta}d^{n}_{0m}(\theta) \tag{8.18}$$

$$\boldsymbol{B}_{mn}(\theta) = \hat{\theta}\frac{d}{d\theta}d^{n}_{0m}(\theta) + \hat{\phi}\frac{im}{\sin\theta}d^{n}_{0m}(\theta) \tag{8.19}$$

$$\boldsymbol{P}_{mn}(\theta) = \hat{r}\cdot d^{n}_{0m}(\theta) \tag{8.20}$$

$$d_n = \left[\frac{2n+1}{4\pi n(n+1)}\right]^{1/2} \tag{8.21}$$

Here $d^n_{om}(\theta)$ represents Wigner d functions. For incident waves (Eq. 8.14), $j_n(kr)$ are spherical Bessel functions, and for scattered waves, $j_n(kr)$ denote Hankel functions of the first kind.

In a linear optical system, the expansion coefficients a, b, p, and q for the two sets of series are linearly correlated by the T-matrix. Following the conventions in ref. 30, the T-matrix is defined by

$$p_{mn} = \sum_{m'n'} T^{(11)}_{mnm'n'}a_{m'n'} + T^{(12)}_{mnm'n'}b_{m'n'} \tag{8.22}$$

$$q_{mn} = \sum_{m'n'} T^{(21)}_{mnm'n'}a_{m'n'} + T^{(22)}_{mnm'n'}b_{m'n'} \tag{8.23}$$

The vector spherical wave function expansion formalism used in the T-matrix representation is identical to that of the Mie theory, although in the classical Mie theory spherical particles have diagonal T-matrices, whereas in the general case of non-spherical particles the T-matrix has non-vanishing off-diagonal elements. According to the T-matrix formalism, a particle of arbitrary size and geometry can be associated with its unique T-matrix independent of the incident electromagnetic fields. Therefore, when the particle is placed in different locations of an optical trap, the T-matrix remains invariant and only the field expansion (Eqs. 8.14 and 8.15) needs to be repeated, an advantage of the method over full-scale modeling techniques such as FDTD or FEM.

It is worth pointing out that while mature algorithms exist for decomposing an incident plane wave or Gaussian beams commonly used in free-space optical tweezers into vector spherical wave functions,[29] the expansion of an arbitrary incident optical field (in particular, localized guided modes in integrated photonics or plasmonics) is non-trivial.[31] A notable piece of work along this line by Moreira et al. has identified analytical solutions to expand an arbitrary incident field into a set of vector spherical wave functions.[32]

8.2.2 Optical Trapping Using Micro- and Nanophotonic Structures

Integrated photonic and plasmonic structures are uniquely poised for nanoparticle trapping for several reasons. First, the strong optical confinement in these devices leads to large optical field gradient that facilitates stable trapping down to a single molecule. Second, a diverse cross section of device designs including resonant or slot structures are available on these platforms either to further enhance trapping force or to offer new capabilities (e.g., sensing and sorting). Last but not least, the integrated photonic or plasmonic devices are readily amenable to integration with other micro- and nano-systems such as microfluidics or microelectronics to augment the functionalities of lab-on-a-chip devices. The latest experimental implementations of nanoparticle trapping using integrated photonic and plasmonic structures will be reviewed in Section 8.3, while in this section we provide a brief overview of the theory behind optical trapping capitalizing on the strong light-particle interactions in these structures and highlight unconventional phenomena that arises in the cavity-enhanced trapping regime.

8.2.2.1 Trapping and transport on optical waveguides

When particles are in placed close proximity to a planar optical waveguide, a propagating mode in the waveguide generates a scattering force that propels the particle along the waveguide. At the same time, the decaying evanescent field of the mode provides a gradient force that traps the particle on the waveguide surface. Both components can be modeled with excellent accuracy using the dipole approximation (Eq. 8.4) for nanoparticles with radii up to $\sim 0.1\lambda$,[18] and therefore here we will exclusively focus on analytical results obtained within the framework of the dipole approximation. We consider the simple case of particle trapping by a transverse electric (TE) polarized mode in a 1-D dielectric slab waveguide whose core, substrate and top cladding indices are denoted by n_1, n_2, and n_3, respectively (Fig. 8.2a). The modal effective index n_{eff} is related to the propagation constant β through $\beta = n_{\text{eff}} \cdot k_0$, where k_0 is the free space wave vector. The evanescent modal field profile in the cladding is thus give by

$$E_3 = E_{13} \cdot \exp(k_x x) = E_{13} \cdot \exp\left(k_0 x \cdot \sqrt{n_{\text{eff}}^2 - n_3^2}\right) \tag{8.24}$$

In Eq. 8.24, k_3 represents the transverse wave vector in the cladding region and E_{13} is the modal electric field at the core/cladding interface. The spatially dependent gradient force induced by the evanescent field is

$$\begin{aligned} F_{\text{grad}}(x) &= -\frac{1}{4}\varepsilon_0 n_3^2 \text{Re}(\alpha) \cdot \nabla E_3^2 \\ &= -\frac{1}{2}\varepsilon_0 n_3^2 k_0 E_{13}^2 \,\text{Re}(\alpha)\sqrt{n_{\text{eff}}^2 - n_3^2} \cdot \exp(2k_0 x \cdot \sqrt{n_{\text{eff}}^2 - n_3^2}) \end{aligned} \tag{8.25}$$

Figures 8.2b,c plot the waveguide modal profile (λ = 1064 nm) as well as the gradient trapping force for three core index n_1 values while keeping the other waveguide parameters constant. We can see from the figures that the optical force is tightly confined within a few hundred nanometers on the waveguide surface due to the field's evanescent nature. In addition, increased waveguide index contrast leads to rapid decay of the evanescent wave and hence enhanced gradient trapping force. This conclusion, while deduced from the simple 1D slab case here, also applies to 2D channel or ridge waveguides.[18]

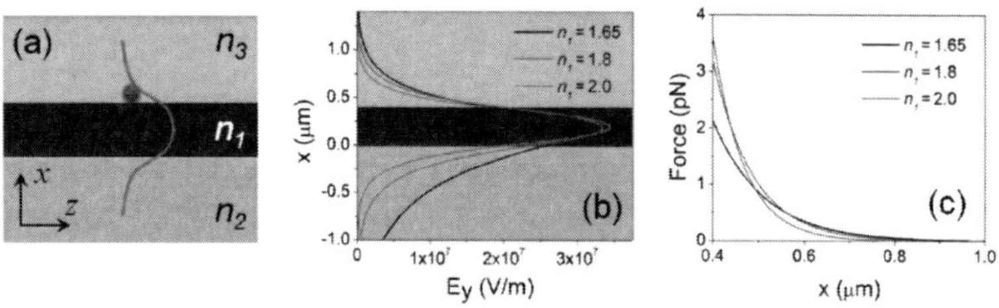

Figure 8.2 (a) Schematic waveguide configuration: in our calculations n_2 = 1.5, n_3 = 1.33 (water), and n_1 takes three values: 1.65, 1.8, and 2.0. The slab thickness is fixed at 400 nm, and the wavelength λ = 1064 nm. The energy flux per unit slab length is 1 W/μm. (b) TE polarized optical mode profile (E_y component) of the waveguide. (c) spatially dependent gradient force on a polystyrene bead with radius r = 50 nm calculated using the dipole approximation.

Slot waveguides, where field localization in a low-index medium is realized by leveraging surface-normal electric field discontinuity at material boundaries, can further strengthen optical force compared to traditional channel dielectric waveguides. It has been shown that in such nano-slot structures the classical dipole approximation still holds for particles with sizes up to 100 nm.[33]

8.2.2.2 Plasmonic trapping

Plasmonic trapping generally takes two forms: trapping by a propagating plasmonic mode (e.g., surface plasmon polariton, SPP, which is a guided mode supported at a metal/dielectric interface) or trapping by a localized surface plasmon resonance (LSPR) mode confined in all three dimensions in a metal nanostructure. Unlike SPP modes, which are confined only in 1D and thus have a continuous frequency spectrum, LSPR modes are discrete in the frequency domain due to the 3D optical confinement.

The former case is similar to trapping by a dielectric waveguide and the gradient trapping force is thus also specified by Eq. 8.25 with n_{eff} now denoting the SPP mode effective index given by

$$n_{\text{SPP}} = \sqrt{\frac{\varepsilon_{\text{dr}}\varepsilon_{\text{mr}}}{\varepsilon_{\text{dr}} + \varepsilon_{\text{mr}}}}, \tag{8.26}$$

where ε_{dr} and ε_{mr} are the *relative* permittivity of the metal and dielectric materials, respectively. In the special case of $\varepsilon_{\text{dr}} + \varepsilon_{\text{mr}} \sim 0$ (the resonant condition), the denominator in Eq. 8.26 vanishes, resulting in subDiffraction limit field localization near the metal surface. Substituting Eq. 8.26 into Eq. 8.25 yields the following trapping force expression in the dielectric side:

$$F_{\text{grad}}(x) = -\frac{1}{2}k_0\varepsilon_0\varepsilon_{\text{dr}}^2 \,|E_{\text{d}}|^2\, \text{Re}(\alpha)\sqrt{-\frac{1}{\varepsilon_{\text{dr}} + \varepsilon_{\text{mr}}}}\cdot\exp\left(2k_0\varepsilon_{\text{dr}}x\cdot\sqrt{-\frac{1}{\varepsilon_{\text{dr}} + \varepsilon_{\text{mr}}}}\right) \tag{8.27}$$

Unlike dielectric waveguide mode, which can either be transverse electric or transverse magnetic (TM) in nature, the SPP mode is TM polarized with three non-vanishing field components: E_x, E_z, and H_y (Fig. 8.3a). As shown by the SPP field profile in Fig. 8.3b, E_x is discontinuous at the metal–dielectric boundary. In Eq. 8.27 E_{d} denotes the total electric field at the metal–dielectric boundary on the dielectric side.

As an example, Figs. 8.3c,e plot the trapping force by the SPP mode supported by a gold/water interface at the wavelengths λ = 1064 nm and λ = 470 nm, respectively. The wavelength-dependent optical constants of gold are quoted from Johnson et al.[34] The figures show that most of the SPP mode is confined in the dielectric (water) side due to the small penetration depth in metal. We also note that the gradient force at λ = 1064 nm is similar to that induced by a slab mode, since the wavelength is far away from the resonant condition and therefore the propagation constants of the SPP mode (Eq. 8.26) and the dielectric slab mode are comparable. In the second scenario, $\mathrm{Re}(\varepsilon_{dr} + \varepsilon_{mr}) \sim 0$ at the wavelength λ = 470 nm. Consequently strong spatial localization of the SPP mode occurs and large gradient force enhancement is observed, as shown in Fig. 8.3e. In this case, the SPP field confinement is primarily limited by the non-vanishing imaginary part of $\varepsilon_{dr} + \varepsilon_{mr}$.

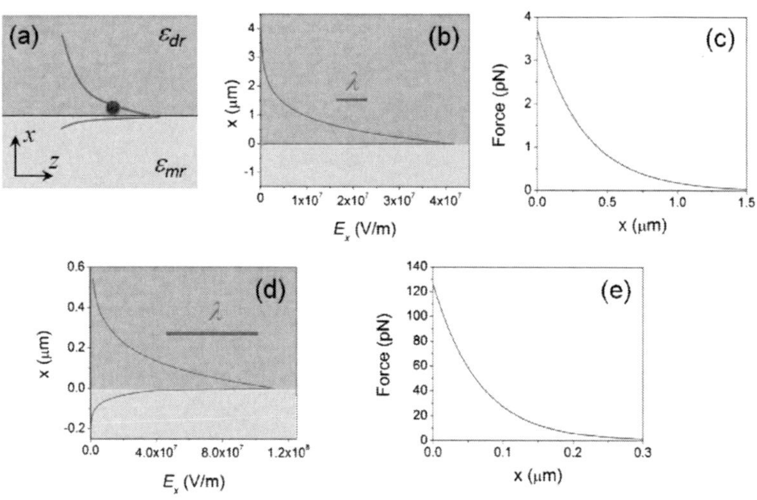

Figure 8.3 (a) Schematic surface plasmon polariton trapping configuration on a gold–water interface. The energy flux per unit slab length is 1 W/μm. (b) SPP mode profile (E_x component); the scale bar represents the free space wavelength λ = 1064 nm. (c) Spatially dependent gradient force on a polystyrene bead with radius r = 50 nm calculated using the dipole approximation for the case in (b). (d) (b) SPP mode profile (E_x component) at wavelength λ = 470 nm. (e) Gradient force on a polystyrene bead with radius r = 50 nm for the case in (d).

In addition to the energy concentration effect, plasmonic trapping can further enhance optical force by overcoming the classical diffraction limit. This is made possible by leveraging tightly confined LSPR modes that arise from the strong capacitive coupling between adjacent metal nanostructures.[35] These nanostructures serve as optical antenna that funnels incident electromagnetic radiation into LSPR modes, leading to sub-diffraction limit optical confinement in the dielectric gap between adjacent metal structures. While the complex geometries of the metal nanostructures generally makes analytical solutions of LSPR modes rather elusive, these localized modes can be numerically solved using FEM or FDTD methods, thus permitting optical force calculations using the dipole approximation. The finite mode volume of LSPR modes given their 3D optical confinement, however, can lead to non-negligible perturbation in the presence of trapped particles that invalidates the classical dipole approximation. The particle-induced perturbation effect in resonant structures will be discussed in the following sections.

8.2.2.3 Resonant cavity enhancement

In an optical resonant cavity, the optical field is significantly amplified compared to the incident excitation field as a consequence of the extended photon lifetime. Such field enhancement, coupled with spatial confinement of fields inside a resonant cavity, makes resonant devices and structures a preferred choice for nanoparticle trapping. The resonant cavity enhancement factor scales with Q/V (quality factor divided by mode volume),[36] suggesting that high-Q nano-cavities with a small mode volume are particularly attractive for optical trapping applications.

Theoretical modeling of optical trapping inside a resonant cavity usually follows one of the following two approaches: the Maxwell stress tensor analysis, or the dipole approximation. As previously discussed, the former approach can be computationally intensive, especially in the case of a high-Q cavity. On the other hand, field perturbation due to the introduction of a particle also becomes strongly enhanced in a nano-cavity with a large Q/V ratio given the high field concentration, which can invalidate the classical dipole approximation.

To account for the cavity perturbation effect, a modified dipole approximation method was implemented that provides optical force predictions with excellent agreement with the much more complicated Maxwell stress tensor simulations.[37] The method applies cavity perturbation theory to calculate the resonance detuning $d\lambda$ due to the entrance of a particle into a cavity:

$$\lambda_{\mathrm{r}} = \lambda_{\mathrm{r0}} + \frac{\lambda_{\mathrm{r0}} \left|E_{\mathrm{N}}\right|^2}{2\int_{\mathrm{cavity}} \varepsilon_{\mathrm{c}} \left|E_{\mathrm{N}}\right|^2 dV} \cdot \varepsilon_2 \operatorname{Re}(\alpha) \tag{8.28}$$

Here ε_{c} denotes the permittivity distribution of the resonant cavity, the field E_{N} is the dimensionless eigenmode of the unperturbed resonant cavity normalized so that $\int_{\mathrm{cavity}} |E_{\mathrm{N}}|^2 dV = 1$, and λ_{r0} and λ_{r} are the cavity resonant wavelengths before and after the particle introduction. The integral in Eq. 8.28 is performed over the entire cavity volume. The resonance detuning (Eq. 8.28) modulates the photon number inside the cavity, which linearly scales with the wavelength-dependent optical coupling coefficient T:

$$T(\lambda) = \frac{\lambda_{\mathrm{r}}^2}{\lambda_{\mathrm{r}}^2 + 4Q^2 \cdot (\lambda - \lambda_{\mathrm{r}})^2}, \tag{8.29}$$

where λ is the excitation wavelength. The gradient optical force is given by

$$F(P,\lambda) = \frac{\lambda_{\mathrm{r}}^2 Q}{\lambda_{\mathrm{r}}^2 + 4Q^2 \cdot (\lambda - \lambda_{\mathrm{r}})^2} \cdot \left[\frac{\varepsilon_2 \lambda \operatorname{Re}(\alpha) \cdot P}{8\pi c_0} \cdot \nabla |E_{\mathrm{N}}|^2\right] \tag{8.30}$$

P denotes the incident excitation optical power and c_0 is the speed of light in vacuum. The term in the bracket is identical to the gradient force term in the classical dipole approximation, and the factor outside the bracket corresponds to the resonant cavity field enhancement ($\propto Q$) after taking into account perturbation induced by the trapped particle ($\propto T$). When the coupling coefficient T approaches unity, i.e., when the following condition is satisfied:

$$\varepsilon_2 \operatorname{Re}(\alpha) \cdot Q \cdot \frac{|E_{\mathrm{N}}|^2}{2\int_{\mathrm{cavity}} \varepsilon_{\mathrm{c}} |E_{\mathrm{N}}|^2 \, dV} \approx \frac{\operatorname{Re}(\alpha) \cdot Q}{V_{\mathrm{c}}} << 1, \tag{8.31}$$

where V_{c} gives the cavity mode volume, Eq. 8.30 reduces to the classical dipole approximation. Based on Eq. 8.31, the condition

$\text{Re}(\alpha) \cdot Q/V_c < 0.1$ can be defined as a semi-empirical criterion for the validity of the classical dipole approximation. Notably, unlike the conventional applicability condition of dipole approximation $r \ll \lambda$, Eq. 8.31 does not explicitly incorporate wavelength dependence. Therefore, large deviation from the dipole approximation may arise in a high-Q nano-cavity with a small mode volume, even for particles much smaller than the wavelength. For instance, in a water-filled photonic crystal cavity with a moderate Q = 10,000 and a mode volume V_c = 1 μm^3 operating at 1550 nm wavelength, a spherical dielectric particle with a refractive index of 1.5 needs to have a radius r of 46 nm ($\lambda/r \sim 33$) or smaller to meet the condition Eq. 8.31. By comparison, it has been found that dipole approximation holds up to $\lambda/r \sim 10$ with high accuracy in an evanescent trapping configuration.[18]

The cavity perturbation effect predicts unique optical trapping behavior of resonant structures different from that of free space optical tweezers. One example is the so-called "self-induced trapping" first analyzed on a photonic crystal cavity.[38] Figure 8.4 shows the simulated field distribution inside the "empty" cavity and in the presence of a trapped particle. In the "empty" cavity without a particle, the resonant wavelength is detuned from the

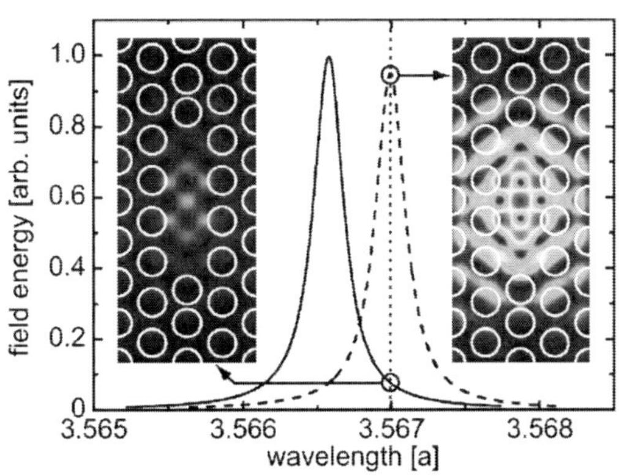

Figure 8.4 Principle of self-induced trapping: field energy inside an "empty" photonic crystal cavity (solid line) and in the presence of a bead with diameter of 0.5a (dashed line), where a is the photonic crystal lattice constant. The inset shows the corresponding electric field distribution. The particle induces resonant detuning that aligns the cavity resonant wavelength with the excitation wavelength and thereby generates large optical force (image courtesy of ref. 38).

excitation wavelength, and therefore the cavity enhancement effect is insignificant. Introduction of the particle shifts the cavity resonance to align with the external excitation. The perturbation leads to resonantly enhanced, strong local field near the cavity and hence large optical force. Since the resonance detuning scales with particle polarizability (Eq. 8.28), the self-induced trapping effect is also sensitive to trapped particle size. Such size-selective trapping characteristics have been quantitatively analyzed and applied to design single nanoparticle optical tweezers.[37]

8.2.2.4 Self-induced back-action optical trapping

Another recent example of optical trapping technique that benefits from optical perturbation effects is self-induced back-action (SIBA) optical trapping demonstrated by the seminal work by Juan et al.[39] The original SIBA configuration utilizes a nanoaperture perforated in a metal film to trap dielectric nanoparticles. The normal incident excitation wavelength is chosen to be slightly above the transmission cut-off of the nanoaperture. The transmission characteristics of the aperture become very sensitive to the position of trapped particles in this configuration: when a particle is situated at the center of the aperture, optical transmission through the aperture reaches a maximum value. As the particle moves away from the equilibrium position, the optical transmittance drops which is accompanied by a decrease of the rate of photon momentum traveling through the aperture. According to Newton's third law of motion, a force in the opposite direction thus acts on the particle in response to the momentum rate change, which pushes the particle back to the equilibrium position. Since SIBA trapping relies on perturbation induced by the particle to generate enhanced optical force, the classical dipole approximation does not apply, and full Maxwell stress tensor analysis was employed to quantify the SIBA optical force.

SIBA is ideally suited for nanoparticle trapping since it can significantly boost the trapping force. Juan et al. showed that the SIBA mechanism contributes to up to sixfold force enhancement compared to the classical gradient force.[39] More important, unlike resonant cavity trapping or plasmonic trapping where force enhancement results from field concentration, SIBA offers large trapping force at comparatively low field intensity, which minimizes optical damage of trapped species.

8.3 Single Nanoparticle/Molecule Trapping Platforms

8.3.1 Introduction

Conventional far field optical tweezers using a tightly focused laser beam face difficulties in trapping single nanoparticle. High optical power is needed to provide sufficient optical gradient force and overcome the Brownian motion of the nanoparticles. On the other hand, high power causes heating or optical damage to the trapping particles, which poses a severe challenge for trapping biological specimens. The diffraction-limited focal spot size also make it hard to precisely trap and position a single nanoparticle.

Near field optical trapping based on plasmonic structures or photonic micro-cavities have been employed to trap and monitor single nanoparticles. As discussed in Section 8.2.2, near field optical trapping by confining light to a sub-wavelength region provides large force enhancement and precise position control. Trapping of nanoparticles smaller than 100 nm have been reported using a broad variety of plasmonic and photonic nanostructures. We will review the experimental demonstrations of plasmonic optical trapping in Section 8.3.2 and photonic micro-cavity–based optical trapping in Section 8.3.3, respectively.

8.3.2 Plasmonic Optical Trapping

Optical trapping using surface plasmon polaritons has been demonstrated using metal films or gold stripes.[40–42] In this case, gradient force along the out-of-plane direction provide a potential well on the surface of the metal film, while the scattering force pushes the trapped particles to move along the surface. While sub-diffraction limit optical confinement is possible as shown in Section 8.2, this traveling wave–based trapping approach cannot offer large enough force enhancement for nanoparticle trapping and also fails to provide a three-dimensional potential well. Therefore, plasmonic tweezers based on LSPR represent a preferred choice for nanoparticle manipulation.

Plasmonic tweezers based on metallic nanostructures including microdisks,[43] micropillars,[44,45] and dipole antennas[46–48] have been reported by several groups. Usually the metallic nanostructures are fabricated by a lift-off process on a dielectric substrate and exhibit a LSPR mode at visible or near infrared frequency.[43,47] The plasmonic resonance mode can be excited from free space using a high NA objective lens or a prism. Unlike high-*Q* photonic cavities, LSPR modes usually feature much wider bandwidth and thus do not need tunable lasers as the excitation source.

Careful thermal management is critical to plasmonic tweezers. The large optical adsorption of metal structures readily causes heating, which limits the maximum power that can be used for trapping. Local heating also potentially complicates the trapping scheme by introducing thermal related forces including thermophoresis and convection. Garces-Chavez et al. have demonstrated that trapping with optical-induced thermal force.[49] There are also debates on the respective contributions of optical force and thermal related forces in plasmonic tweezers.[50]

Recently, Wang et al. reported trapping of particles with a size of 110 nm using a gold nanopillar with a heat sink.[45] Figure 8.5a shows a schematic of the gold nanopillar sitting on top of a gold film. The nanopillar has a diameter of 280 nm and a height of 130 nm. A copper layer is sandwiched between the gold and silicon substrate to provide good thermal conductivity and thereby circumventing the heating problem. This heat sink design reduced the temperature rise by 100 times; therefore, larger power can be used to trap smaller particles. The gold nanopillar supports a localized surface plasmon resonance mode at the wavelength of 974 nm as shown in Figs. 8.5b,c. Two field maxima located on both sides of the pillar depending on the polarization state of the exciting field. Compared with other plasmonic structures directly sitting on a dielectric substrate, this design supports a mode maxima located near the top of the nanopillar, which is accessible to the nanoparticles. The field intensity in the center of the hotspots has been enhanced by 490 times comparing to the normal incident excitation field. The large field enhancement together with larger optical power made possible by the heat sink potentially enables trapping of nanoparticles smaller than 100 nm: the current platform is mainly limited by the detection capability of nanoparticles.

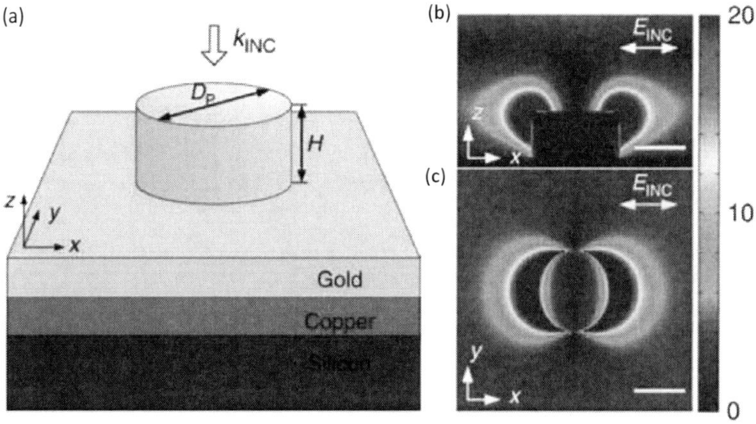

Figure 8.5 (a) Schematic of gold nanopillar with a heat sink underneath. (b,c) Normalized field intensity distribution on the vertical and horizontal cross sections. Image courtesy of ref. 45.

Plasmonic dipole antenna formed by placing two nanostructures close to each other can dramatically enhance the field intensity in the nano-gap. Optical trapping of nanoparticles has been reported by several groups using this type of structures.[46,47,51] As an example, Zhang et al. reported trapping of gold nanoparticles with a radius of 10 nm using a plasmonic dipole antenna with a pair of metal nano-rods separated by a nanometric gap as shown in Fig. 8.6. The nano-gap provides a two-orders-of-magnitude field enhancement compared to a single nano-rod design. The dipole antenna resonance frequency can be tuned by changing the length of the nano-rod. The resonance frequency is also extremely sensitive to perturbation in the nano-gap, a feature used to detect trapped nanoparticles by measuring the scattering spectrum as shown in Fig. 8.6b. White light illumination combined with trapping laser launched from the transparent substrate excites the LSPR mode, and the scattering spectrum is measured using a charge-coupled device (CCD) camera. The trapping events can therefore be detected in real time by monitoring the LSPR peak.

Figures 8.6c,d show a sample measurement of the Rayleigh scattering spectra and the resonance wavelength shift when a 10 nm Au nanoparticle enters and gets trapped in the nano-gap. Each nanoparticle causes a 5 nm resonance shift. The trapping and sensing system eliminates the need for fluorescently labeled nanoparticles and high resolution imaging system.

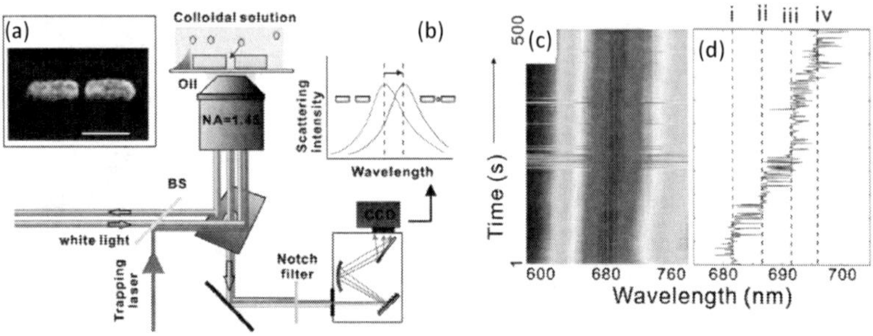

Figure 8.6 (a) SEM image of the gold dipole antenna, (b) optical trapping and sensing setup, (c) Waterfall plot of the Rayleigh scattering spectra, and (d) the peak wavelength shift as a function of the collection time of trapping 10 nm Au nanoparticles using a dipole antenna with a gap of 15 nm. Image courtesy of ref. 47.

The aforementioned plasmonic optical tweezers are still based on gradient force, which is proportional to the polarizability of the particles and therefore scales with the volume of the particles. The field enhancement of the LSPR mode reduces the required incident power but still relies on high field intensity in the hotspot to trap nanoparticles. Self-induced back action optical trapping demonstrated by Juan et al. using nano-hole in a metallic film provides a low power approach to trap nanoparticles.[39] As discussed in Section 8.2.3, the particle to be trapped strongly modifies the LSPR mode and facilitates the trapping. Numerical simulation has indicated that the SIBA-based optical force does not scale with the particle volume, which makes it a compelling platform for single nanoparticle trapping with reduced optical damage.

Trapping of 50 nm polystyrene particles has been demonstrated using an aperture of 310 nm on a gold film with a power of only 1.9 mW based on the SIBA mechanism. Figure 8.7a shows a schematic of the trapping platform. The aperture is fabricated on a 100 nm-thick gold film using focused-ion-beam milling. The gold film is then inserted into a microfluidic chamber facing down. Trapping laser beam above the cut-off frequency of the aperture is injected from the bottom through a high NA objective lens. Numerical simulation based on the Maxwell stress tensor method confirms that the SIBA force is an order of magnitude larger than the perturbative optical force using the conventional gradient force formula (Eq. 8.4). The particle-induced transmitted power change can be used to monitor the trapping events. Since the particle plays an active role in the transmission of light through the aperture, the optical transmission is very sensitive to the size and properties of the particle. Unlike the dipole antenna case, a photodetector is capable of detecting the trapping events.

Later, shaped apertures have been investigated to provide further force enhancement by the same group. A double-hole aperture as shown in Fig. 8.7c has been used to trap 12 nm polystyrene particles,[52] and single protein molecules.[53] The low optical power reduces the risk of thermal and optical damage to the trapped nanoparticles, which makes it an ideal platform for performing the biological analysis such as protein binding demonstrated in ref. 53.

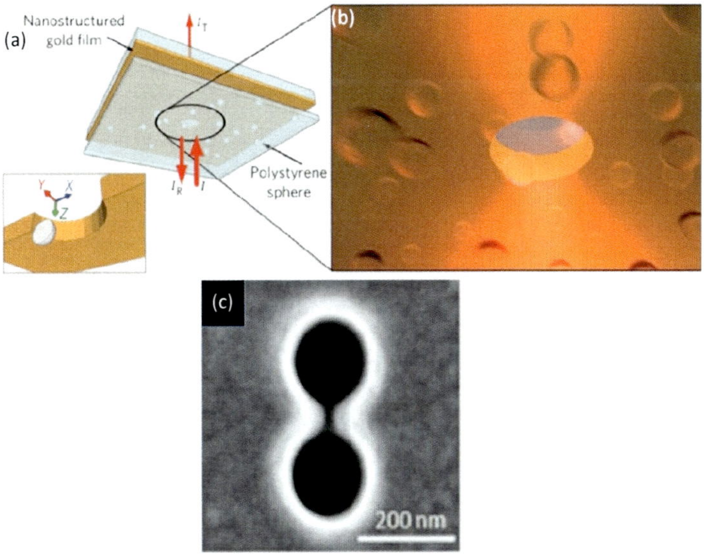

Figure 8.7 (a) Schematic representation of the SIBA optical trapping system using a nanoaperture on a gold film. (b) Schematic illustration of trapping of a polystyrene particle (images courtesy of ref. 39). (c) SEM image of the double-hole aperture (image courtesy of ref. 53).

8.3.3 Photonic Cavity–Based Optical Trapping

Optical trapping of particles using gradient force induced by the tightly localized field on optical waveguides has been reported.[54,55] To trap nanoparticles with sizes below 100 nm, microcavities, which offer large field enhancement, have been extensively investigated as a preferred trapping platform.[56–59] The first experimental demonstration of optical trapping based on photonic microcavities was reported by Arnold and coworkers[56] in 2009. Whispering-gallery mode of a glass microsphere was used to trap particles with sizes down to 140 nm. Light is coupled from a tapered fiber to the glass microsphere and build up the intensity around the equator of the microsphere. Particles coming into the vicinity of the microsphere are attracted to and circulate around the microsphere equator due to the enhanced gradient and scattering force. The cavity-induced optical force enhancement has reduced the minimum trapping power by an order of magnitude comparing to waveguide trapping.

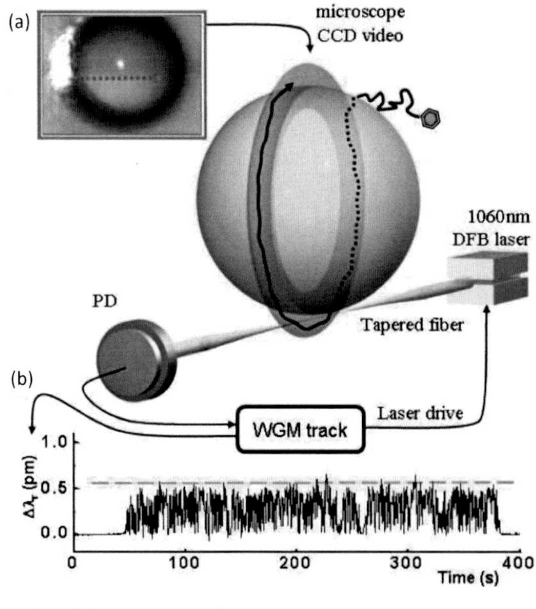

Figure 8.8 (a) Schematic of the optical trapping using a microsphere. Inset: microscope image of the microsphere. (b) Resonance shift as a function of time while one particle trapped and orbiting around the microsphere. Image courtesy of ref. 56.

The trapped particle induces perturbation to the cavity mode and causes resonance shift, which is used to monitor the trapping event and quantify the size of the trapped particles. The paper demonstrated that the same microsphere can successfully trap and detect particles down to 140 nm in size. Trapping smaller nanoparticles like viruses is achievable by increasing the quality factor (Q) of the cavity and/or increasing the optical power.

In the experiment, the glass microsphere possesses a high Q-factor of 10^7. A tapered fiber is used to evanescently couple incident light, which makes integration with other devices or microfluidic channels difficult. After the first demonstration of the cavity-based optical trapping, multiple groups have demonstrated optical trapping using silicon or silicon nitride–based planar micro-cavities, including the microrings,[57,60] microdisks[58] and photonic crystal cavities.[59,61,62] Light can be easily coupled into these planar cavities via a monolithically integrated bus waveguide fabricated using mature semiconductor processes. The configuration is also amenable to integration with microfluidic channels for analyte delivery. These features enable advanced on-chip optical micromanipulation systems for particle trapping, delivery, detecting, and sorting.[63,64]

Among the silicon-based microcavities, microrings and microdisks support whispering gallery modes similar to those of microspheres. As mentioned above, the field enhancement in the cavity is proportional to Q/V, where V is the mode volume. Compared to glass microspheres, silicon-based planar cavities can maintain smaller radius due to their high refractive index contrast and therefore reduce its mode volume. However, Q-factors of planar cavities are generally lower than microspheres due to side wall roughness scattering. Limited force enhancement as well as the presence of large scattering force that propels particles along the trapping track limits the capability of precise trapping and positioning nanoparticles smaller than 100 nm using whispering gallery microcavities.

Photonic crystal cavities, on the other hand, can achieve small mode volume and high Q simultaneously. The mode volume of a photonic crystal cavity is usually in the order of $\sim(\lambda/n)^3$. With appropriate designs, a Q-factor of up to 10^7 is attainable.[65,66] Therefore, photonic crystal cavity in principle can provide much larger field enhancement comparing to whispering gallery cavities, which enables trapping of nanoparticles with sizes down to the single molecule level (<10 nm).

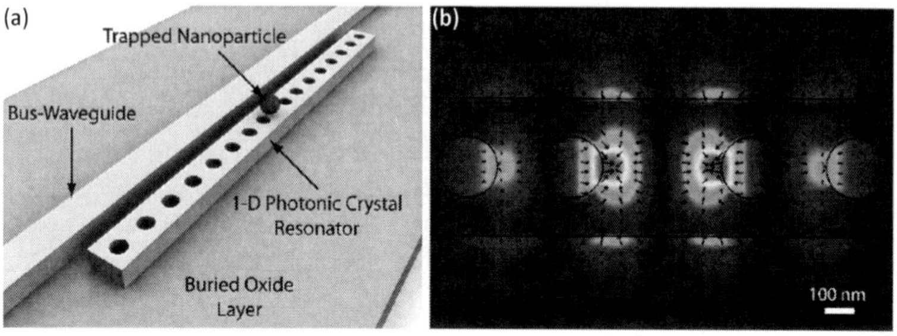

Figure 8.9 (a) Schematic diagram of optical trapping using 1D photonic crystal cavity. (b) field distribution near the cavity center. Image courtesy of ref. 59.

Mandal et al. first reported the trapping of nanoparticles with a size of 48 nm using a one dimensional silicon photonic crystal cavity as shown in Fig. 8.9. Periodic holes are etched into a silicon channel waveguide with a defect in the middle by increasing the gap between the two holes in the center. A straight bus waveguide is used to couple light to the cavity. The simulated field distribution near the center of the microcavity in Fig. 8.9b shows that the optical field is tightly confined in the center of the cavity. Optical gradient forces calculated using the dipole approximation are indicated by the black arrows pointing toward the field maxima. Similarly, field gradient in the out-of-plane direction exerts enhanced gradient force on the nanoparticles and traps them onto the surface of the cavity. The trapping stiffness of a 200 nm particle is estimated to be 2.86 pm/W/nm from the displacement of a trapped particle recorded by microscope, several orders of magnitude larger than conventional far field optical tweezers. In addition to the large force enhancement, photonic crystal cavities also trap and precisely position the trapped particle in the field maxima due to its standing wave nature.

Another way to enhance the optical field of a photonic structure is to leverage the slot effect.[29] Yang et al. demonstrated optical trapping and delivery of 75 nm nanoparticles and λ-DNA molecules using a silicon slot waveguide.[55] Such a slot waveguide with a low index region sandwiched by high index materials possesses large electrical field discontinuity and support a quasi-

TE mode with large field enhancement in the sub-wavelength slot region. Similarly, slot effect has also been incorporated in the recently demonstrated photonic crystal cavity to further enhance the optical force. We first proposed a nanoslotted 1D photonic crystal cavity by integrating a nano-slot in the center of a 1D photonic crystal cavity. Numerical simulation based on 3D FDTD shows that in addition to the 250 times field intensity enhancement from the photonic crystal cavity with a Q of around 1800, the nanoslot provides an additional fivefold force enhancement. Chen et al. later reported an experimental demonstration based on a similar structure on a silicon nitride platform.[62] Figure 8.10a shows a top view SEM image of the cavity fabricated on a Si_3N_4 waveguide on top of SiO_2 substrate. The simulated field intensity distribution in Fig. 8.10b indicates the strong field enhancement inside the nanoslot region. The cavity with a Q of ~5,000 and a mode volume of 4.4 $(\lambda/n)^3$ can successfully trap 22 nm polystyrene beads with an incident power of 11 mW. Trapping of streptavidin-coated CdSe/ZnS quantum dots with diameter of 10–15 nm and Wilson disease proteins have also been achieved using the same structure.

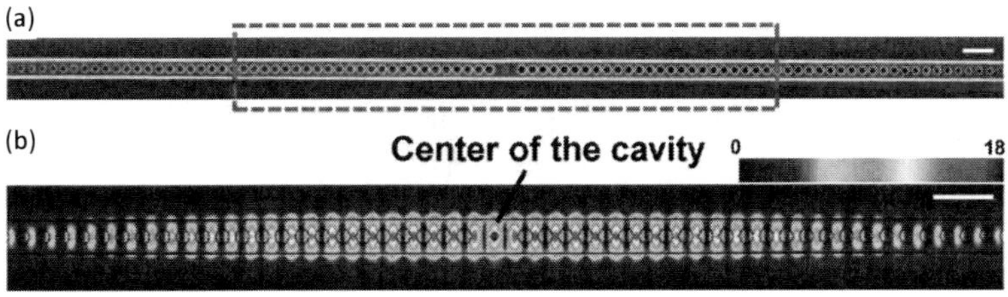

Figure 8.10 (a) Top view SEM image of the nanoslotted 1D photonic crystal cavity. (b) Electrical field intensity distribution of the 1D photonic crystal cavity by 3D FDTD simulation. Image courtesy of ref. 62.

Comparing to the plasmonic structures, dielectric photonic crystal–based devices have low optical absorption, and therefore no significant heating was observed. Chen et al. reported a temperature increase of 0.3 K at 11 mW incident power. Note that wavelength of 1064 nm is used in this demonstration which has low absorption in water. Si_3N_4 has been chosen as the material platform given its low absorption in this wavelength range. The minimal temperature increase makes it suitable for trapping objects that are temperature sensitive such as biological specimens.

In sum, single nanoparticle trapping based on both localized surface plasmon resonator and photonic crystal cavity have been demonstrated. The plasmonic tweezers usually have broader working wavelength range and the LSPR mode can be easily excited from free space using a laser beam. The simplicity of the system make it ideal for quick and disposable trapping and analysis applications. However, careful design with good thermal management is required to prevent the thermal related forces and thermal damage to the trapped objects.

In comparison, photonic tweezers have low optical absorption and the issue of heating is thus mitigated. However, they usually involved coupling light into a sub-wavelength waveguide using an optical fiber, which requires high precision alignment systems. Meanwhile, high-Q cavities are required for trapping nanoparticles, which means only a very narrow wavelength range near the resonance condition can be used. The resonance frequency of the cavity is sensitive to temperature and refractive index of the surrounding liquid. Even the trapped particles themselves can shift the resonance frequency due to cavity perturbation. For practical applications, stabilization of the environmental conditions or dynamical locking the incident wavelength to the resonance is likely mandated.

Particle induced resonance shift may be used to monitor the trapping events on both platforms. However, similar to the optical force, the resonance shift roughly scales with the third power of the particle radius. Detecting the resonance shift for small nanoparticles is difficult especially in a fluidic environment, where the observed resonance drift can be easily dominated by the change of local temperature or refractive index of the surrounding medium. Stabilization of the environmental conditions including careful thermal management and fluid regulation will be critical for nanoparticle optical trapping and sensing applications.

8.4 Concluding Remarks

While optical tweezers have been a well-established technology, extending the trapping capabilities of optical tweezers to the nanoparticle regime, however, is not a trivial challenge. The high optical power required as well as the ensuing optical damage present a major technical barrier to nanoparticle trapping. Novel designs based on plasmonic or photonic nanostructures have shown competitive advantages over traditional free-space optical tweezers, and are promising platforms to overcome the optical power requirement taking advantage of the large optical force enhancement due to strong optical confinement. New trapping schemes, in particular the SIBA mechanism, may further mitigate the optical damage issue. The transition to these nanophotonic or nanoplasmonic trapping configurations also requires revisiting the classical dipole approximation for force calculation, as perturbative effects by the trapped particles become pronounced at the small length scale. The perturbative effects can also provide an effective method for detecting trapped particles and quantifying their optical properties, as demonstrated by several recent reports. We anticipate that the continuing growth of this dynamic field will provide ample opportunities for scientific investigation and technology development spanning the areas of photonics, nanotechnology, and bioengineering.

References and Further Reading

1. A. Ashkin (1970). Acceleration and trapping of particles by radiation pressure. *Phys. Rev. Lett.,* **24**(4), 156.
2. A. Ashkin, et al. (1986). Observation of a single-beam gradient force optical trap for dielectric particles. *Opt. Lett.,* **11**(5), 288–290.
3. A. Ashkin (2000). History of optical trapping and manipulation of small-neutral particle, atoms, and molecules. *Selected Top. Quantum Electron. IEEE J.,* **6**(6), 841–856.
4. D. McGloin (2006). Optical tweezers: 20 years on. *Philos. Trans. R. Soc. A: Math. Phys. Eng. Sci.,* **364**(1849), 3521–3537.
5. K. C. (2004). Neuman and S. M. Block, Optical trapping. *Rev. Sci. Instrum.,* **75**(9), 2787–2809.
6. J. E. Molloy and M. J. Padgett. Lights (2002). Action: Optical tweezers. *Contemp. Phys.,* **43**(4), 241–258.
7. J. R. Moffitt, et al. (2008). Recent advances in optical tweezers. *Biochemistry,* **77**(1), 205.
8. K. C. Neuman and A. Nagy (2008). Single-molecule force spectroscopy: Optical tweezers, magnetic tweezers and atomic force microscopy. *Nat. Methods,* **5**(6), 491.
9. K. Svoboda and S. M. Block (1994). Biological applications of optical forces. *Ann. Rev. Biophys. Biomol. Struct.,* **23**(1), 247–285.

10. P. C. Ashok and K. Dholakia (2012). Optical trapping for analytical biotechnology. *Current Opin. Biotechnol.*, **23**(1), 16–21.
11. H. Zhang and K.-K. Liu (2008). Optical tweezers for single cells. *J. R. Soc. Interface*, **5**(24), 671–690.
12. K. Dholakia and T. Čižmár (2011). Shaping the future of manipulation. *Nat. Photon.*, **5**(6), 335–342.
13. L. Tong, R. Kerstin and K. Mikael, Optical tweezers for Raman spectroscopy. *Raman Spectroscopy for Nanomaterials Characterization*. Springer Berlin Heidelberg, 2012. 507–530.
14. See, for example, J. Vanderlinde. *Classical Electromagnetic Theory*. Vol. 145, Section 8.2.2. Springer, 2006.
15. The readers should be careful with the definition of polarizability α. According to our convention $P = \varepsilon_2 \alpha E$, and thus polarizability has the SI unit of cubic meter (volume).
16. L. Novotny and B. Hecht. *Principles of Nano-Optics*. Cambridge university press, 2012.
17. P. C. Chaumet and M. Nieto-Vesperinas (2000). Time-averaged total force on a dipolar sphere in an electromagnetic field. *Opt. Lett.*, **25**(15), 1065–1067.
18. S. Gaugiran, et al. (2005). Optical manipulation of microparticles and cells on silicon nitride waveguides. *Opt. Express*, **13**(18), 6956–6963.
19. M. Meier and A. Wokaun (1983). Enhanced fields on large metal particles: Dynamic depolarization. *Opt. Lett.*, **8**(11), 581–583.
20. C. F. Bohren and D. R. Huffman (1983). Absorption and scattering by a sphere. *Absorption Scattering of Light Small Particles*, 82–129.
21. E. M. Purcell and C. R. Pennypacker (1973). Scattering and absorption of light by nonspherical dielectric grains. *Astrophys. J.*, 186, 705–714.
22. B. T. Draine and P. J. Flatau (1994). Discrete-dipole approximation for scattering calculations. *JOSA A*, 11(4), 1491–1499.
23. C. E. Dungey and C. F. Bohren (1991). Light scattering by nonspherical particles: A refinement to the coupled-dipole method. *JOSA A*, **8**(1), 81–87.
24. M. A. Yurkin and A. G. Hoekstra (2007). The discrete dipole approximation: An overview and recent developments. *J. Quant. Spectrosc. Radiative Transfer*, **106**(1), 558–589.
25. P. C. Chaumet, et al. (2005). Efficient computation of optical forces with the coupled dipole method. *Phys. Rev. E*, **72**(4), 046708.
26. A. G. Hoekstra, et al. (2001). Radiation forces in the discrete-dipole approximation. *JOSA A*, **18**(8), 1944–1953.

27. S. H. Simpson and S. Hanna (2011). Application of the discrete dipole approximation to optical trapping calculations of inhomogeneous and anisotropic particles. *Opt. Express,* **19**(17), 16526–16541.

28. A compiled list of DDA computation codes in the public domain: http://en.wikipedia.org/wiki/Discrete_dipole_approximation_codes, retrieved on 08/10/2014.

29. T. A. Nieminen, et al. (2007). Optical tweezers computational toolbox. *J. Opt. A: Pure Appl. Opt.,* **9**(8), S196.

30. T. A. Nieminen, et al. (2001). Numerical modelling of optical trapping. *Comput. Phys. Commun.,* **142**(1), 468–471.

31. G. Gouesbet and J. A. Lock (2013). List of problems for future research in generalized Lorenz–Mie theories and related topics, review and prospectus [Invited]. *Appl. Opt.,* **52**(5), 897–916.

32. W. L. Moreira, et al. (2010). Expansion of arbitrary electromagnetic fields in terms of vector spherical wave functions. *arXiv preprint arXiv,* **1003**(2392).

33. A. H. J. Yang, T. Lerdsuchatawanich, and D. Erickson (2009). Forces and transport velocities for a particle in a slot waveguide. *Nano Lett.,* **9**(3), 1182–1188.

34. P. B. Johnson and R. W. Christy (1972). Optical constants of the noble metals. *Phys. Rev. B,* **6**(12), 4370.

35. M. L. Juan, M. Righini, and R. Quidant (2011). Plasmon nano-optical tweezers. *Nat. Photon.,* **5**(6), 349–356.

36. K. J. Vahala (2003). Optical microcavities. *Nature,* **424** (6950), 839–846.

37. J. Hu, et al. (2010). Optical trapping of dielectric nanoparticles in resonant cavities. *Phys. Rev. A,* **82**(5), 053819.

38. M. Barth and O. Benson (2006). Manipulation of dielectric particles using photonic crystal cavities. *Appl. Phys. Lett.,* **89**(25), 253114.

39. M. L. Juan, et al. (2009). Self-induced back-action optical trapping of dielectric nanoparticles. *Nat. Phys.,* **5**(12), 915–919.

40. G. Volpe, et al. (2006). Surface plasmon radiation forces. *Phys. Rev. Lett.,* **96**(23), 238101.

41. K. Wang, E. Schonbrun, and K. B. Crozier (2009). Propulsion of gold nanoparticles with surface plasmon polaritons: Evidence of enhanced optical force from near-field coupling between gold particle and gold film. *Nano Lett.,* **9**(7), 2623–2629.

42. K. Wang, et al. (2010). Scannable plasmonic trapping using a gold stripe. *Nano Lett.,* **10**(9), 3506–3511.

43. M. Righini, et al. (2007). Parallel and selective trapping in a patterned plasmonic landscape. *Nat. Phys.*, **3**(7), 477–480.

44. K. Wang and K. B. Crozier (2012). Plasmonic trapping with a gold nanopilla. *Chem. Phys. Chem.*, **13**(11), 2639–2648.

45. K. Wang, et al. (2011). Trapping and rotating nanoparticles using a plasmonic nano-tweezer with an integrated heat sink. *Nat. Commun.*, **2**, 469.

46. M. Righini, et al. (2009). Nano-optical trapping of Rayleigh particles and Escherichia coli bacteria with resonant optical antennas. *Nano Lett.*, **9**(10), 3387–3391.

47. W. Zhang, et al. (2010). Trapping and sensing 10 nm metal nanoparticles using plasmonic dipole antennas. *Nano Lett.*, **10**(3), 1006–1011.

48. B. J. Roxworthy, et al. (2012). Application of plasmonic bowtie nanoantenna arrays for optical trapping, stacking, and sorting. *Nano Lett.*, **12**(2), 796–801.

49. V. Garcés-Chávez, et al. (2006). Extended organization of colloidal microparticles by surface plasmon polariton excitation. *Phys. Rev. B*, **73**(8), 085417.

50. M. Ploschner, et al. (2010). Optical forces near a nanoantenna. *J. Nanophoton.*, **4**(1), 041570–041570.

51. A. N. Grigorenko, et al. (2008). Nanometric optical tweezers based on nanostructured substrates. *Nat. Photon.*, **2**(6), 365–370.

52. Y. Pang and R. Gordon (2011). Optical trapping of 12 nm dielectric spheres using double-nanoholes in a gold film. *Nano Lett.*, **11**(9), 3763–3767.

53. A. A. Al Balushi, A. Zehtabi-Oskuie, and R. Gordon (2013). Observing single protein binding by optical transmission through a double nanohole aperture in a metal film. *Biomed. Opt. Express*, **4**(9), 1504–511.

54. S. Kawata, and T. Tani (1996). Optically driven Mie particles in an evanescent field along a channeled waveguide. *Opt. Lett.*, **21**(21), 1768–1770.

55. A. H. J. Yang, et al. (2009). Optical manipulation of nanoparticles and biomolecules in sub-wavelength slot waveguides. *Nature*, 457(7225), 71–75.

56. S. Arnold, et al. (2009). Whispering gallery mode carousel—a photonic mechanism for enhanced nanoparticle detection in biosensing. *Opt. Express*, 17(8), 6230–6238.

57. S. Lin, E. Schonbrun, and K. Crozier (2010). Optical manipulation with planar silicon microring resonators. *Nano Lett.,* **10**(7), 2408–2411.

58. H. Cai, and A. W. Poon (2011). Optical manipulation of microparticles using whispering-gallery modes in a silicon nitride microdisk resonator. *Opt. Lett.,* **36**(21), 4257–4259.

59. S. Mandal, X. Serey, and D. Erickson (2009). Nanomanipulation using silicon photonic crystal resonators. *Nano Lett.,* **10**(1), 99–104.

60. A. H. J. Yang and D. Erickson (2010). Optofluidic ring resonator switch for optical particle transport. *Lab on a Chip,* **10**(6), 769–774.

61. S. Lin, et al. (2009). Design of nanoslotted photonic crystal waveguide cavities for single nanoparticle trapping and detection. *Opt. Lett.,* **34**(21), 3451–3453.

62. Y.-F. Chen, et al. (2012). Controlled photonic manipulation of proteins and other nanomaterials. *Nano Lett.,* **12**(3), 1633–1637.

63. K. Grujic, et al. (2005). Sorting of polystyrene microspheres using a Y-branched optical waveguide. *Opt. Express,* **13**(1), 1–7.

64. S. Lin and K. B. Crozier (2011). Planar silicon microrings as wavelength-multiplexed optical traps for storing and sensing particles. *Lab on a Chip,* **11**(23), 4047–4051.

65. P. B. Deotare, et al. (2009). High quality factor photonic crystal nanobeam cavities. *Appl. Phys. Lett.,* **94**(12), 121106.

66. Y. Akahane, et al. (2003). High-*Q* photonic nanocavity in a two-dimensional photonic crystal. *Nature,* **425**(6961), 944–947.

67. V. R. Almeida, et al. (2004). Guiding and confining light in void nanostructure. *Opt. Lett.,* **29**(11), 1209–1211.

Chapter 9

Rainbow Trapping Effect in Horizontal and Vertical Directions

Qiaoqiang Gan, Dengxin Ji, Haifeng Hu, and Xie Zeng

Electrical Engineering Department, University at Buffalo, 230M Davis Hall, Buffalo, NY 14260, USA

qqgan@buffalo.edu

9.1 Introduction

Optics has inherent physical limitations (e.g., velocity and diffraction limit) that make manipulation of light difficult. There is a pressing need to find new mechanisms that would improve our ability to manipulate light. Designs for slowing light employing ultracold atom clouds,[1] photonic crystals,[2] and optical fibers[3] have been reported recently and generated considerable attention. However, physically realizing low values of the group velocity in solid-state materials at room temperature, especially slow light on a chip, remains a great challenge.[4]

It is known that light can be slowed down in the vicinity of resonances in dispersive materials.[5] In order to reduce the group velocity (v_g) of light coherently, there are primarily two major approaches employing either electronic or optical resonances. In the electronic scheme, drastic slowing down and complete

Integrated Nanophotonic Resonators: Fundamentals, Devices, and Applications
Edited by Ya Sha Yi

ISBN 978-981-4613-78-1 (Hardcover), 978-981-4613-79-8 (eBook)
www.panstanford.com

stopping of light pulses can be accomplished by converting optical signals into electronic coherences.[6–9] The use of electronic states to coherently store the optical information, however, imposes severe constraints in the scheme, including narrow bandwidth, limited working wavelengths and strong temperature dependence. While promising steps have been taken towards slowing light in solid-state media and semiconductor nanostructures operating at room temperature,[10] it still remains a great challenge to implement such schemes on-a-chip incorporating optoelectronic devices.[9] Besides, only a few special and delicate electronic resonances available in nature possess the required properties. As a result, there has been great interest in pursuing alternative approaches utilizing optical resonances in photonic structures, such as microcavities,[11] photonic crystals,[12] semiconductor waveguide ring resonators,[13] etc. In 2007, researchers in UK proposed a tapered structure based on metamaterial with negative refractive index.[14] Theoretically, this structure should be able to slow light to a standstill. The material would halt each frequency of light, and hence color, at slightly different places, to make so called "trapped rainbow" (see Fig. 9.1), which opened a new and attractive approach to stopping light.

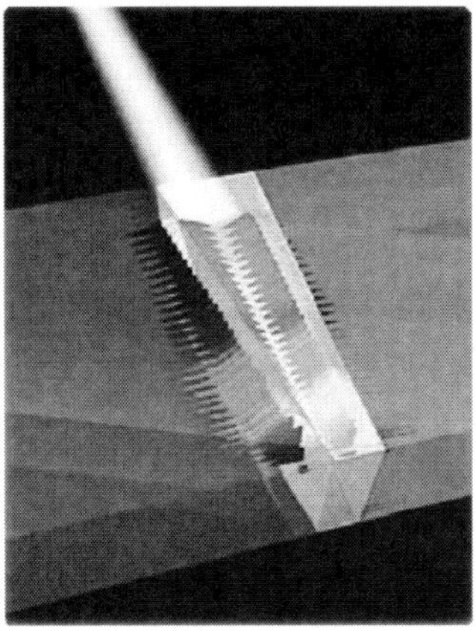

Figure 9.1 A schematic of the trapped "rainbow" storage of light in a tapered metamaterial waveguide. Reproduced from *Nature*, **450**, 330 (2007).

In this chapter, two different types of rainbow trapping structures will be discussed. The first type is based on surface plasmonic structure, which can trap surface wave on the horizontal plane. The other one is based on structured hyperbolic metamaterial (HMM) film, which is able to realize the trapped "rainbow" effect along the vertical direction.

9.2 Rainbow Trapping Effect along Horizontal Directions

Plasmonic structures offer advantages for on-chip integration of optical circuits, data storage and compact biosensing, etc.[15] What makes plasmonic structures particularly interesting is their potential for spatial confinement of electromagnetic (EM) energy within subwavelength dimensions over a wide spectral range.[16,17] Moreover, their abilities to support surface plasmon polaritons (SPP) make them an attractive candidate for slowing light in a solid-state system.[18] In this section, we study a graded surface grating plasmonic system, which is able to slow multiple wavelengths simultaneously at arbitrary temperatures (including room temperature). Compared with traditional slow light schemes, such surface plasmonic structures constitute a nanophotonic platform capable of simultaneous operation over a wide range of wavelengths. Moreover, it is not limited in its working temperature, which makes it particularly suited for practical applications.

9.2.1 Ultra-Wideband Surface Wave Trapping at THz Frequencies

9.2.1.1 Confining THz waves at 1D metallic gratings with a constant depth

For simplicity, we first analyze the dispersion curves of the metallic surface with a constant grating depth, which consists of a 1D groove array engraved in the metal surface, with a depth h, width d, and lattice constant p. In most previous studies of SPPs in metallic structures for THz and GHz domains, the metals were treated as perfect electrical conductors (PECs) due to the fact that the depth of the wave penetration into the metal is much shorter than the corresponding wavelength.[16,18] Under this assumption, the dispersion relations for TM-polarized (E_x, E_z, and H_y) waves propagating in the x axis can be obtained to a first-order approximation from the expression[19]

$$k_x = \frac{\omega}{c}\sqrt{\frac{d^2}{p^2}\tan^2\left(\frac{\omega}{c}h\right)+1}, \tag{9.1}$$

where c is the light velocity in vacuum. The dispersion relations for the 1D groove arrays with various parameters (h, d, p) can be obtained by solving Eq. (9.1). Using Eq. (9.1), we can optimize the structures by properly choosing the values for h, d, and p that can support THz surface modes. For such cases, most of the energy carried by each mode is guided above the metallic surface. One can see that the dispersion curves significantly deviate from the light line, indicating that the metallic grating has dramatically modified the dispersions of the THz wave. In addition, for each groove depth the corresponding curve asymptotically approaches a specific cutoff frequency above which the SPPs can be no longer supported by the metallic grating. In addition, for the fixed values of d and p, the larger is h, the lower is the cutoff frequency. It should be noted that by introducing an eigenvalue equation in the corrugated conducting plane of a periodic system and taking higher-order scattering components into account, the dispersion curve can also be plotted, which is in a good agreement with the curves shown in Fig. 9.2. One can refer to ref. [21] for the details of the numerical procedures of this approach.

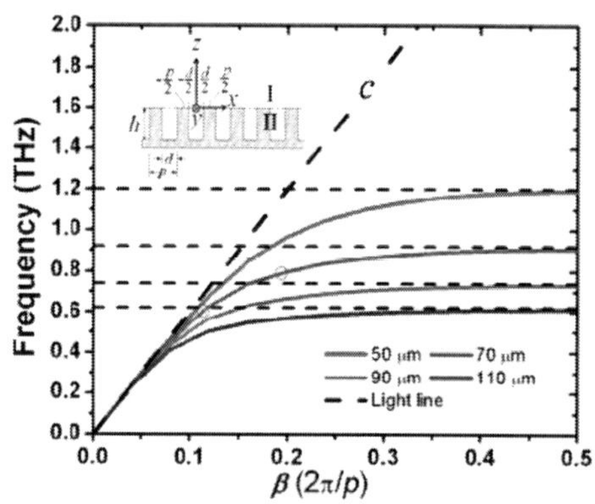

Figure 9.2 Solid curves—dispersion curves calculated for d = 20 μm and p = 50 μm with different depths of the grooves. Horizontal dashed lines—corresponding cutoff frequencies. Dashed line labeled by c—light line. Inset: structure of 1D metallic grating having a constant depth of grooves (copyright info @ IEEE).[21]

Now, we investigate how this 1D metallic grating can be used to spatially confine and guide THz waves. Based on the surface dispersion properties of SPP modes, the larger k_x is (corresponding to β), the stronger is the confinement of the EM wave by the structure.[22] As illustrated in Fig. 9.2, for the grating with h = 70 μm, the cutoff frequency is approximately 0.9 THz. The value of k_x for the SPP mode at 0.8 THz is larger than that at 0.6 THz (see the two circles in the dispersive relation for h = 70 μm). Consequently, the confinement for the 0.8-THz wave is stronger than that for the 0.6-THz wave.

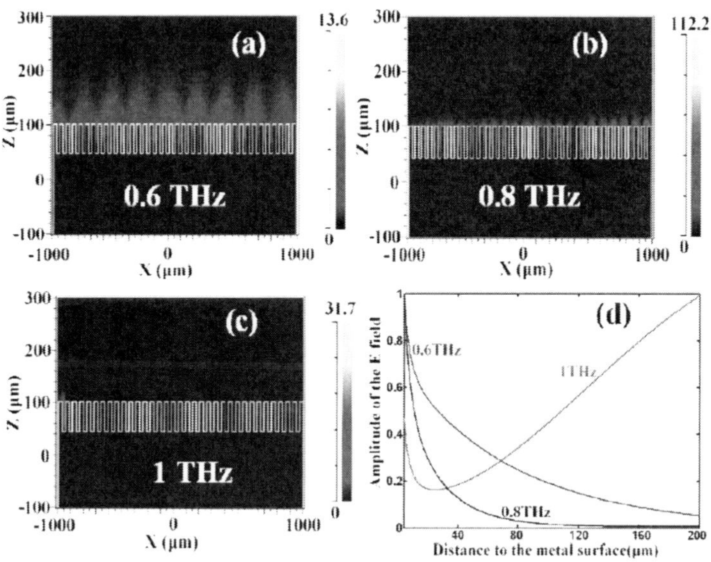

Figure 9.3 (a), (b), and (c) correspond to the 2D distributions of the field intensities ($|E|^2$) at 0.6 THz, 0.8 THz, and 1 THz, respectively, obtained by using 2D FDTD simulations. While conducting these simulations, the thickness of the metal is set to 400 μm and the period, width, and depth of the grating are 50, 20 and 70 μm, respectively. (d) Amplitude of the electric field vs. distance from the metallic surface at different incoming frequencies obtained from (a), (b), and (c) (copyright info @ IEEE).[21]

To support the results obtained above, we investigate the THz SPPs using a 2D FDTD simulation tool.[23] We set the period, width, and depth of the periodic grating to 50, 20 and 70 μm, respectively. As one can see from Figs. 9.3a–c, the THz SPPs are strongly confined within a subwavelength region near the surface of the metallic grating for the 0.6-THz and 0.8-THz waves (Figs. 9.3a,b). However, there are no SPP waves propagating along the surface at 1 THz (Fig. 9.3c), since this frequency is already beyond the cutoff frequency of 0.9 THz. In addition, according to Figs. 9.3a,b, the confinement of the SPPs is much stronger at 0.8 THz than that at 0.6 THz. Indeed, as shown in Fig. 9.3d, the confinement depth of the THz wave, defined by the distance to the metal surface for which the *E* field is reduced to a half, is 7.5 μm at 0.8 THz vs. 24.5 μm at 0.6 THz. The confinement depth decreases by a factor of 3.3 when the frequency is increased from 0.6 THz to 0.8 THz. These results illustrate that a THz wave with a certain frequency can be effectively guided by a uniform surface grating, and that the confinement depth can be significantly smaller than the wavelength of the THz wave. On the other hand, the 1-THz wave is scattered to the far field.

9.2.1.2 Metallic surface grating having graded depth

(A) *Dispersion properties of the graded gratings*

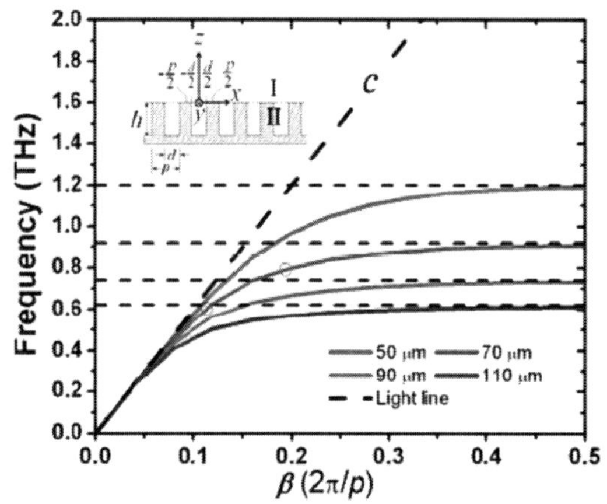

The dispersion curves in Fig. 9.2 and the FDTD simulation results in Fig. 9.3 provide clues to the confinement and waveguiding of THz waves by metallic grating surfaces when the groove depth is properly chosen. The dispersion curves also indicate that the group velocity of the guided SPP mode, which is given by the slope of the tangent line at the point on the dispersion curve, decreases as the frequency of the input wave is increased.[14,24] In particular, as the SPP frequency approaches the cutoff frequency, the group velocity approaches zero.

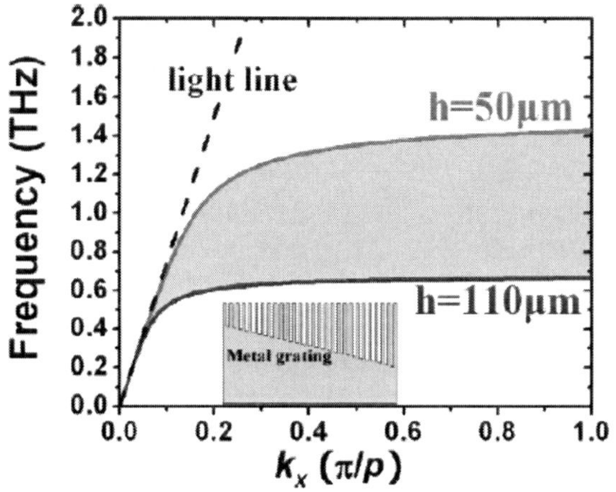

Figure 9.4 Dispersion curves calculated for d = 20 μm and p = 50 μm with different groove depths (h = 50 and 110 μm). Inset: Schematic illustration of the graded grating structure (copyright info @ APS).[25]

Now in this section, we consider a graded grating as shown by inset in Fig. 9.4, where the depth of the grating is now graded linearly from left to right, and the corresponding dispersion curves differ from those for a constant grating depth (see Fig. 9.2). For example, considering that the depths of the grating in the left-hand side and the right-hand side are 50 and 110 μm, respectively, the dispersion curves between the two ends now lie in the gray region between the two lines in Fig. 9.4. Obviously, the cutoff frequency of the structure increases from 0.6 to 1.4 THz approximately. The v_g for the surface waves may be derived from Eq. (9.1) and expressed as

$$v_g = \frac{d\omega}{dk_x} = \frac{c}{A + \dfrac{2d^2 h\omega}{p^2 cA}\tan\left(\dfrac{\omega}{c}h\right)\sec^2\dfrac{\omega}{c}h}, \tag{9.2}$$

where $A=\sqrt{\frac{d^2}{p^2}\tan^2\left(\frac{\omega}{x}h\right)+1}$. From Eq. (9.2), the group index for the SPP modes can be calculated to be $\frac{c}{v_g}=\frac{c}{d\omega/dk_x}=A+\frac{2d^2h\omega}{p^2cA}\tan\left(\frac{\omega}{c}h\right)\sec^2\frac{\omega}{c}h$. One can see from Fig. 9.5 that the values of v_g at the frequencies of 0.6, 0.7, 0.8, and 0.9 THz are reduced significantly from the speed of light in vacuum. If the grading of these grating depths is sufficiently small, the v_g at these four frequencies can in principle be reduced significantly. This implies that the EM waves are "trapped" at the different locations along the surface grating, depending on the frequencies of the incoming EM waves. A broadband EM wave within a frequency band of 0.6–1.4 THz can be dramatically slowed down in a single graded grating depth structure, showing the ultra-wideband capability of this approach. This structure has clear advantages compared to approaches based on the cold or warm atomic gases or semiconductor nanostructures and devices.[2,3]

(B) *Trapped "rainbow" storage of THz light on the graded gratings*[25]

It is worth noting that the different frequencies are localized at various positions corresponding to different grating depths as shown in Fig. 9.5. For instance, the EM wave at the frequency of 0.9 THz is almost "stopped" at the location having the depth of approximately 83.335 μm, whereas that at 0.6 THz is localized at the grating depth of 125.000 μm. The process of stopping light for the graded grating depths was simulated using a two-dimensional (2D) FDTD model. Simulation settings are similar to those for Fig. 9.3. The period and width of the grating structure are 50 and 20 μm, respectively. In Fig. 9.6a, the depth of the grating structure linearly changes from 50 μm to 150 μm. The total length of the grating structure is approximately 2 mm (t_{pass} = 2 mm/c), and the calculation time is set to $T = 7.5t_{pass}$ = 15 mm/c. An EM wave propagating in free space should be able to propagate from the left-hand side to the right-hand side of the entire simulation region within the calculation time.

2D FDTD simulation results indicate that different frequencies within the range of 0.6–0.9 THz are localized at different locations along the surface after a calculation time, $T_1 = 7.5t_{pass}$ (see Figs. 9.6a–d). When the calculation time is longer, e.g., $T_2 = 15t_{pass}$,

$50t_{pass}$, and $100t_{pass}$, the locations of these frequencies more or less stay unchanged (data not shown here), confirming that the values of v_g of the SPP modes have been reduced drastically. These characteristics are also reflected in the 1D steady-state wave intensity at a location of 2 μm above the metal surface (see Fig. 9.6e). These results clearly show that the graded grating depths can be used to slow down the EM wave packet and even stop it at different locations on the surface depending on the frequency of the incoming EM wave. Such a structure could be used to disperse different frequency components of the incoming EM pulse, i.e., to function as a spectrometer. If our surface grating structure is incorporated into a chip, the wavelength or frequency of the EM wave can be readily analyzed by monitoring the localized evanescent waves. If different waves at different locations can be coupled out for THz communications, a novel plasmonic wavelength-division multiplexing is also realizable. Interestingly, if the grade of the grating depths is changed, the separations between the adjacent frequencies can be tuned freely. For example, if the depth of the grating structure changes linearly from 50 to 130 μm from left to right across the structure, the peak locations of the different frequencies will be expanded horizontally (as shown in Fig. 9.6f).

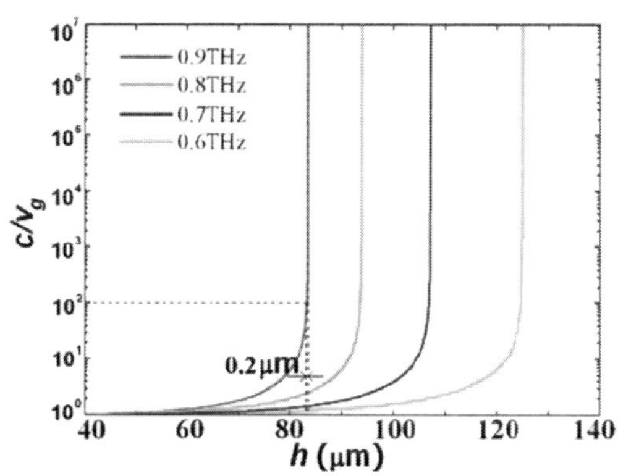

Figure 9.5 Reciprocal of the v_g of the SPP modes are calculated according to the dispersion curves in the gray region in Fig. 9.4. One can see that the v_g of the SPP modes is sensitive to the depth of the gratings: A depth change about 0.2 μm will cause the v_g to change from 10^7c to 10^2c at 0.9 THz (copyright info @ APS).[25]

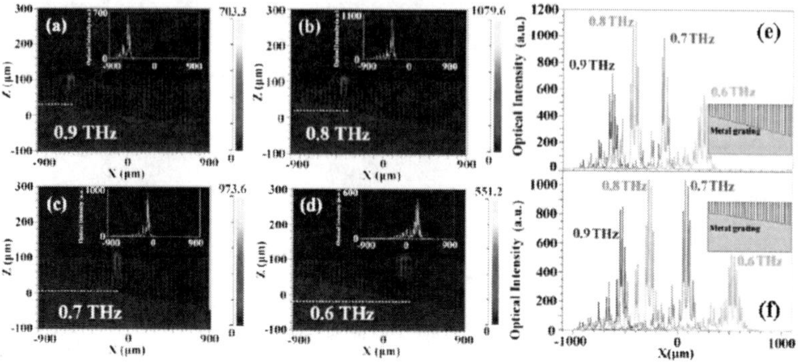

Figure 9.6 Results obtained by using 2D FDTD simulations: (a–d) corresponds to the 2D field distribution at four different frequencies of 0.6, 0.7, 0.8, and 0.9 THz. (e) and (f) represent 1D optical intensity ($|E|^2$) distribution 2 μm above the structured metal surface. The depth of the grating structure in (e) and (f) changes from 50 to 150 μm linearly [the inset in (e)] and from 50 to 130 μm linearly [the inset in (f)] from the left-side end to the right-side end, respectively. In these simulations, the period and width of the grating structure are 50 and 20 μm, respectively. The simulation time T = 15000 μm/c, where c is the light velocity in vacuum (copyright info @ APS).[25]

9.2.2 Rainbow Trapping and Releasing in Telecommunication Domain

After the reports on "trapped rainbow" storage of THz light in metamaterials[14] and plasmonic graded structures,[25] a debate arose as to whether light can be "stopped" in realistic metamaterials[26,27] when loss is properly considered in the modeling. In addition, since plasmonic structures and devices operating in the visible and near infrared region have the advantage for merging photonics and electronics within nanoscale dimensions,[15] it is of value to incorporate the dispersion properties of the metal and scale the operating frequencies of these structures from THz to the telecom domain, or even to visible frequencies to trap a rainbow visible to the human eye.[28,29] In this section, we discuss the investigations on "rainbow" trapping at telecommunication wavelengths.

9.2.2.1 Dispersion properties of metallic nanogratings at telecommunication frequencies

Scaling laws[30] indicate how the cutoff frequency in the dispersion relations of periodic gratings can easily be scaled to other frequency regimes. In this Section, we explore decreasing the feature sizes of metal gratings (see the inset in Fig. 9.7) to nanoscale dimensions to extend operational frequencies to the visible and infrared domains, and examine the dispersion properties of the metal nanostructures in detail. We employ the complex refractive indices of the metal, in this case silver,[31] and determine the dispersion relations of the gratings by means of the FDTD simulations.[23] As can be shown by the results in Fig. 9.7, the cutoff frequencies of the silver gratings with structure 1 (h = 230 nm, d = 200 nm, p = 400 nm) and structure 2 (h = 140 nm, d = 200 nm, p = 400 nm) are now both in the telecom domain [1.7808×10^{14} Hz (λ = 1.6846 μm) for structure 1 and 2.26637×10^{14} Hz (λ = 1.3237 μm) for structure 2, respectively].

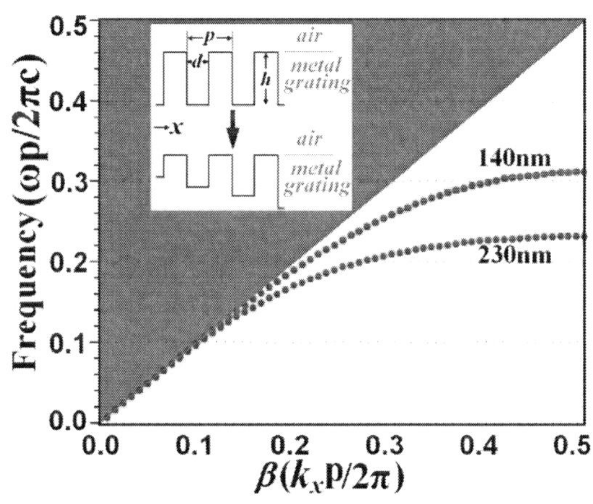

Figure 9.7 Dispersion curves calculated for d = 200 nm and p = 400 nm with different groove depths (h = 140 and 230 nm). Inset: a schematic of grating structures with a constant depth and graded depth. In this figure, the x axis is the reduced propagation constant with a unit of $k_x p/2\pi$; the y axis is the reduced frequency with a unit of $\omega p/2\pi c$, here c is the light velocity in vacuum (copyright info @ APS).[28]

It is known that the group velocity of SPP modes, which is given by the slope of the tangent line at a given point on the dispersion curve, decreases significantly with increasing the frequency as the cutoff frequency is approached.[21,25] Consequently, if the light at the cutoff frequency is coupled into the grating structure, its group velocity becomes quite low. However, it remains a considerable challenge to overcome the large momentum mismatch in the first Brillouin zone and to directly couple the light into one of the SPP modes around the cutoff frequency. In a recent study, the group velocity of SPP modes at telecom frequencies was reported to be approximately half that of the light velocity,[32] which implies that the working frequency is still far away from the cutoff frequency.

To address the issues of optical coupling and momentum mismatch, we employ the graded grating structure illustrated in Fig. 9.7 to gradually couple light into SPP modes possessing a very low group velocity. Assuming that the grade is small enough, a graded grating could be approximated by a series of many small gratings, each with a constant groove depth. In this case, the SPP dispersion relations vary with position along the grating. The dispersion relations for such a non-uniform grating are expected to change gradually along the x axis as the groove depth is increased. If a wave propagates along the surface of such a grating, the group velocity of the SPP mode gradually decreases along the propagation direction. By properly choosing the grating depths, the group velocity for an incoming wave can be greatly reduced at a specific location. For example, let us assume that the groove depth changes gradually from 140 to 230 nm over the 25 μm grating length. For this structure, the dispersion relations will vary as a function of position along the grating, roughly changing from the dispersion curve shown for 140 nm in Fig. 9.7 to that for 230 nm. At the input port, the shallow end of the grating, we only need to overcome a relatively small momentum mismatch to couple the light into the SPP mode. Subsequently, this mode will couple, step by step, as it enters into regions of increasing grating depth. This series of steps was introduced to help overcome the large momentum mismatch and gradually couple light into SPP modes with a large value of k_x and a very small group velocity. It should also be noted that as the grating depth changes, so does the cutoff frequency, so that incoming waves at different frequencies will be "trapped" at different positions along the grating, i.e., leading to the so-called "trapped rainbow" storage of light.[14,25]

9.2.2.2 Trapped "rainbow" at telecommunication frequencies

The concept described above can be validated by two-dimensional FDTD simulations. Figure 9.8 illustrates how light waves at four different wavelengths in the telecom domain, i.e., 1.33, 1.45, 1.55 and 1.65 μm, are "trapped" at different positions along the graded grating structures. These four wavelengths correspond to the cutoff frequencies associated with the grating depths at these positions. According to Fig. 9.7, the cutoff frequencies for the gratings with constant depths of 140 and 230 nm are about 226.6 THz (i.e., a wavelength of about 1.32 μm) and 178.1 THz (a wavelength of about 1.68 μm), respectively. These values are in good agreement with those obtained from the FDTD simulations (see Figs. 9.8a,d): As shown in Fig. 9.8a, the 1.33 μm wave is "trapped" at x = 13 μm, where the grating depth is about 150 nm. Similarly, Fig. 9.8d shows how the 1.65 μm wave is "trapped" at about x = 20.5 μm, where the grating depth is about 225 nm. In this way, it should be possible to trap a rainbow that is directly observable at visible wavelengths.

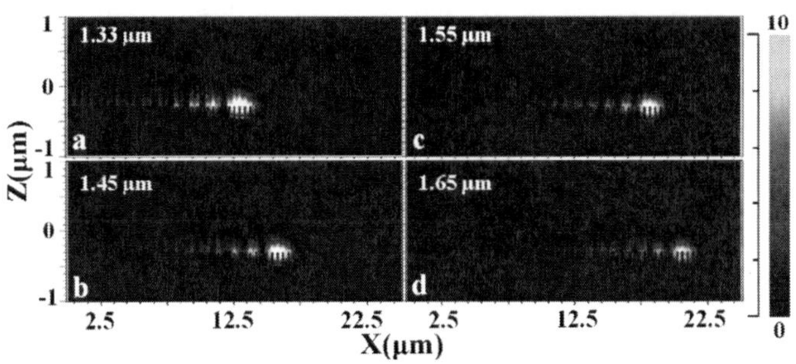

Figure 9.8 Trapped rainbow for telecom wavelengths obtained by using 2D FDTD simulations: (a–d) correspond to the 2D field distribution at four different wavelengths of 1.33, 1.45, 1.55 and 1.65 μm. The depth of the grating structure changes from 20 to 270 nm linearly in a 25 μm region. In these simulations, the period and width of the grating structure are 400 and 200 nm, respectively. Nonuniform mesh sizes are employed in this modeling: The edge grid sizes are Δx = 10 nm and Δz = 2 nm, and body grid sizes are Δx = 20 nm and Δz = 20 nm. The simulation time T = 2000 μm/c, here c is the light velocity in vacuum (copyright info @ APS).[28]

9.2.3 Experimental Verification of "Rainbow" Trapping Effect in Visible Domain

In previous sections, we have discussed the "rainbow" trapping effect in the THz and telecommunication domains. With a graded grating structure, the dispersion relations are expected to change gradually, as are the SPP properties, so that incoming waves at different frequencies can be "trapped" at different positions along the propagation direction on the grating, i.e., the "trapped rainbow" storage of light is feasible.[14,25,28] In this section, we discuss the experimental efforts to validate the "rainbow" trapping effect in visible domain.

9.2.3.1 Surface dispersion engineering using 1D metallic grooves with various depths and periods

Before this remarkable effect in plasmonic structures is verified, two theoretical predictions have to be confirmed experimentally: (1) The dispersion properties of the metallic surface and v_g of the SPP modes could be engineered; and (2) the photonic bandgap could be observed.

It was theoretically predicted that the dispersion curves of 1D metallic grating structures with various geometric parameters could be freely tuned.[28] To show the tunability of the dispersion curves of the nanopatterned surfaces, the angular-dependent reflection spectra of silver grating surfaces were calculated using RCWA method and their corresponding dispersion curves are plotted in Fig. 9.9. In this plot, the period and width of these grooves are 520 and 150 nm, respectively. One can see that the dispersion curve in the first Brillouin zone is obviously bent downward. Consequently, v_g for the SPP modes is reduced as the groove depth is increased. Our previous modeling reveals that the shape of the dispersion diagrams could be controlled by the geometric parameters (see Fig. 9.4). In addition, it was discussed that surface photonic bandgaps could be created via periodic nanopatterns.[33] This interesting feature can be also seen in this plot. One can see that a bandgap is broadened centered at the wavelength of approximate 540–560 nm, indicating that SPP modes at this frequency region cannot be supported by this nanopatterned surface. Such groove structures have been employed in the design of bidirectional plasmonic splitters.[33]

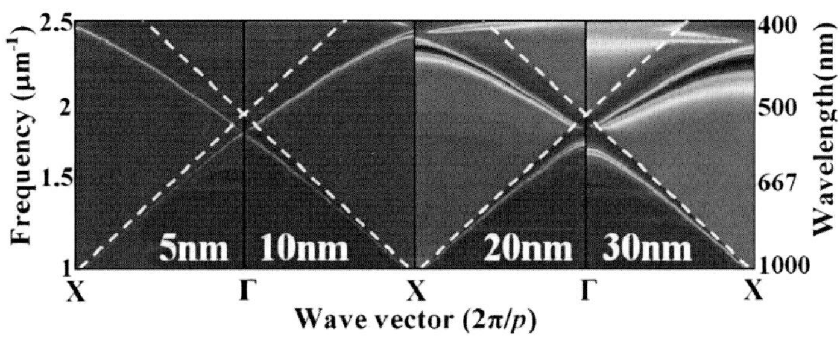

Figure 9.9 (a) Dispersion diagrams of the silver surface groove structures with the depth of 5, 10, 20 and 30 nm calculated with RCWA method. The dotted lines are light lines in the free space (copyright info @ 2014 National Academy of Sciences).[34]

9.2.3.2 Metallic graded gratings[34]

With these adiabatic changes to the surface dispersion properties of the nanograting structures, now we discuss the experimental verification of the "rainbow" trapping effect. A graded grating is fabricated by using FIB milling technology. As shown in Fig. 9.10a, the period and the width of the grooves are approximately 475 and 150 nm, respectively. The milling time for each groove is increased steadily to control the depth. According to the AFM characterization in Fig. 9.10a, the depth of the grooves increases approximately 1.2~1.6 nm by each step. A nanoslit is fabricated at 13 μm from the left edge of the graded grooves to function as an SPP coupler. It is illuminated from the back side to couple the free-space light to SPP modes at the top surface, and the emission light from the nano-grooves could be observed from the top side. With a white light illumination filtered by a multi-band filter (Transmission bands centered at 542 and 639 nm), a colorful emission in red and green is observed in Fig. 9.10b, reflecting the "rainbow" trapping effect. Here blue light is not observed because 464 nm is close to the cutoff frequency of the SPP dispersion curve and the propagation loss is therefore quite large. In addition, the wavelength at 464 nm is weak in our white light source. We then used two bright line filters at 546 nm (bandwidth 6 nm) and

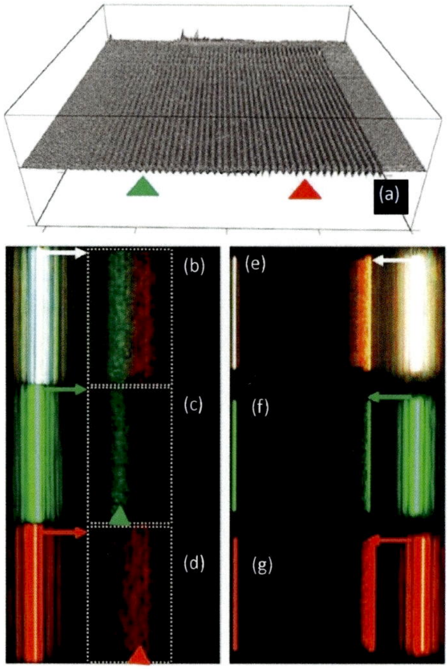

Figure 9.10 (a) The AFM image of a graded grating. The period and the width of the grooves are approximately 475 and 150 nm, respectively. The groove depth increases from approximately 6 to 100 nm. (b–d) Images of the emission from the structure with different filters, respectively. (e–g) Emission images with the SPP modes coupled and propagating in a direction opposite to that for (b–d). (copyright info @ 2014 National Academy of Sciences).[34]

655 nm (bandwidth 12 nm) to control the illumination wavelength. According to our theoretical predictions (results not shown here), the waves in green at 546 nm and in red at 655 nm should be trapped at the grooves with a depth of approximately 60 and 30 nm, respectively (indicated by the triangles in Fig. 9.10a). One can see from Figs. 9.10c,d that the SPP emissions obviously extended to different positions along the graded grating structures. The central positions of the green and red emission spots fit reasonably well with the theoretical predictions (see triangles in Fig. 9.10b). However, the bright regions are relatively wide because of the following reasons: (1) The transmission bands of the filters employed in the measurement are relatively broad. (2) For a real

device in Fig. 9.10a, one can see obvious roughness at the bottom of the nanogrooves, leading to a larger scattering loss. (3) As shown by the numerical band diagrams in Fig. 9.9, the dispersion curve is obviously broader at the photonic band edge because of the larger propagation loss, which is consistent with our previous numerical simulations on trapped "rainbow" in the visible region.[28,35] All these three reasons may lead to the broader emission regions for different wavelengths. If the SPP modes are guided from the opposite direction, they will reach to the deepest groove first, which is approximately 80 nm in depth. One can predict that the SPP modes should be scattered at once without propagating in the waveguide. To confirm this prediction, we fabricate another nanoslit at 13 μm from the right edge of the graded grooves to couple the light to SPP modes in the opposite direction. As shown in Figs. 9.10e–g, it is obvious that the emissions scattered at the deepest grooves without further extension compared with the images shown in Figs. 9.10 b–d.

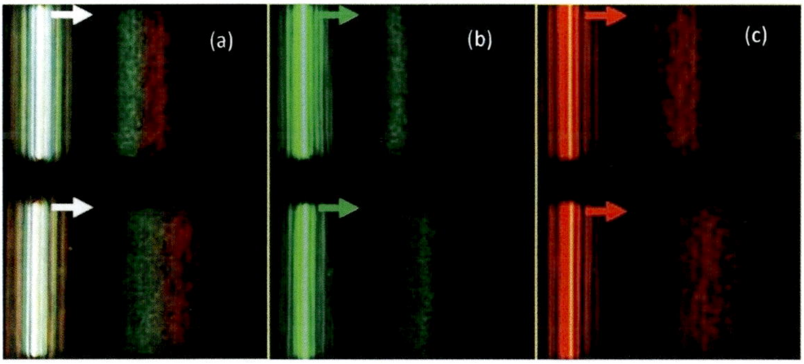

Figure 9.11 Microscope images of graded gratings with different gradients. In this measurement, three filters are employed: (a) A multi-band filter (Semrock, transmission bands centered at 542 and 639 nm). (b) A green filter centered at 546 nm (bandwidth 6 nm). (c) A red filter centered at 655 nm (bandwidth 12 nm) (copyright info @ 2014 National Academy of Sciences).[34]

It is interesting to note that, if the grade of the grating depths is changed, the trapped positions of different wavelengths should be tuned freely, according to our previous modeling result.[25] To validate this prediction, we fabricated another graded grating structure, whose period and width are fixed at 475 and 150 nm, respectively. The milling time for each groove is reduced to a half, whereas the number of the grooves is doubled. The input coupler nanoslit is still fabricated at 13 μm from the edge of the graded grooves. As shown in Fig. 9.11, we put the microscope images of the two graded grating structures together for comparison. One can see that the center positions of the green and red emissions are shifted slightly. In addition, the vertical widths of the SPP emission areas are broadened for the second structure with a smaller grade, which is consistent with the numerical modeling for THz waves in ref. [25], indicating that the local dispersion properties of the graded gratings could be tuned in a similar way from THz waves to telecommunication and visible waves.[28,35]

9.2.4 Summary

The idea of "trapped rainbow" storage of light[14,25,28] promises a potential to simultaneously slow down multiwavelengths in a broad band in a single structure. Our results verify that that the surface dispersion can be engineered by changing the structure of nanopatterned metallic surfaces. By using a graded surface grating structure, different wavelengths could be "trapped" at different positions on this structure. Although fabrication error and roughness introduce uncertainty, recently developed nanofabrication techniques that produce ultrasmooth patterned metals can be utilized to further validate this effect, and therefore to allow the improvement in design and fabrication of nanopatterned metal surface.[36] These new "metamaterials" offer the ability of storing photons insides solid-state structures at room temperature, which is promising for realizing novel optical components, integrated photovoltaic devices and biosensors on a chip.

9.3 Rainbow Trapping Effect in Vertical Direction[37]

In surface rainbow trapping effect discussed in the previous section, the white light SPP modes were launched into the on-chip chirped surface gratings through nanoslits,[34,38] leading to a relatively weak total coupling efficiency. According to previous theoretical and experimental investigations,[26,39] only the light impinging on the nanoslit area can be coupled to SPP modes. Most energy of the incident white light beam cannot be utilized. In addition, the 1D graded grating structure is polarization dependent. Consequently, this surface rainbow trapping platform is difficult to utilize in developing applications for photon harvesting or enhanced light–matter interactions, which require efficient coupling from the incident free-space light to slow SPP modes.

To overcome this intrinsic limitation, insulator-negative-index-insulator (INI),[14,40] insulator-metal-insulator (IMI),[41] and metal-insulator-metal (MIM) waveguide tapers[42–44] are proposed to achieve "rainbow trapping" effect. In a recent theoretical investigation, the rainbow trapping performance of INI, IMI and MIM waveguide tapers were analyzed and compared.[44] It was concluded that the MIM structure can trap the first order (and higher order) transverse magnetic (TM) modes, which makes them the best practical candidate for rainbow trapping among the three tapered waveguide structures (i.e., INI, IMI and MIM). However, even for the MIM structure, there are theoretical constraints to achieve the rainbow trapping conditions.[44] Consequently, trapped "rainbow" schemes based on INI, IMI or MIM waveguide tapers remain largely in the theoretical domain due to intrinsic constraints and difficulties in fabrication.

In this section, we will discuss a HMM waveguide taper structure, which can realize a highly efficient trapping effect in the vertical direction. This structure is not limited by those severe theoretical constraints required by INI, IMI, and MIM waveguide tapers, and therefore representing a significant promise to realize the rainbow trapping structure practically. The experimental realization of such HMM waveguide taper array will also be discussed.

9.3.1 Ultra-Wideband Light Wave Trapping on HMM Waveguide Taper Array

9.3.1.1 Hyperbolic metamaterials

HMM is referred to an artificial medium with subwavelength features whose iso-frequency surface is a hyperboloid.[45–49] For such a metamaterial with a diagonal form of the permittivity tensor, ε = diag (ε_x, ε_y, ε_z), its diagonal elements have different signs (i.e., $\varepsilon_x < 0$, $\varepsilon_z > 0$), leading to the hyperbolic iso-frequency surface, i.e., $\omega^2/c^2 = k_x^2/\varepsilon_z + k_y^2/\varepsilon_z + k_z^2/\varepsilon_x$.[45,46] Intriguingly, the photonic density of states in this kind of HMM is infinite due to the open (hyperboloid) iso-frequency surface.[47] Therefore, it is also called indefinite medium,[45] which is promising for a variety of new applications including three-dimensional indefinite cavities,[49] spontaneous emission enhancement[48] and active nanoplasmonic devices,[50] etc. In this work, we will employ an HMM constructed by alternating metal-dielectric multi-layers to design a tapered waveguide structure for efficient and practical rainbow trapping.

9.3.1.2 Effective medium theory model

To explain the fundamental mechanism, we first analyze the waveguide modes supported in an air/HMM/air planar waveguide by solving the eigenequation analytically based on the effective medium theory. Considering an HMM waveguide in the inset of Fig. 9.12a with a dielectric cladding layer ($\varepsilon_1 = 1$) and an anisotropic core layer (e.g., $\varepsilon_{2x} = -25 + 0.25i$, $\varepsilon_{2z} = 5 + 0.05i$), the propagation constant b of the TM modes can be obtained by solving Eq. (9.3),

$$\frac{\varepsilon_{2z}\gamma_1}{\varepsilon_1\gamma_2} = \tan(\gamma_2 w/2), \quad \text{for even modes} \tag{9.3a}$$

$$\frac{\varepsilon_{2z}\gamma_1}{\varepsilon_1\gamma_2} = -\cot(\gamma_2 w/2), \quad \text{for odd modes,} \tag{9.3b}$$

where $\gamma_1 = (\beta^2 - k_\omega^2\mu\varepsilon_1)^{1/2}$, $\gamma_2 = (k_\omega^2\mu\varepsilon_{2z}\beta^2 - \varepsilon_{2z}\beta^2/\varepsilon_{2x})^{1/2}$. $k_\omega = \omega/c$ is the vacuum wave vector, w is the width of the core layer, and μ is the permeability (for nonmagnetic materials, ($\mu = 1$). Here we first analyze the lossless case by neglecting the imaginary part of the permittivity of the HMM core layer, i.e., $\varepsilon_{2x} = -25$, $\varepsilon_{2z} = 5$. Using Eq. (9.3), the geometric dispersion curves of TM_0, TM_1, and TM_2 modes are plotted in Fig. 9.12a. One can see that an HMM core layer with a given width can support two different propagation constants for each TM mode. As the core layer width increases, a degeneracy point can be obtained as indicated by empty circles. To distinguish the optical properties of these two modes with different propagation constants, we calculate the normalized power flow in Fig. 9.12b, i.e., where S_z is the z-component of the Poynting vector. For modes with smaller propagation constants below degeneracy points in Fig. 9.12a (see solid lines), the power flow is parallel to the propagation direction (which is defined as |f> mode), i.e., $P > 0$ as plotted by solid lines in Fig. 9.12b; while for modes with larger propagation constants above degeneracy points shown in Fig. 9.12a (see dotted lines), the power flow is anti-parallel to the propagating direction (which is defined as |b> mode), i.e., $P < 0$ as plotted by dotted lines in Fig. 9.12b. One can see that these two modes are getting close to each other and be degenerate finally as the width of the core layer increases, which is similar to the TM_0 mode behavior in IMI waveguide (when $|\varepsilon_{\text{core}}| < \varepsilon_{\text{cladding}}$).[43] Interestingly, at the degeneracy point (see empty circles), P_{norm} is 0 indicating that these TM modes will be trapped in the proposed HMM waveguide structure. To validate this prediction, we then model the electric-field (|E|-field) distribution guided in a tapered HMM waveguide using finite element method (FEM). In the numerical simulation shown in Fig. 9.12c, the guided TM_0, TM_1, and TM_2 modes are directly launched into the waveguide taper from the narrow end. For instance, the core layer width of the narrow end is 0.1λ in the upper panel of Fig. 9.12c, corresponding to two propagation constants of $1.077k_\omega$ and $9.180k_\omega$ for the TM_0 mode in Fig. 9.12a. Since the smaller β (i.e., $1.077k_\omega$) is very close to the vacuum wave vector, k_ω, it should be easy to launch the |f> mode practically using a free-space beam. As the |f> mode is guided along the waveguide taper with an increasing width, it will be trapped at the degeneracy point finally (i.e., $w = 0.173\lambda$, indicated by white arrows in Fig. 9.12c), agreeing very well with the FEM modeling. Similarly, |f> modes for TM_1 and TM_2 modes cannot propagate beyond their corresponding degeneracy points at $w = 0.380\lambda$ and 0.591λ, as show in the middle and lower panels of Fig. 9.12c, respectively. Due to the power flow of the modes decreases to 0, the incident energy is trapped at these degeneracy points. In addition, the tangent line at the degeneracy point is parallel to the vertical direction, indicating that the group velocity is zero and therefore revealing an encouraging promise to realize rainbow trapping in these HMM waveguide tapers.

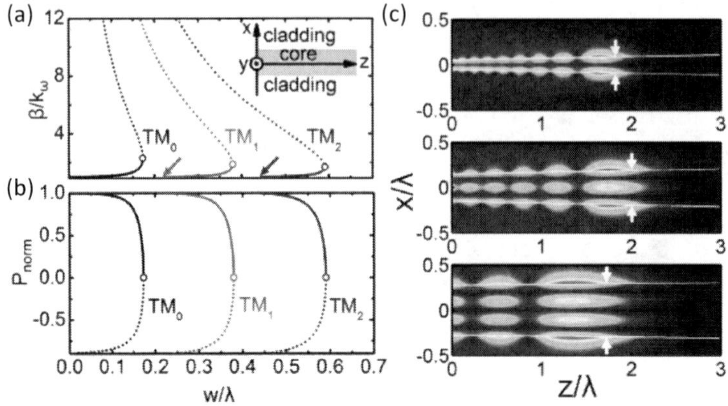

Figure 9.12 (a) Propagation constants of TM_0, TM_1, and TM_2 modes, β, as a function of the HMM waveguide width. The two branches for each mode are separated by a degeneracy point, represented by empty circles. (b) Normalized power flows of TM_0, TM_1 and TM_2 modes, P_{norm}, as a function of the HMM waveguide width. (c) shows $|\mathbf{E}|$-field distribution of TM_0, TM_1 and TM_2 modes in HMM waveguide tapers. In the length of 3λ along the z-direction, the tapered width increases from 0.1λ to 0.22λ in the upper panel, from 0.3λ to 0.42λ in the middle panel, and from 0.55λ to 0.62λ in the lower panel, respectively (copyright info @ NPG).[37]

In previously reported simulation results exploring the feasibility of realizing rainbow trapping in INI, IMI and MIM waveguide tapers,[41–44] a rainbow trapping condition parameter for having the degeneracy point, $\sigma_\varepsilon = |\varepsilon_{core}/\varepsilon_{cladding}|$, has been analyzed and compared. According to ref. [44], this condition for the relatively practical MIM waveguide taper structure is still challenging (e.g., for TM_1: $1 < \sigma_\varepsilon < 1.3510$), which requires a very large permittivity of the dielectric core layer (e.g., GaP) so that the rainbow trapping can be realized in a very limited spectral range for TM_1 modes. It will be demonstrated that the proposed HMM waveguide structure is not limited by these severe theoretical constraints required by INI, IMI, and MIM waveguide tapers. Here we first define three parameters, $V = \frac{\pi\omega}{\lambda}[\mu\varepsilon_{2z}(\varepsilon_1/|\varepsilon_{2x}|+1)]^{1/2}$, $U = w\gamma_2/2$, and $\sigma = \sqrt{|\varepsilon_{2x}|\varepsilon_{2z}}/\varepsilon_1$, to simplify the characteristic Eq. (9.3) as follows:

$$\sigma\sqrt{U^2 - V^2}/U = \tan U, \quad \text{for even modes} \tag{9.4a}$$

$$\sigma\sqrt{U^2 - V^2}/U = -\cot U, \quad \text{for odd modes} \tag{9.4b}$$

In this equation, V and σ are constants for a given waveguide structure. U is related to the propagation constant of the waveguide mode. The solution to this equation can be obtained by graphically determining the intersection points between curves of $W = \tan U$ or $W = -\cot U$ (i.e., the term on the left-hand side of Eq. (9.4)) and $W = \sigma\sqrt{U^2 - V^2}/U$ (i.e., the term on the right-hand side of Eq. (9.4)) in the (U, W) space. Based on this graphical analysis, conditions for supporting |f> and |b> branches of TM_0, TM_1, and TM_2 modes are listed in Table 9.1. Particularly, the TM_1 and TM_2 |f> modes exist in the ranges of $\pi/2 \le V \le V_1$ and $\pi \le V \le V_2$, respectively, corresponding to ranges from cutoff points for TM_1 and TM_2 modes (indicated by arrows in Fig. 9.12a) to their degeneracy points. In order to realize the rainbow trapping for a given wavelength, degeneracy points have to be realized, which can be met by tuning the core layer width (w) of the HMM waveguide. Consequently, as the incident TM_0 mode propagates along the HMM waveguide taper from the narrow end to the broad end, the |f> and |b> modes will get close to each other and degenerate finally. Noticeably, during this process, only the hyperbolic condition (i.e., $\varepsilon_{2x} < 0, \varepsilon_{2z} > 0$) of the core layer is needed. Any reported metamaterials with hyperbolic iso-frequency surfaces (e.g., refs. [45–49]) can be used to enable the degeneracy condition, and therefore representing a significant promise to realize the rainbow trapping practically.

Table 9.1 Conditions for existence of TM_0, TM_1 and TM_2 modes in the HMM waveguide.

	TM_0	TM_1	TM_2
|f>	$0 < V \le V_0$	$\pi/2 \le V \le V_1$	$\pi \le V \le V_2$
|b>	$0 < V \le V_0$	$0 < V \le V_1$	$0 < V \le V_2$

Note: Consider that $V = \frac{k_\omega}{2}[\mu\varepsilon_{2z}(\varepsilon_1/|\varepsilon_{2x}|+1)]^{1/2}$, the corresponding thickness of the HMM core layer to support |f> and |b> modes can therefore be determined. For instance, when $\varepsilon_1 = 1$, $\varepsilon_{2x} = -25$, $\varepsilon_{2z} = 5$, the thickness of the core layer is $w/\lambda = 0.14$ V. Therefore, one can control the value of V to support different TM modes by choosing the value of w/λ. At the degeneracy conditions for different order TM modes, $V_0 = 1.24$, $V_1 = 2.72$ and $V_2 = 4.23$. Consequently, when $w = 0.1\lambda$, $V \approx 0.714$ indicating the existence of TM_0 |f> and |b> modes, TM_1 |b> and TM_2 |b> modes in this HMM waveguide. When $w = 0.2\lambda$, $V \approx 1.429$ indicating the existence of TM_1 |b> and TM_2 |b> modes only.

In the discussion presented above, the loss of the HMM material was neglected. This ideal but impractical assumption was usually employed to approximately predict optical behaviors of plasmonic structures and metamaterials. For example, the metal loss was neglected in previous theoretical designs of trapped rainbow of THz waves in metamaterial[51] and plasmonic surface grating structures,[14] which resulted in a debate on the feasibility of the proposed "stop" light effect.[26,27] In recent years, many simulation results have shown that adiabatic metallic metamaterials or plasmonic structures do not permit the complete "stopping" of light waves even under the assumption of lossless materials due to the fundamental nonadiabaticity near the degeneracy point.[44] More specifically, the adiabaticity condition for the graded core width variation,[52] $dw/dz << wk_{\omega}\Delta n/\pi$ where Δn is the effective index difference between eigenmodes (i.e., |f> and |b>), cannot be met at the degeneracy point since $\Delta n = 0$. However, this criterion was still omitted occasionally in recent theoretical designs.[53] It should be noted that a complete stop of light is also not achievable in the lossless HMM waveguide tapers although the group velocity at the degeneracy point is 0 in principle. The incident |f> mode will convert to |b> mode and escape from the "trapped" position. In the next paragraph, we will consider the loss of HMM materials to explore this mode conversion process and evaluate the rainbow trapping performance further.

To reveal the difference by considering the absorption (i.e., $\varepsilon_{2x} = -25 + 0.25i$, $\varepsilon_{2z} = 5 + 0.05i$ in our modeling), we plot the real and imaginary parts of the propagation constants, β_r and β_i, as shown in Figs. 9.13a,b, respectively. One can see from Fig. 9.13a that the absorption breaks the degeneracy point connecting |f> and |b> modes in the lossless case. The geometric dispersion curve for TM_0 modes will extend to the right side of the lossless degeneracy point with significantly larger loss as shown in Fig. 9.13b. The TM_0 mode behaves as an evanescent wave in this "cutoff" region, due to the large attenuation coefficient, which is defined as β_i/k_{ω}. On the other hand, the propagation loss for the |b> mode is much larger than that of the |f> mode, revealing a key difference before and after the mode conversion. To further evaluate the field confinement, the effective modal area is defined by $A = \int |\mathbf{E}|^2 dx / \max\{|\mathbf{E}|^2\}$, as plotted in Fig. 9.13c. Generally, the |b> mode indicated by the dotted line has a smaller modal area than the |f> mode indicated by the solid line. The model areas of these two eigenmodes match approximately near the degeneracy point as the mode conversion occurs. To reveal the behavior of the mode propagation in the tapered lossy HMM waveguide, we perform the FEM simulation in Fig. 9.13d. One can see that the incident |f> mode cannot propagate beyond the degeneracy point position indicated by the vertical dotted line. An obvious oscillation field distribution is observed due to the interference between the incident |f> mode and the reflected |b> mode.

To provide a quantitative understanding on this mode conversion and light trapping mechanism, the mode expansion method is employed to calculate the amplitude of the local guide modes in the tapered waveguide. The total field can be expressed as the linear combination of all the possible modes supported by the waveguide structure in Eq. (9.5).

$$E_{x,\text{total}} = \sum a_\sigma E_{x,\sigma} = (a_{f+} + a_{f-})E_{x,f+} + (a_{b+} + a_{b-})E_{x,b+} + \sum_{\rho \neq f,b} a_\rho E_{x,\rho} \tag{9.5a}$$

$$H_{y,\text{total}} = \sum a_\sigma H_{y,\sigma} = (a_{f+} - a_{f-})H_{y,f+} + (a_{b+} - a_{b-})H_{y,b+} + \sum_{\rho \neq f,b} a_\rho H_{y,\rho}, \tag{9.5b}$$

Here the signs "+" and "–" denote the forward- and backward-propagation directions; ρ represents the high order modes or the radiation modes. In the lossy waveguide, we use the unconjugated general form of orthogonality condition,[52,54] $[H_{y,\sigma}|E_{x,\sigma'}] = \delta_{\sigma\sigma'}[H_{y,\sigma}|E_{x,\sigma}]$, to obtain the amplitudes of |f+>, |f–>, |b+> and |b–> TM_0 modes in the tapered waveguide, as shown in Fig. 9.13f. One can see that the incident |f+> mode converts to the |b+> mode near the degeneracy point, while the amplitudes of the other two modes, |f–> and |b–>, are negligible in this structure. Consider that the power flow of the |b+> mode is antiparallel to the propagation direction (the power flow directions for these two eigenmodes are indicated by arrows in Fig. 9.13f), the incident energy therefore escapes from the trapped position through the mode conversion process. To distinguish the mode areas of these two eigenmodes, the $|E|$-field for a f+|f+> and a b+|b+> are plotted in Fig. 9.13e. When the TM_0 |f> mode is launched from the narrow end of the structure, the mode width of the $|E|$-field distribution is indicated by white dots, showing that the guided modal area shrinks towards the degeneracy point (see the upper half panel). After the mode conversion from |f> to |b> near the degeneracy point, the |b> mode is squeezed into the HMM waveguide taper further due the much larger propagation loss (see the lower half panel in Fig. 9.13e). Due to this decreasing modal area from the |f+> mode to the |b+> mode with a significantly enhanced propagation loss, the tapered HMM waveguide is therefore promising to develop applications for super absorbers[55,56] based on the intriguing rainbow trapping effect.

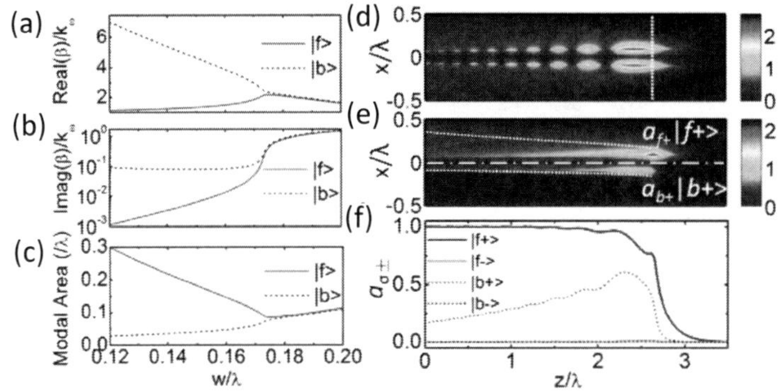

Figure 9.13 The width-dependence of (a) the real part of the propagation constant, β_r, (b) the attenuation coefficient, β_i, and (c) the effective modal area of the fundamental mode (TM_0) in the lossy HMM waveguide taper. (d) The |*E*|-filed distribution in the HMM waveguide taper for the TM_0 mode. (e) The |E|-filed distribution for the |f+> mode (the upper half panel) and |b+> mode (the lower half panel), respectively. (f) The local modal amplitudes of |f+>, |f−>, |b+> and |b−> modes along the *z*-direction (copyright info @ NPG).[37]

In the next section, we discuss the fabrication and characterization to realize the spectrally tunable on-chip broadband super absorptive hyperbolic metafilm from near-IR to mid-IR spectral regions.

9.3.2 Experiment Realization of HMM Waveguide Taper

To demonstrate the broad band absorption property based on the proposed HMM waveguide taper, we first deposited four-pair Ag/SiO_2 layer on top of a 150 nm-thick Al film in a multi-target electron-beam evaporation system.[57] The thickness of each Ag/SiO_2 layer was controlled at 30 ± 5 nm. Focused ion beam lithography was employed to fabricate tapered structure on Al/SiO_2 layers. Due to the optically opaque ground Al film, the absorption, A, can be characterized by $A = 1 - R$ where R is the reflection intensity. Compare to a one-pair perfect absorber with period and width of the top Ag pattern of 950 and 720 nm, whose absorption peak at 3.5 μm as shown in Fig. 9.14a (sample 1), the full width at half maximum (FWHM) of the absorption band of four-pair HMM waveguide taper as shown in Fig. 9.14a (sample 2) was broadened from 2.5 to 4.7 μm. To further broaden the absorption band, an eight-pair HMM waveguide taper array with the period, top width and bottom width of 1.35 μm, 480 nm, 1.14 μm as shown in Fig. 9.14a (sample 3) was fabricated. According to the measurement and modeling results shown by solid blue curves in Figs. 9.14b,c, the FWHM of the absorption band can be extended to 2 to 6.53 μm and 2.70 to 5.52 μm, respectively. Importantly, this broad absorption band is tunable by changing the geometric parameters. For instance, for the four-pair HMM waveguide taper array, when the period, top width and bottom width were tuned to 1.57, 850, 1.35 μm, respectively (see sample 4 in Fig. 9.14a), the FWHM of the absorption band was tuned to 3.80–6.03 μm (measured result) and 4.05–6.00 μm (modeled result). As shown in Fig. 9.14f (measured result) and Fig. 9.14g (modeled result), we also tuned the absorption band to 0.90–2.00 μm by adjusting the period, top width and bottom width to 300, 110, 270 nm based on the eight-pair multi-layers (see sample 5 in Fig. 9.14a). Furthermore, by extending the 1D tapered structure into the two-dimensional (2D) pyramidal pattern array (see sample 6 in Fig. 9.14a), a polarization insensitive absorption band can be realized for normal incident light, as shown by the black dotted (i.e., y-polarization) and blue solid curves (i.e., x-polarization) in Fig. 9.14h (measured result) and the black curve in Fig. 9.14i (modeled result).

However, an obvious challenge to realize the proposed hyperbolic metafilm pattern is the quality control and improvement for the multi-layered metal/dielectric films. For instance, 20 and 15 pairs of metal/dielectric thin films were required to realize the broadband absorption band from mid-IR[53] to near-IR spectral regime,[58] respectively, which is extremely challenging to maintain the flat and continuous films in practice. Here we combine the multi-unit pattern array proposed for single-paired perfect absorber and the multi-layered HMM waveguide taper in a single structure

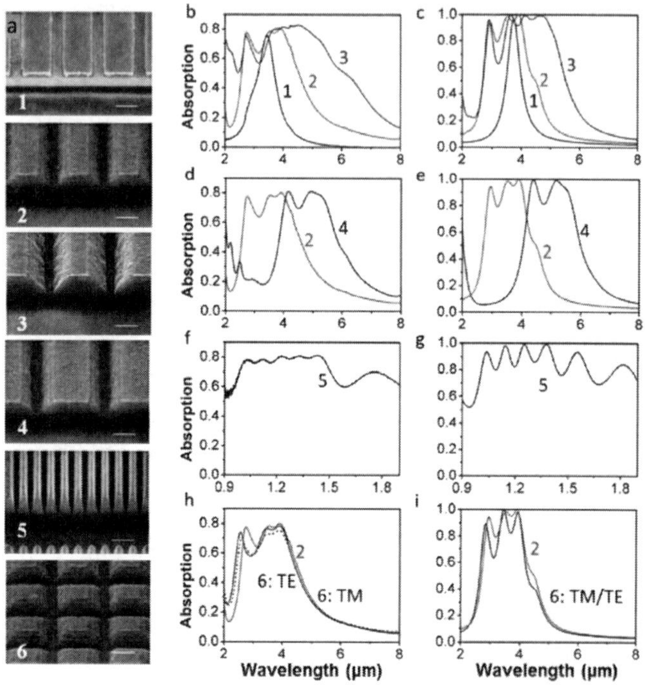

Figure 9.14 (a) 54°-tilted SEM images of 6 samples of super absorptive metafilms. The period and width of the top Ag pattern of sample 1 are 950 nm and 720 nm, respectively. For samples 2–6, the period, top width and bottom width parameters are (1.17 μm, 500 nm, 950 nm) for sample 2, (1.35 μm, 480 nm, 1.14 μm) for sample 3, (1.57 μm, 850 nm, 1.35 μm) for sample 4, (300 nm, 110 nm, 270 nm) for sample 5 and (1.17 μm, 500 nm, 950 nm) for sample 6, respectively. The scale bar is 500 nm. (b, d, f, h) and (c, e, g, h) show measured and modeled absorption spectra of these 6 samples (indicated by sample number), respectively (copyright info @ NPG).[57]

to minimize the required number of layers and realize an ultra-broad absorption band. As shown in Fig. 9.15a (sample 7), a two-unit HMM waveguide taper array was fabricated on the four-paired Ag/SiO_2 film. The period of the pattern unit is 2.26 μm, and the top and bottom widths of two units increase from 580 nm to 860 nm and from 790 nm to 1.2 μm, respectively, similar to the width range of the tapered structure fabricated on the eight-paired Ag/SiO_2 film shown in Fig. 9.14a (sample 3, from 480 nm to 1.14 μm). One can see that the FWHM of the absorption band ranges from 2.50 to 5.57 μm in experiment (see the solid black curve in Fig. 9.15b) and from 2.90 to 5.31 μm in modeling (see the black curve in Fig. 9.15c), which is equivalent to the one achieved by sample 3 in Fig. 9.14a. Furthermore, the polarization dependence of these 1D pattern arrays can be overcome by introducing 2D distributed multi-unit patterns, as shown in Fig. 9.15a (sample 8). In this fabrication, the periods along two directions are both 2.26 μm, and the square pyramidal widths of the four units increase from 630–860 nm (i.e., $a_1 - a_2$) and 790 nm–1.20 μm (i.e., $b_1 - b_2$), respectively. In this case, a polarization insensitive absorption band is obtained with an FWHM from 2.60 to 5.56 μm. The required number of metal/dielectric layers is reduced by half, therefore simplifying the sample preparation and experimental realization of the on-chip broadband super absorptive metafilms significantly.

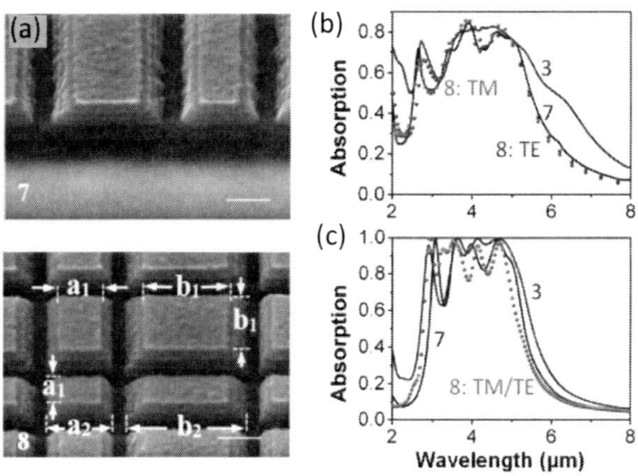

Figure 9.15 (a) 54°-tilted SEM images of super absorptive metafilms with multiple patterns in a single period. The scale bar is 500 nm. Images (b) and (c) show measured and modeled absorption spectra of these two samples, respectively. For comparison, the measured and modeled absorption spectra of sample 3 with eight-pair Ag/SiO_2 layers are plotted by solid red curves (copyright info @ NPG).[57]

9.3.3 Summary

In summary, we have explored the feasibility of rainbow trapping based on a multi-layered metal-dielectric film stack supporting broadband slow light resonances. Due to the eigenmode conversion between |f> and |b> modes with significantly different optical properties, the HMM waveguide taper constructed by metal and dielectric films can be used to develop practical on-chip optical super absorbers with a tunable absorption band. Consequently, we experimentally realized a patterned hyperbolic metafilm with engineered and freely tunable absorption band from near-IR to mid-IR spectral regions based on multilayered metal/dielectric HMM waveguide taper. By cascading resonant metal-insulator-metal perfect absorber elements with gradually tuned widths along the vertical direction, the absorption band of the patterned HMM film is extended significantly.

9.4 Outlook

This chapter discussed fundamental understanding of light/thermal management in structured surfaces/materials by including coupled theoretical and experimental studies. The spatial control of the localized dispersion properties of patterned plasmonic surface and hyperbolic metafilm can provide a practical platform with engineerable refractive index, which will pave the way for future investigations on a broad range of optical/thermal and energy technologies. For instance, if the absorption band can be tuned to mid- and far-IR spectral regions, it will enable the further exploration of improved on-chip microbolometers and thin-film thermal collectors for energy recycling after sunset and using the waste heat from industrial plants.[59,60] In addition, as a widely used light source, filament lamps convert heat to light with very low efficiencies. A lot of energy-efficient light sources (e.g., LED) have been developed to replace this conventional heat-to-light converter. Remarkably, if high-temperature materials (e.g., tungsten) can be implemented to realize a high index metafilm emitter, it may significantly enhance the heat-to-light conversion efficiency and therefore create renewed research opportunities to revolutionize this inefficient technology. These promising and transformative applications enabled by the rainbow trapping effect will have enormous long-term impact on our nation's energy, environmental and sustainability needs.

References

1. L. V. Hau (2008). *Nat. Photon.,* **2**, 451.
2. T. Baba (2008). *Nat. Photon.,* **2**, 465.
3. L. Thévenaz (2008). *Nat. Photon.,* **2**, 474.
4. R. Boyd and D. Gauthier (2009). *Science*, **326**, 1074.
5. L. Brillouin, *Wave Propagation and Group Velocity* (Academic, New York, 1960).
6. A. Kasapi, M. Jain, G. Y. Yin, and S. E. Harris (1995). *Phys. Rev. Lett.*, **74**, 2447.
7. L. V. Hau, S. E. Harris, Z. Dutton, and C. H. Behroozi (1999). *Nature* (London) **397**, 594.
8. C. Liu, Z. Dutton, C. H. Behroozi, and L. V. Hau (2001). *Nature* (*London*), **409**, 490.
9. A. V. Turukhin et al. (2002). *Phys. Rev. Lett.*, **88**, 023602.
10. C. J. Chang-Hasnain, P. C. Ku, J. Kim, and S. L. Chuang (2003). *Proc. IEEE*, **91**, 1884.
11. Y. Yamamoto and R. E. Slusher (1993). *Phys. Today*, **46**, 66.
12. M. F. Yanik and S. H. Fan (2004). *Phys. Rev. Lett.*, **92**, 083901.
13. F. Xia, L. Sekaric, and Y. Vlasov (2007). *Nat. Photon.,* **1**, 65.
14. K. L. Tsakmakidis, A. D. Boardman, and O. Hess (2007). *Nature* (*London*), **450**, 397.
15. W. L. Barnes, A. Dereux, and T. W. Ebbesen,(2003). *Nature* (*London*), **424**, 824; E. Ozbay (2006). *Science*, **311**, 189; C. Genet and T. W. (2007). Ebbesen, *Nature* (*London*), **445**, 39.
16. J. B. Pendry, L. Martin-Moreno, and F. J. Garcia-Vidal (2004). *Science*, **305**, 847.
17. A. P. Hibbins, B. R. Evans, and J. R. Sambles (2005). *Science*, **308**, 670.
18. Z. Ruan and M. Qiu (2007). *Appl. Phys. Lett.*, **90**, 201906.
19. F. J. Garcia-Vidal, L. Martin-Moreno, and J. B. Pendry (2005). *J. Opt. A: Pure Appl. Opt.*, **7**, S94.
20. K. Zhang, D. Li, *Electromagnetic Theory for Microwaves and Optoelectronics* (Springer-Verlag, Berlin Heidelberg 1998).
21. Z. Fu, Q. Gan, Y. Ding, and F. Bartoli (2008). *IEEE J. Sel. Top. Quant. Electron.*, **14**, 486.

22. H. Raether, *Surface Plasmons on Smooth and Rough Surfaces and on Gratings*, Springer Tracts in Modern Physics Vol. **111** (Springer-Verlag, Berlin, 1988).
23. *Fullwave* 6.1, Rsoft Inc (2008).
24. J. T. Shen, P. B. Catrysse, and S. H. Fan (2005). *Phys. Rev. Lett.*, **94**, 197401.
25. Q. Gan, Z. Fu, Y. Ding, and F. Bartoli (2008). *Phys. Rev. Lett.*, **100**, 256803.
26. A. Reza, M. M. Dignam, and S. Hughes (2008). *Nature*, **455**, E10.
27. K. L. Tsakmakidis, A. D. Boardman, and O. Hess (2008). *Nature*, **455**, E11.
28. Q. Gan, Y. J. Ding, and F. J. Bartoli. (2009). *Phys. Rev. Lett.*, **102**, 056801.
29. L. Chen, G. Wang, Q. Gan, and F. Bartoli. (2009). *Phys. Rev. B*, **80**, 161106.
30. J. D. Joannopoulos, R. D. Meade, and J. N. Winn, *Photonics Crystals: Modeling the Flow of Light* (Princeton University, Princeton, NJ, 1995).
31. E. D. Palik, *Handbook of Optical Constants of Solids* (Academic, Orlando, FL, 1985), vol. 1, p. 366.
32. M. Sandtke and L. Kuipers (2007). *Nat. Photon.*, **1**, 573.
33. Q. Gan, F. Bartoli (2010) *Opt. Lett.*, **35**, 4181.
34. Q. Gan, Y. Gao, K. Wagner, D. Vezenov, Y. J. Ding, and F. J. Bartoli (2011). *Proc. Natl. Acad. Sci.*, **108**, 5169.
35. L. Chen, G. Wang, Q. Gan, and F. Bartoli (2009). *Phys. Rev. B*, **80**, 161106.
36. P. Nagpal, N. C. Lindquist, S. H. Oh, and D. J. Norris (2009). *Science*, **325**, 594.
37. Q. Gan, and F. Bartoli (2011). *Appl. Phys. Lett.*, **98**, 251103.
38. H. Hu, D. Ji, X. Zeng, K. Liu, and Q. Gan (2013). *Sci. Rep.*, **3**, 1249.
39. P. Lalanne, J. P. Hugonin, and J. C. Rodier (2005). *Phys. Rev. Lett.*, **95**, 263902.
40. S. He, Y. He, and Y. Jin (2012). *Sci. Rep.*, **2**, 583; DOI:10.1038/srep00583.
41. J. Park, K.-Y. Kim, I.-M. Lee, H. Na, S.-Y. Lee, and B. Lee (2010). *Optics Express*, 18(2), 598.
42. W. T. Lu, Y. J. Huang, B. D. F Casse, R. K. Banyal, and S. Sridhar (2010). *Appl. Phys. Lett.*, **96**, 211112.

43. B. Prade, J. Y. Vinet, and A. Mysyrowicz (1991). *Phys. Rev. B*, **44**, 13556–13572.

44. M. S. Jang, and H. Atwarter (2011). *Phys. Rev. Lett.*, **107**, 207401.

45. D. R. Smith, P. Kolinko, and D. Schurig (2004). *J. Opt. Soc. Am. B*, **21**, 1032.

46. L. Alekseyev, and E. Narimanov (2006). *Opt. Express*, **14**, 11184–11193.

47. Z. Jacob, L. V. Alekseyev, and E. Narimanov (2006). *Opt. Express*, **14**, 8247.

48. H. N. S. Krishnamoorthy, Z. Jacob, E. Narimanov, and I. Kretzschmar (2012). *Science* **336**, 205.

49. X. Yang, J. Yao, J. Rho, X. Yin, and X. Zhang (2012). *Nat. Photon.*, **6**, 450.

50. Hess, O., et al. (2012). *Nat. Mater.*, **11**, 573–584.

51. R. Boyd and D. Gauthier (2009). *Science,* **326**, 1074.

52. A. Snyder, and J. D. Love, *Optical Waveguide Theory* (Chapman and Hall, New York, 1983).

53. Y. Cui, K. H. Fung, J. Xu, H. Ma, Y. Jin, S. He, and N. X. Fang (2012). *Nano Lett.*, **12**, 1443.

54. P. Lalanne, J. P. Hugonin, H. T. Liu, and B. Wang (2009). *Surf. Sci. Rep.*, **64**, 453.

55. K. Aydin, V. E. Ferry, R. M. Briggs and H. A. Atwater (2011). *Nat. Commun.*, **2**, 517.

56. Teperik, T. V., et al. (2008). *Nat. Photon.*, **2**, 299–301.

57. D. Ji, et al. (2014). *Scientific Reports* 4.

58. Q. Liang, T. Wang, Z. Lu, Q. Sun, Y. Fu, and W. Yu (2013). *Adv. Opt. Mater.,* **1**, 43–49.

59. R. F. Service (2008). *Science*, **320**, 1585.

60. D. K. Kotter, S. D. Novak, W. D. Slafer, and P. Pinhero (2008). Proceedings of ES2008, ES2008-54016.

Index